ADVANCE PRAISE FOR

Ecocities

While most green city books are at best sustainability-light,
Richard Register's *Ecocities: Rebuilding Cities in Balance with Nature* steers a course
to deep green. His hard-hitting solutions are on a scale commensurate with the problem.
Read it and work quickly to implement his many suggestions if you want a
tolerable planet upon which to house your home.

— RANDY HAYES, Executive Director, International Forum on Globalization
& Founder of Rainforest Action Network

Ecocities takes you on an exploratory journey into the cities of the future.
You will wander along streets flanked by clear streams and flowering shrubs where
there is no polluting traffic, or traverse bridges linking gardens and fruit trees many stories
above the ground. Yet this is no half-baked dream, but rather a reality based on technologies,
architectural designs and functioning buildings that actually exist in several places
around the globe, and concepts that have been approved by down-to-earth city planners.
I want everyone, especially high school and college students, to read and think about
Richard Register's book. A copy of *Ecocities* should be in every school library.

— DR. JANE GOODALL, DBE, Founder, the Jane Goodall Institute
& UN Messenger of Peace www.janegoodall.org

That we are entering a new world, one of rising temperatures and the
coming of peak oil production is no longer debatable. What most people miss is
that the largest things human beings create — the built community of cities, towns and
villages — is the cause of and potential solution to these largest of all problems
facing humanity and future life on Earth.

Ecocities places us all right there, well prepared with a vision, enthusiasm, and powerful, practical tools. This book enables us to challenge the doom-sayers and doom-makers to this race for a healthy, sane, compassionate future. We have to win, and, explains the author, here's how.

— LESTER R. BROWN, President of the Earth Policy Institute and author of
Plan B 2.0: Rescuing a Planet Under Stress and a Civilization in Trouble.

There are few people who know and understand cities in the way Richard Register does and even fewer who are prepared to stand up and be counted as genuine friends of the city. Richard Register is one of those rare authors who can speak and write about cities from both the 'head and the heart', who combines solid technical knowledge and skills with a vision of the beauty and ecological possibilities of the city.

In this book Richard Register provides a wonderful account of how we can transform cities into places that are not only sustainable, but which are a joy and inspiration to live in. This book needs to be read not only for its intellectual contribution to future city development, but also for how it speaks to the heart about the true meaning of the city.

— JEFFREY KENWORTHY, Professor in Sustainable Cities, Institute for Sustainabilty and Technology Policy, Murdoch University, Perth, Western Australia

As an environmental activist I have long admired Richard Register and Ecocity Builders. It's so clear that with politicians and the private sector focused on bottom lines dictated by inter-election intervals or quarterly dividends, long-term ecological issues tend to lie out of sight.

By taking the perspective of decades and working for incremental change in the right direction, enormous changes can happen in the end. Richard gave me a powerful insight when he showed the rate of renewal within cities and how inculcating ecological principles could be integrated and have a huge effect over time.

— DAVID SUZUKI, The David Suzuki Foundation

ecocities

ecocities

Rebuilding Cities in Balance with Nature

REVISED EDITION

Richard Register

NEW SOCIETY PUBLISHERS

Cataloging in Publication Data:
A catalog record for this publication is available from the National Library of Canada.

Cover design by Diane McIntosh. Illustration by Richard Register.
All interior illustrations by Richard Register.

Printed in Canada.
First printing May 2006.

Paperback ISBN-10: 0-86571-552-1
Paperback ISBN-13: 978-0-86571-552-3

Inquiries regarding requests to reprint all or part of *Ecocities* should be addressed to New Society Publishers at the address below.

To order directly from the publishers, please call toll-free (North America) 1-800-567-6772, or order online at www.newsociety.com

Any other inquiries can be directed by mail to:

New Society Publishers
P.O. Box 189, Gabriola Island, BC, V0R 1X0, Canada
1-800-567-6772

New Society Publishers' mission is to publish books that contribute in fundamental ways to building an ecologically sustainable and just society, and to do so with the least possible impact on the environment, in a manner that models this vision. We are committed to doing this not just through education, but through action. We are acting on our commitment to the world's remaining ancient forests by phasing out our paper supply from ancient forests worldwide. This book is one step toward ending global deforestation and climate change. It is printed on acid-free paper that is **100% old growth forest-free (100% post-consumer recycled)**, processed chlorine free, and printed with vegetable-based, low-VOC inks. For further information, or to browse our full list of books and purchase securely, visit our website at: www.newsociety.com

NEW SOCIETY PUBLISHERS   www.newsociety.com

To Nancy, Bill, Sylvia and Kirstin for years of sustained friendship and help,
in the spirit and substance that made this book possible

Ecocites Become Us

It's a secret now, but can't be for long.
We are of three bodies:
One created for us, given at birth,
One created by us, labored through life,
One created with us, eternity's rebirth.
One the individual, of flesh, blood and soul,
One the community of wood, stone and fire,
One the universe, of stuff, time and changing.
This is who we are, where we are from, are, and going.
The secret is out. We are building ourselves.
We are the universe, painful joyous work unfolding.

Contents

Acknowledgments

Without the financial and material assistance of the following people this book, either would not have been completed or would have taken much longer: Charlotte Rieger, Ron Chilcote and the Foundation for Sustainability and Innovation, Marci Riseman, Walt Christiansen, Ross Jackson, Hamish Stewart and Gaia Trust, Jerry Mander, Doug Thompson, Ralph Kratz, the Foundation for Deep Ecology, Patrick Kennedy, Pat and Graeme Welch, Diana and Arjun Divecha, and Marco Vangelisti. For editing that's indispensable — and I should think very difficult when an artist/activist like myself dutifully tries to write a book, something that's not necessarily natural to his inclinations and skills — thanks to Barbara Metzer for a big ever so thorough job and, in the second edition, to Ingrid Witvoet at New Society, and to my editor Michael, and to Kirstin Miller for many good suggestions and reflections. For the moral support, encouragement, and clear reflection over the years on ideas and issues that make up so much of this book, I have to give considerable credit to Nancy Lieblich, Bill Mastin, Susan Felter, Sylvia McLaughlin, and Huey Johnson. For particularly important help with Ecocity Builders, Urban Ecology, and Arcology Circle local projects and international conferences that provided grist for this book's mill, my great appreciation to Rus Adams, Ray Bruman, Jim Caid, Paul Downton, Chérie Hoyle, Joan Bokaer, Rusong Wang, Will Wright, and Joell Jones. My ecocity debate team members were mainly Ken Schneider, Thomas Berry, Brian Swimme, E.F. Schumacher, Hazel Henderson, Jerry Brown, Ernest Callenbach, Walt Anderson, Lynne Elizabeth, Joseph Epes Brown, and, providing crucial ideas and inspiration at a more remote distance, Jane Jacobs, Ian McHarg, E.O. Wilson, and — for the ecocity *is* the city to make peace on Earth and with Earth — Mahatma Gandhi. Special thanks to Paolo Soleri for his pivotal insights, his willingness to share them with me, and his courage to build.

And finally, the key link in getting any book out from the struggling writer's brain to the world at large: thanks so much to the publisher of the first edition, John Strohmeier, and his fine company, Berkeley Hills Books, and to Chris and Judith Plant and the other folks at New Society, one of those rare publishers with a powerful vision, whose books lead to a better world in action as well as contemplation.

FOREWORD

By Hazel Henderson

As Richard Register says, "the quality of life depends largely on how we build our cities. The higher the density and diversity of a city, the less dependent it is on motorized transport; and the fewer resources it requires, the less impact it has on nature."

This book is a treasure trove of such insights summarizing the role of cities in human evolution. Register reminds us that our cities are living systems that we shape and which then shape us, our lives, and even our evolutionary possibilities. As someone who has pondered the role of cities, as an environmentalist and a futurist, I must say that *Ecocities* is the most satisfying and comprehensive new volume on the subject. Register embraces every aspect of cities and the urban experience — advocating and showcasing the best visions, concepts, designs, and working solutions from all over the world.

As I read this deeply engrossing book, I thought of my childhood. I grew up in the seaside town of Clevedon, on the Bristol Channel in the west of England in the 1940s.

Nestled into limestone cliffs, most of its houses and public buildings were constructed of this same gray stone. Like most European towns, Clevedon was compact, with a density that allowed its some 4,000 inhabitants to walk and bicycle everywhere. There were few cars, plentiful and frequent bus services, and a rail line connecting Clevedon to Bristol, the nearest city, fifteen miles away. The air was clean, because not only were cars unnecessary, but also gasoline was rationed during World War II. Food was also rationed; so most people grew vegetables and kept hens in their gardens or on their "victory plots."

We walked or cycled to the pier and bought the day's fresh catch from the regular fishers. We also cycled to the nearby farms on the town's edge to buy our milk, butter, and produce. The High Street shops were stocked with local meats, including wild rabbit — a delicacy — and a wide array of local, seasonal fruits and vegetables. People shopped every day because most had no refrigerators. Like most children, we walked to our schools and ran errands for our parents and helped with the chores and gardening. My biggest thrills were going down into the basement room where we grew our mushrooms, marveling at how they appeared almost magically, and collecting the still-warm eggs from under the sitting hens. My mother grew our food without pesticides and volunteered at the local "cottage hospital" and its well-baby clinic. My three siblings and I were born at home with a midwife in attendance. I also experienced the air raids and bombs, the loss of friends, seeing the night skies ablaze as the city of Bristol was bombarded. Yet my overall experience of life in Clevedon was richly tactile, with close relationships and a cohesive community.

It is hard to believe that all this was only fifty years ago. I believe, along with many other systems thinkers like Richard Register, that industrial societies are in a classic "overshoot." We have overshot the optimum in cars, suburbs, and sprawl and their attendant patterns of energy waste, pollution, and environmental destruction. We have overshot the mark in losing community and identity among thousands of acres of huge tract homes in former family farms — with even more demand for more roads, concrete, parking lots, and strip malls.

Fast-forward to the 1960s. I found myself living in a high-rise apartment in midtown Manhattan. I studied these overshoots already in progress, hollowing out the city's residen-tial boroughs and businesses. New York is one of the most complex and challenging cities in the world. I was excited by the constant activity, the contrasts. The density allowed one to live car-less, but the acres of squalid housing, the noise, and the polluted air appalled me. My response was civic engagement and environmental activism in co-founding Citizens for Clean Air in 1964. This accelerated my education and led me to the works of Lewis Mumford, Buckminster Fuller, Ian McHarg, and Jane Jacobs and to my association with E.F. Schumacher who wrote the Foreword to my first book, *Creating Alternative Futures*. I became associated with Fritjof Capra through contributing to his book, *The Turning Point,* my ideas about how faulty economics was causing many of these "overshoots," which are now leading to global climate change. I joined the Lindisfarne Association founded by meta-historian William Irwin Thompson, together with Paul Hawken, Amory and Hunter Lovins, Stewart Brand, Lynn Margolis, James Lovelock, John and Nancy Todd — many of them cited in *Ecocities.*

Many years later, I visited Paulo Soleri's Arcosanti, which is described in *Ecocities.* I absorbed Soleri's philosophy and experienced his vision of "arcologies" as an expression of humanity evolving consciously toward Pierre Teilhard de Chardin's "noosphere." Yet, as Register points out, we humans are slow learners. Solari's Arcosanti, a brave and important experiment in urban design costing a fraction of that of an aircraft carrier, has never received any funding from business, government, or foundations — and survives by selling its cast bronze bells and wind chimes.

I ponder that we humans have spent 98 percent of our collective history together as hunter-gatherers in roving bands. Yet we now comprise a six billion-person human family, living largely in huge mega cities, like Sao Paulo, Mexico City, Shanghai, and Tokyo, with very little experience of managing our affairs at such a scale. We are consuming some 25 percent of the entire planet's primary biomass production and 40 percent of its land-based biomass production. This is accelerating the rate of extinction of our fellow species on which we are dependent, as we have migrated to the ends of this Earth.

No one has been tracking all these dilemmas as they relate to cities better than Richard Register. *Ecocities* recognizes all these dilemmas and opportunities and the new realities of the 21st century, from rising atmospheric carbon dioxide levels, shrinking water tables, and loss of agricultural land to sprawl — and the energy-wasting dead end of the automobile/highway/fossil-fueled industrial complex. Yet Register also richly describes and brilliantly advocates steps of the transition to sustainable communities and cities, which we must make if we humans are to survive. I will enjoy and learn from this passionate book for years to come.

Hazel Henderson
St. Augustine, Florida
August 2001

Preface to the Second Edition

WHAT A DIFFERENCE THREE YEARS MAKE! How does our dream of literally building a better future grow? Are we moving more rapidly toward ecologically healthy cities, towns, and villages — or in the opposite direction?

I thought the original *Ecocities*, the first edition, was about as positive an approach to meeting the challenges of our times as one could find. A book dealing with a fast-changing subject needs updating, but what has rather suddenly happened amounts to a more basic shift that calls for more than a few simple additions. Something profoundly disturbing is happening, coming at us from several directions at once, and, depending on the way we play it, the nature of our future as a species hangs in the balance. Suddenly everything is accelerating and a crisis of linked crises is getting close to the other side of the door. As positive as the ecocity vision is and as good as steps in that direction have been in many ways, there are growing threats against such a vision.

The first shock hit just as I completed the first edition of this book: the mind-boggling attacks of 9/11. Then followed, almost immediately, the invasion of Afghanistan and shortly thereafter that of Iraq. Could oil, without which our cities would grind to a halt, be involved?

Meantime, the consensus among climate scientists that the Earth was indeed warming became conspicuous beyond doubt when a planet's worth of people (minus a few in the White House) noticed that glaciers on every continent and mountain range were shrinking and disappearing, when a heat wave killed 35,000 people in Europe in the summer of 2003, and when this writer, for one, climbed up Mount Blanc in the Alps to see for himself and — yes, ladies and gentlemen, the glaciers there are melting away for sure. You have my word. I've been there. Checked it out. Took pictures.

Something else has been creeping up on us for years, noticed merely by those who pay attention to such things, something which has only in the last year become known much more broadly, clustered around the phenomenon embracing the whole Earth called "peak oil." That's the situation in which oil hits peak production and then, as time goes by, inevitably declines — forever. It's a serious concern when the resource that begins declining is the one that holds our cities' transportation systems together, heats and cools our buildings, provides fertilizers, insecticides, herbicides, and tractor fuel for our food production, makes fibers for our clothes and plastics for thousands of uses, and paves our streets and covers the roofs of our homes and workplaces with asphalt shingles and roofing tar. Transportation, indoor climate, food, clothing, shelter — practically everything depends on oil.

Oil is a finite resource, so there is no debate that this crucial commodity will peak and then decline — the only debate is as to exactly when we can expect it to do so. The members of the Association for the Study of Peak Oil and the sudden swarm of authors writing books on peak oil in the last two years are largely retired oil geologists and associated scientists, people no longer on the payrolls of the oil companies. All of them are expecting peak oil to arrive long before alternatives are sufficiently available to prevent a major economic, human, and ecological crisis.

More to the point, especially regarding neglected solar and wind energy, which together constitute barely one thousandth of the energy used by Americans, the debate is about whether we will have anything remotely as convenient and powerful in place to continue propping up our gigantic human population and enormous levels of consumption when the supply of oil begins sliding away. Quantity will go down and price will go up — forever.

What that means in terms of the economy, the ability to make products of raw resources, and the continuation of life systems on the planet as a whole became suddenly a headline concern. *National Geographic* stepped right into the middle of these controversial issues in 2004 with a cover article on the end of oil and another on climate change — and its editor said he knew they were going to create controversy and lose some subscribers. But he felt compelled to tackle these most important survival issues as a matter of the needs of the times. "I can live with some cancelled subscriptions," said Bill Allen. "I'd have a harder time looking at myself in the mirror if I didn't bring you the biggest story in geography today."[1]

"Forever" is one of those big words that casts it ominous shadow when we notice it lurking behind other words, like "extinction." That's right — that an animal or plant species is not going to become rare or confined to some distant place you and I might visit someday. It will be gone from the planet, and nothing will ever be able to bring it back. Your kids can skip the idea of seeing it, learning from it, enjoying it, maybe eating it, or utilizing its potential products or healing properties — nothing. In addition, if that species was a major player in its ecological niche, the environment it was part of will, because of its absence, also be gone forever.

The truly disturbing thing about peak oil is that it appears likely to come much sooner than the oil companies say they expect. They have been exaggerating their reserves to maintain investor confidence, which means they are claiming we have more time to solve the problem than is likely to actually be available. When the cheap energy just isn't there anymore, what might that imply? In this book I am promoting the extremely low-energy city, a redesigned city that can function on very little energy. But the hour is getting late. It will take energy to build such a city, and, I have begun realizing in the last three years, the investment has not been made.

Here's another sobering thought. Since finishing the first edition of *Ecocities* I have been hosted in North America, Asia, Africa, and Europe by dozens of organizations full of caring people. I have been to China four times to witness the world's largest experiment in city building first hand, and I can tell you precious little progress on ecocity development has transpired anywhere while enormous "progress" has been made on behalf of the car/sprawl/freeway/cheap energy way of building. Automobile factories are being rapidly constructed along with thousands of miles of highway in China while Americans are buying ever more anachronistically, bizarrely, *insanely* large cars. The world is becoming more dependent on cheap fuel, not less. *Much* more, every year, and just when we should know better.

Regarding cars, only in the last few months have I noticed the obvious, which seems to have escaped practically everyone else as well: only one out of ten of us drives one. There are more than 600 million cars on the planet and 6,500 million people. That means nine out of ten are disenfranchised from the dream of mobility — making it an impossible dream and a vast social injustice. In fact, since those one out of ten who drive are causing most of the planet-wide catastrophes of the day and, via climate change and extinctions, disasters to last millions of years, the dream turns into a genuine nightmare come true. One could find nothing more destructive of human opportunity and environmental health. Period. The energy-efficient car, you begin to realize as you look such numbers in the eye and remember that the supply of cheap energy is ending starting right now, is a delusional attempt at a remedy. The Prius — the energy-conserving automotive darling of many American environmentalists who are hanging on to the hope that they can continue driving forever — will not save us. It will only convince a few well-meaning people to delay longer finally dealing with reshaping our built environment — cities, towns, and villages — for human beings, not cars. But there is no time to delay solutions; time is running out. Forever starts today.

And not just ecologically speaking, because the finality of that "forever — nevermore" concept is popping up in the cultural context as well. Jane Jacobs writes *Dark Age Ahead*[2], chronicling the incivility of a general and growing malaise seemingly unrelated to ecology, with failing families and schools credentialing instead of teaching, subverting cultural contributions instead of encouraging them. The worst thing about cultures slipping into their "dark ages," she says, is that they forget and the contributions and lessons of that culture disappear utterly. Jared Diamond writes *Collapse*[2], comparing ancient and recent collapses from the central Pacific islands to the Soviet Union, describing the interrelated causes of collapse and the means by which some cultures survive, which, when they do survive, always include a strong element of respect for ecology and resources — otherwise they are not on the list of survivors.

Societies develop cultural blind spots to ecological realities, spots so blind as to baffle people from outside of the culture looking in. How could the Easter Islanders in the most remote corner of the ocean cut down their last tree and with it eliminate the fuel for their fires and the building materials for their houses on a cold, windy island? How could

they do that, thereby making boat building impossible and eliminating their main means to stay warm, cook, harvest food, and connect with the outside world? How could we be blind to something so gigantic and obvious as the sprawling city of cars with its direct impact on climate, war, and the "cutting down" of the last "tree" of finite oil? How could the people of the world's most oil-addicted country have a President, a Vice President, and a Secretary of State who are all oil company executives and believe that conquering the second largest oil field in the world has nothing to do with oil? Oil, that's the word that never passes government lips except in rare disclaimers that it has anything to do with anything. The people don't want to see and the press shuts its eyes and closes its mouth, too. If Easter Islanders look insane to us on the outside, how about us? They could at least dream of replanting tree seeds, rescued by some wise elder or upstart genius appearing like magic among them in an ancient version of the "techno-fix." But oil spawns no seeds.

Such cultural blind spots to ecological and resource realities destroy civilizations, with enormously negative impacts that are most unpleasant to contemplate. Collapse, whether faltering and episodic or a freefall into cannibalism, makes good reading regarding somebody else's culture, but regarding our own?

Despite these dark words made necessary by the events of the last three years, this is a

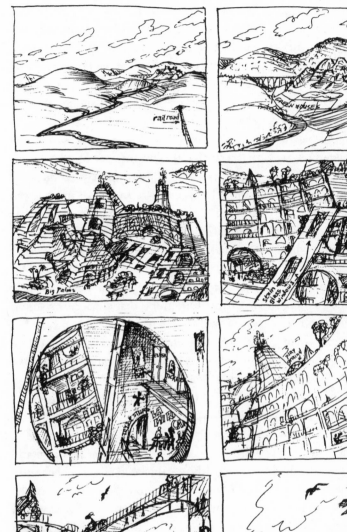

Arcology zoom-through.
My first eye-opening "ecocity" experience was meeting Paolo Soleri, who posited the more three-dimensional, car-free and ecologically healthy city, something like the Indian pueblos of the American Southwest, but updated to modern times. I tried various ways of visualizing these very tall, small-footprint cities, such as this 1970 drawing in which I imagined flying close by one of his cities.

book of many hopeful, practical and, I think you will agree, inspiring concepts and examples. It brims with tools for healthy change and lays out basic principles for ecological designs of cities, towns, and villages. You've no doubt already noticed the many illustrations attesting to the great potential we have before us. But it would be irresponsible to ignore the sudden arrival of bad news or, at least, news that should sound the alarm. We need a wake-up call. Maybe enough people to make a difference are beginning to hear it right now. Maybe that will make us ready — at long last — to think seriously about what we build and how we live in it.

In any case, our environmental awareness is rising (where it hasn't broken the Rapture or Jihad barrier among fundamentalist religions), and generally, major efforts at improvement are afoot. With the possible exception of overfishing and only partially restrained logging, we no longer hunt or harvest species to extermination. We've learned that much about maintaining our biological capital. But what we haven't noticed in the meantime is that something else has quietly raised its gigantically destructive head in the background like a dark cloud — and that 800-pound gorilla nobody is talking about, sitting on the couch over by the corner, is the city itself or, more particularly, the city in the form of sprawl, cars, asphalt, and cheap energy (but it won't be cheap for long). Well, we are the first true Whole Earth Civilization

(with a few pockets of energetic dissent), so it is only natural that there might appear a wholly new reality not recognized or faced by past civilizations. Because we face something new and unprecedented, we have some excuses for a slow wake up — though they are growing thinner every day.

It's coming back to me now, something that clarifies the parameters of this book's content; one of those important thoughts I heard a long time ago. In 1971 I interviewed scientist Aden Meinel, who had recently directed the construction of Kitt Peak National Astronomical Observatory. I went to Arizona to cover his solar energy story for *West Magazine*, the *Los Angeles Times'* Sunday magazine. One thing that stuck in my mind was his warning that we'd need to spend a fair fraction of our fossil fuel endowment putting into place a renewable energy system. We couldn't wait until oil became expensive. We'd need too much of it. His favorite solar "system" was something he liked to call a solar energy "farm," with reflectors to send light to a central boiler to generate steam and turn a turbine, then an electric generator. Another notion was to collect the heat of the sun under panels with big lenses that would magnify the sun's light, to heat to a very high temperature a liquid material like sodium, to pump that to the central plant, and then to produce electricity with the steam boiler, the turbine, and the generator. In any case, he said, we'd have to melt lots of glass and build a massive infrastructure for the renewable energy system because we, not the geology of the Earth over the one hundred to two hundred million years it took to make the fossil fuels, would need to gather and concentrate the energy ourselves. That year, 1971, was coincidentally the same year oil production peaked in the United States, though I had no idea about it at the time.

It is now 35 years later — and no such investment in renewable energy has been made. That's not all. The physical infrastructure of cities, towns, and villages, designed to fit renewable energy systems, has also not been built. It's not even being planned, and only visualized in a few rare places such as this book. The ecocity, if dawning, is still way out over the horizon. We've waited so long I'm beginning to wonder if we will have enough time to build a sustainable, much less a vitally healthy and inspiring civilization at all. But if we do, it will look a lot more like what's offered in this book than what you will see by looking around Yourtown today.

The last three years have been a splash of cold water in the face for me. I'm hoping this book will help wake us all up to not cutting our last tree and building very carefully something that will last. The three years since writing the first edition have emphasized to me just how rare and important the insights for a better future that you will find in these pages really are. Maybe ecocities will be our last — and best — chance.

INTRODUCTION

CITIES ARE BY FAR THE LARGEST CREATIONS of humanity. Designing, building, and operating them has the greatest destructive impact on nature of any human activity. As we are constructing them today, cities also do little for social justice, not to mention the grace and subtlety of human intercourse. Yet our built communities, from village and town to city and megalopolis, also shelter and launch many of our most creative collaborations and cultural adventures, arts and artifacts. When we build the automobile/ sprawl infrastructure, we create a radically different social and ecological reality than if we build closely-knit communities for pedestrians. Contrast American sprawl with traditional European cities. We will go way beyond that comparison soon enough — far enough, in fact, to demonstrate that cities can actually build soils, cultivate biodiversity, restore lands and waters, and make a net gain for the ecological health of the Earth.

Ecocities proposes a fundamentally new approach to building and living in cities, towns, and villages, an approach based on solid principles from deep history and an honest assessment of a troubled future. Our prescription will be replete with examples of wonderful architecture, transit solutions, and new approaches to public and natural open spaces. We'll explore dozens of practical tools for healthy urban transformation, and invent a few.

Given that cities are so large, damaging, and yet potentially beneficial, you'd think we would have long ago devised the science, study, discipline, and art of ecologically healthy city building. Why aren't people trying to systematically think it through and develop it? Well, they are — it's you and I right here in these pages. Welcome to the new frontier in home building.

Philip Shabekoff, long-time environmental writer for the *New York Times,* interviewed hundreds of environmentalists and one of their most common themes, he reports,[1] was this: how is it we can be winning so many battles, yet regarding the really big things we are

1

losing the war? Species extinctions, global warming, climate change, soil loss, the collapse of ocean ecologies, and on and on — all are getting worse every day.

It's no mystery to me. We've never engaged the big battles. We try to make cars better rather than greatly reduce their numbers. We try to slow sprawl development rather than reverse its growth and shrink its footprint. We keep making freeways wider and longer, dreaming of "intelligent highways" rather than removing lanes and replacing them with rails, small country roads, and bicycle paths. We continue to provide virtually every subsidy and support policy the oil companies want. It's no wonder that we're not winning the war. The objective of this book is to lay out an evolving strategy that faces the big problems head on and gives us at least a chance of winning.

Of Cars and Dinosaurs

I'll start my story in the age of dinosaurs when our ancestors, the first mammals, were darting about between thundering footfalls trying to avoid the fangs and claws of those gigantic monsters, the thought of which takes our breath away even at the safe distance of 65 million years. Those furry, warm-blooded creatures managed to survive by staying just below the size of the smallest dinosaur *hors d'oeuvre*, and there they waited, frozen for millions of years in their diminutive forms. Then, in an instant — wham! — the world was slammed into another reality, and a new stage of evolution burst forth in a swirl of asteroid or comet dust. Mammals were given the opportunity to grow, diversify, learn, build, think collectively, here on the Sun's third planet. The oppressor was gone and freedom burst upon the evolution of consciousness like a sunrise.

Cars are the dinosaurs of our time. They are destroying the reasonable and happy structure of cities, towns, and villages. Once communities have been shaped for cars, they remain dependent upon them. Sprawled communities can't function without their speed. Their addiction is a structural addiction built into the physical structure of the

Car city vs. compact city. How to save land, energy and time; in three sequential illustrations.

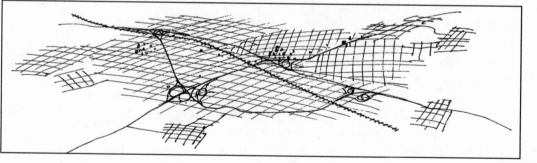

city. Cars are preventing the next step in our now cultural evolution, the step in which we build as if we knew we were evolving.

For the past 40 years I have been an activist, trying to persuade people to add some sane features to their houses, neighborhoods, and cities, to change larger policies, and to adopt tools and techniques for making cities more ecologically healthy. It has been so difficult that I sometimes think it will take a truly colossal disaster to wake us up. This is hardly a unique thought; many people volunteer the notion to me repeatedly. I think that disaster is likely to be the same one that banished the dinosaurs: climate change. This time the cause won't be a wandering asteroid, with instant concussion and fire followed by a dark and icy nuclear winter under dense clouds that turn day into night for weeks, months, or years. This time it's going to be the human-created heating of the planet, the spreading of deserts, the flooding of low lands, and the fury of super storms. The evolutionary leap 65 million years ago was the bursting forth of the mammals. If the fear of the prospect, instead of the thing itself, motivates us, the next opportunity could be humanity collectively learning how to build a healthy future. Inescapably, that means building ecologically healthy cities — ecocities. The difference between doing it now and doing it later is the difference between saving most of what's left of biodiversity — that is, life on Earth and our collective souls — or

losing both, leaving to our descendants a deeply impoverished planet.

I see this book as a kind of overview of something that exists, so far, only in small pieces scattered here and there around the world, throughout history, and in the imaginations of creative thinkers. Some people are becoming very sophisticated about the principles of ecocity building, and I want you to meet them and see what they've been doing. Many more are building systems new and traditional that can fit elegantly into the ecocity, though they may not yet be aware of the potential of that larger context, and I want to suggest how we might tap that potential.

We'll start off looking briefly at cities in their most basic form, with some attention to the rich detail that can confound or contribute to a healthy relationship with nature. Then we'll consider where the city fits in its largest of all contexts, the evolution of matter,

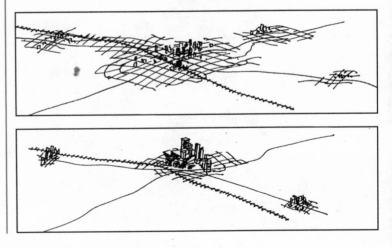

life, and consciousness in the universe. Next we will look at the city in nature and in history, and after that we'll examine the city as it seems to be functioning and malfunctioning right now. Finally, we will explore the direct approach to building ecocities, beginning with basic principles and then moving on to dozens of ideas and projects that have great promise for the task. Throughout, doggedly adhering to the old saw that a picture is worth a thousand words, I'll be making sure that you have enough imagery before you that you can't possibly digest it in one sitting. I want you to come away from the experience saying, "Wow, can we really build something like that?" or "I hadn't thought of that before!" or "That's a place I could enjoy living in!" or "Where did all that life come swarming in from? Does life just appear when you build to make room for it?" That it does.

Elevated plaza with view to ocean and mouth of a stream.

CHAPTER 1

As We Build, So Shall We Live

A S WE BUILD, SO SHALL WE LIVE. The city, town, or village — this arrangement of buildings, streets, vehicles, and planned landscapes that serves as home — organizes our resources and technologies and shapes our forms of expression. It is the key to a healthy evolution of our species and will determine the fate of countless other species as well. The city, in fact, is the cornerstone of the civilization that currently embraces the entire planet. Insofar as our civilization has gone awry, especially with regard to its impacts on the environment, a very large share of the problem can be traced to its physical foundations. Given the crisis state of life systems on Earth — the collapse of whole habitats and the increasing rates of extinction of species — it follows that cities need to be radically reshaped. Cities need to be rebuilt from their roots in the soil, from their con-crete and steel foundations on up. They need to be reorganized and rebuilt upon ecological principles.

Yet most people believe instead that fine-tuning the same old civilization will be enough, that there is no real need for funda-mental change in the way we build and live. But simply failing to notice the crisis in the built environment won't make it go away. Only rebuilding will. Building the ecocity will create a new cultural and economic life in which we can tackle the problems of healthy evolution rather than fighting a rear-guard action aimed at repairing the damage. While it probably won't rid the world of greed, ethnocentrism, and violence, building a nonviolent city that respects other life forms and celebrates human creativity and diversity is consistent with solving those problems.

Cities need to be radically reshaped. Cities need to be rebuilt from their roots in the soil, from their concrete and steel founda-tions on up. They need to be reorgan-ized and rebuilt upon ecological principles.

5

What we build creates possibilities for, and limits on, the way we live. What we build teaches those who live in the city, town, or village about our values and concerns. It says, "This is the way it should be," or at least, "This is the best we could do." The edifice edifies. Children in today's typical car-dominated cities learn that cars are valued so highly that it is worth risking human life and enduring high costs and serious pollution to make way for them. They also learn that people don't care much for public life or nature. In Berkeley, California, where I lived for 30 years until recently, there are no public plazas or pedestrian streets, though you can be sure there are thousands of parking places and cars on every street. Creeks are buried for miles, and the ridgelines from which we could once enjoy the sunset and a view of San Francisco Bay are increasingly being filled with private houses that banish public access. In ways like this each city tends to reproduce in its children the values embodied in its form and expressed in its functioning.

More to the point, if our cities are built for cars, one sixth of us will find jobs in that industry and its support systems, and another sixth will be building and fixing the buildings and infrastructure that go along with the layout automobiles require. But if we build the ecocity, large numbers of people will find jobs involving its building and operation. By shifting steadily toward an ecocity infrastructure we could soon train people to be streetcar and bicycle builders and mechanics, organic gardeners, restorationists, naturalists, "green" designers and builders, and pedal-powered delivery people, all with a minimal impact on nature. As we build, so shall we live.

Even more, as we live, so shall we become. And not only in attitudes, skills, and habits, but also, eventually, in physical form. As Americans spend more and more time sitting motionless before television and computer screens and in ever fatter SUVs, they are becoming increasingly overweight and unhealthy. Another way of looking at this is to see that we build environments that help build us. We are indirectly self-designing in a very significant way, turning ourselves into a species that reinforces its own design by building its environment in particular ways. In ecology and evolutionary biology, we have learned that species shape one another physically and behaviorally. Pollinating insects and birds adjusted their proboscises and beaks to the task, and flowers shaped themselves to cooperate. The birds were fed. The flowers were pollinated. Some birds' nests have evolved over thousands of years, affecting the birds' behavior and even body shape. All creatures respond to and alter their habitats, climate, and even atmospheric and soil chemistry, thus altering themselves over the years. If we want to live an ecologically healthy and responsible life, we need to build so that we can lead such a life.

Each city tends to reproduce in its children the values embodied in its form and expressed in its functioning.

What We Seem to Be Building

Since 1991 I have traveled several times to all the continents but Antarctica talking about ecocities. I am one of a small number of people on this loosely linked lecture circuit who are thinking about redesigning and building whole cities on ecological principles. My experience suggests that a significant and growing number of people around the world are beginning to take a strong interest in urban form and related dysfunction and in the ecologically healthy alternatives that ecocities offer. Cities such as Vancouver, British Columbia; Portland, Oregon; Curitiba, Brazil; and Waitakere, New Zealand, are making ecological progress on a number of fronts, and the international ecovillage movement is steadily growing. We are seeing good work, and that is significant and heartening. But society is still moving overwhelmingly in the opposite direction.

For my 30 years in Berkeley, for example, I had been trying to help reshape the city. With others I cofounded two nonprofit organizations — Urban Ecology in 1975 and Ecocity Builders in 1992 — and these groups have managed to have portions of creeks that had been buried for decades opened, to have a street redesigned as a "Slow Street," to have a bus line established, to have an ordinance written making attached solar greenhouses legal in front yards, to build a few of these greenhouses, to plant and harvest street fruit trees, to have energy-saving ordinances developed, to delay freeway construction, to tear up parking lots to plant gardens and urban orchards, and to affect the course of particular development projects by pointing out their impacts on the city's ecological health. These and other projects have not only created physical features and functions and established laws, but stimulated discussion of these issues, as well.

It has gradually become clear, however, that few people notice. New one-story buildings are being built in the downtown, where taller buildings would place transit and passengers in mutually supportive proximity. Transit service itself is being cut and fares are rising. The neighborhood around one of the major transit stations has been rezoned for reduced development, when a higher-density "transit village" would have helped housing, transit, energy conservation, pollution abatement, and the economy. Large freeway-oriented projects and big parking lots have been built as an outcome of a planning process with considerable public support and input — many people wanted it that way. These projects and parking lots, not only in Berkeley but all over the region, encouraged the widening of Interstate 80 from eight to ten crowded lanes. The University of California built several very large buildings and parking lots in that period with little regard to its surrounding urban and natural environment.

At the national level, despite ever more intense verbal attacks on sprawl, the big news early in the first decade of the third millennium is bigger sport utility vehicles and capitulations to ever more highways, parking lots, bridges, and other car infrastructure. For example, in 1960 one-third of the citizens of the United States lived in cities, one-third in suburbs, and one-third in rural locations. By 1990, well over half lived in suburbs. Between 1970 and 1990, the population of California increased approximately 40 percent while the land area of cities and suburbs went up 100 percent. Between those years the country witnessed what has sometimes been called "the second suburbanization of America," in which, instead of commuting daily from suburb to city center, tens of millions of people began traveling from suburban house to suburban workplace. It's not uncommon for spouses to work thirty to sixty miles apart while their children attend school at the third corner of a geographical triangle, encompassing hundreds of square miles.

Since 1970, giant suburban developments have popped up on a scale and in numbers barely imaginable before. Industrial and "back office" business parks have appeared on farm, range, forest, and filled marsh lands miles from any other development. These office and commercial zones are surrounded by acres of land converted to dead-level asphalt and concrete slabs, sweltering in the summer, pouring car-contaminated waters into the creeks, rivers, and bays when it rains, and draining rubber dust and oily, salty, sooty snowmelt in winter. Privately owned malls, accessible virtually only by car, have largely replaced town and neighborhood centers, shifting social spaces from public to private control. New communities of hundreds, even thousands of families are hiding behind walls and guard posts, "forting up," in the words of the real estate pages, abandoning the inner cities and physically distancing themselves from social accountability. They are attaining this social isolation through the physical isolation of sprawl, cars, asphalt, gasoline, and concrete block walls eight to eighteen feet high. Only expected guests and the electromagnetic waves and wires of radio, television, computers, and telephones dare enter.

In 1972, before the oil embargo and the subsequent energy crisis, cars consumed on the average about 30 percent more energy per mile than they did 15 years later. About a third of the United States' oil was coming from the Middle East. By 1992, after the new wave of suburbanization, the United States was getting approximately 60 percent of its oil from the Middle East. The less fuel it takes to drive about and the cheaper per mile it is, the farther people are willing to drive. The better the mileage, the more the suburbs sprawl out over vast landscapes, the more demand there is for cars and freeways, the more cars are needed to service expanding suburbia, and, ultimately and ironically, the more gasoline is needed. Thus the energy-efficient car creates the energy-inefficient city, the "better" car the worse the city. The car is part of a whole system of complex, necessarily interconnecting parts existing in an interdependent relationship with the total environment it helps create.

The bigger picture — represented by the total commuting time for large populations, the world use of fossil fuels, ocean tanker oil spills, wars for oil in the Middle East, the waste of investment capital in building infrastructure that will go on damaging the world for many decades, and so on — is far from encouraging. China started closing Beijing's streets to bicycles to make way for cars in 1998, and it is currently engaged in a massive highway-building program. It plans enormous shifts of population from rural areas and farming to cities and manufacturing and business, and shifts from rail, bicycle, and

Good place for a freeway. *This drawing I took as something of a joke until ten years later, when I learned the freeway on the far side of the Danube from downtown Vienna was buried for more than two miles — with above-ground air ventilation boxes of identical design.*

pedestrian cities to cities for motor vehicles on rubber tires — a colossal transformation in the wrong direction. In Brazil, Turkey, India, Africa, and Australia, large highway projects are being built as they all emulate America's destructive example. People in these places say quite directly and from their point of view completely reasonably, "You Americans have cars — who are you to say we shouldn't? Now it's our turn."

Getting down to Basics

To move forward from this point, we need to look at the whole system of which we are part, rather than at one small part at a time. Theologian Thomas Berry gets us off on the right track:

> Unaware of what we have done or its order of magnitude, we have thought our achievements to be of enormous benefit for the human process, but we now find that by disturbing the biosystems of the planet at the most basic level of their functioning we have endangered all that makes the planet Earth a suitable place for the integral development of human life itself.
>
> Our problems are primarily problems of macrophase biology ... the integral functioning of the entire complex of biosystems of the planet ....[1]

The complex biosystem that Berry speaks of is also known as the biosphere. It is our home we are wrecking with our "plundering industrial life patterns," and the major engine of our civilization's shortsighted exploitation and destruction is the sprawled city with its flood of traffic, its thirst for fuels, and its vast networks of concrete and asphalt. The car/sprawl/freeway/oil complex reproduces more of itself — a peculiar kind of economic "vitality" — while it paves agricultural and natural land, kills a half million people outright in accidents every year, injures more than ten million, many permanently, and is completely destroying a reserve of complex chemicals that took 150 million years to create, the so-called fossil fuels which should be seen as a fossil chemical treasure chest, an inheritance of the species and a responsibility for carefully restrained use into the deep future. This infrastructure also consumes enormous quantities of steel and energy; in a 1999 television ad, the Ford Motor Company asserted that it uses enough steel to build 700 Eiffel Towers every year. Seven hundred Eiffel Towers of steel racing about the countryside powered by flame, requiring the earth to be paved, and transforming the atmosphere! And that's only one of a dozen major automobile manufacturers.

Today, we have more than 600 million cars worldwide, and we transform into sprawl development several million acres of land every year, removing it from agriculture, nature, and more pedestrian-oriented traditional towns and villages. Already, in 2004, barely providing a sip for the thirst of American cars for fuel, ten million acres of corn were taken from feeding people to feeding cars.[2]

The car/sprawl/freeway/oil complex is destroying habitat, and directly and indirectly animals and plants through road kill, noise disruptions, poisonous water runoff, contaminated air, and climate change. Repetitive small car-dependent buildings scattered over vast areas, often made of wood from shrinking forests, not only require enormous quantities of gasoline to maintain, but share

walls with no one and lose their energy from heating or cooling to the surrounding air after a single use. Thus scattered, small-building development wastes energy not only for transportation but also for space heating and air conditioning. This four-headed monster of the twentieth-century Apocalypse — cars, sprawl, freeways, and oil — is what we are building and in what we are committing the next generation to live. It is what is largely defining our jobs and many other life activities. It commits us to one of the most expensive and dangerous common activities

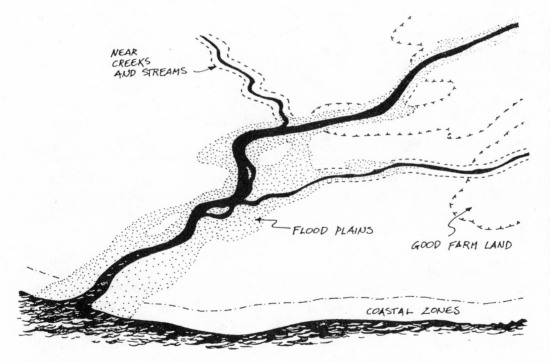

NEAR CREEKS AND STREAMS

FLOOD PLAINS

GOOD FARM LAND

COASTAL ZONES

Where not to build. *Land uses are the foundation of city design. Avoid flood plains and best farm land, celebrate coastal and waterside with special features. Maintain maximum permeability of surfaces, and to do this, the compact city is the most important solution.*

allowed, namely driving. The car/sprawl/ freeway/oil infrastructure has enormous arms in the form of shipping routes and pipelines, both subject to accidental disasters, and in the form of military forces that maintain constant pressure and occasionally go to war to keep petroleum cheap and flowing.

Because this monster is a "whole system" structure, we can effectively attack it by taking on any one of its four main components. Working against cars, sprawl development, freeways and paving, or oil dependence will help bring down the whole destructive edifice. Even better, we know that if we provide positive alternatives to any of these components, we will be starving the system and nurturing another "whole system" creation: the pedestrian/three-dimensional system. We will be building a whole new infrastructure for a new civilization.

Paul and Anne Ehrlich[3] have suggested summarizing the situation in a simple formula: Humanity's impact (I) on the world's environment is roughly equal to population (P) multiplied by affluence (A) multiplied by technology (T): $I = PAT$. The baseline is population, which for human beings on Earth is now around 100 times larger than what is within the Earth's carrying capacity should we be hunter gathers still scouring the landscape like other animals.[4] But of course we have agriculture and industry, wealth and tools to support a far larger population than would otherwise be possible. A large number of peo- ple with low consumption and benign technologies might have the same impact as a far smaller number of people with higher affluence and/or people with more damaging technologies. These three variables give us a way of thinking about the dynamics of damage to the Earth and its resources and life systems.

However, something is missing from the formula: the anatomy of the built community. Land use and infrastructure, understood as a single created item or human "artifact," constitute the home to the population, the engine of most of the affluence, and shelter and organize the technologies so that they can function. Unless we take into account the city's physical structure and organization, we won't be able to solve the all-too-physical problems of a disintegrating biosphere and a rapidly degrading resource base for human sustenance. We can see cities, towns, and villages as whole, functioning, potentially-healthy living systems, or we can get lost in the details as Thomas Berry warns us we may. The manner in which the city is laid out and organized is the foundation for virtually everything else. Without understanding urban anatomy we will fail to understand how population, affluence, and technology relate to each other and to the environment. Impact (I) = population (P) times land use and infrastructure (L) times affluence (A) times technology (T): $I=PLAT$. One last refinement: "impact" is usually understood as negative. "Effect" (E) might be a better, more neutral word, but we could even imagine a benefit (B) arising from the healthy interrelation of P, L, A, and T.

The urban planner Kenneth Schneider adds depth to this line of reasoning by saying, rather straightforwardly, that dealing with only the major issues rather than the entire system, can preclude attaining the desired results.

"Despite the statements of many urbanists and environmentalists to the contrary, the central issues are not clean air and water, endangered species or environments, more money for housing and urban renewal, or even energy; certainly not in their separate capacities. These issues are relevant, perhaps necessary, but not basic. What is basic is the structure of the human environment, the city. Building a good city — a framework for all separate things to work harmoniously — is essential in order to alleviate each of the separate issues of development." Considering separate concerns separately merely traps us into building an even more destructive environment.[5]

It helps to see that the city is a particular creation, entity, organism, apparatus, meta-machine — whatever the name, it is an artifact with certain characteristics and purposes. One of the clearest descriptions of the city's purpose comes from David Engwicht, an Australian activist and planning consultant. According to Engwicht, "the city is an invention for maximizing exchange and min-

Nature center.
A proposed nature center for Mills College on their revitalized small lake, with solar orientation and greenhouse, meeting space and glass wall for viewing the lake.

imizing travel." What is exchanged? Ideas, goods, food, money, friendship, hopes, fears, genes — virtually everything at the core of human culture and economy, genetics and evolution. The quality and the content of that exchange, as well as its influence on the environment in which the city is placed, is very largely determined by the physical form and arrangement of the city's parts. Seen in this way, the city is something like a ship for a journey into humanity's future. Better build it well!

A New Synthesis Architecture

At the European Eco Logical Architecture Congress in Stockholm and Helsinki in August 1992, there was much discussion about a number of imaginatively designed "ecological buildings." These buildings featured renewable energy systems, built-in recycling, non-toxic building materials, interior greenhouse planting, and occasional rooftop gardens. Many had effective — even beautiful — natural lighting. One, the NMB Bank Building in Amsterdam, featured dynamically sloped walls to bounce outside motor vehicle sound away from people inside and to reduce wind effects. One feature, spectacular in its smallness, was a thin cataract of fresh water running down grooves in banisters. But with the exception of my own ecocity slide show, Paolo Soleri's proposed cities, and some traditional village architecture from desert regions around the Mediterranean described by the French architect Jean Bouillot, each building stood like an independent entity, separate from the other functions of society and the economy. Some of Joaquim Eble's buildings in the six-story range, with tall solar greenhouses and rooftop sod and flowers, came close to real "whole city" designs in that they were at least located on major transit lines.

"Green buildings" have received increasing interest in the United States. In particular, those designed in the offices of Sim Van der Ryn, David Orr, and William McDonough — respectively the headquarters and display rooms of Real Goods in Hopland, California, the Center for Environmental Studies at Oberlin College in Oberlin, Ohio, and the Gap Headquarters at San Bruno, California — are attractive and energy-efficient, since they are all largely passively heated and cooled by sun and shade and use a number of low-energy, nontoxic, and recycled building materials and furnishings. The Real Goods Building in Hopland features straw bale worked into recycled wooden posts and beams in a particularly pleasant environment facing to the sun on the south. It has a grape arbor that provides shade in the summer when the leaves intercept most of the sunshine and, when the leaves fall off, lets light through in the winter to warm the building. It has a pond and a constructed-marsh waste system full of plants and buzzing with dragonflies. But these American buildings called "green," "healthy," or "ecological" are only one story high (Real Goods) or two (Oberlin's Environmental Studies Center and the Gap headquarters). Thus they address neither higher-density urban problems nor transit and pedestrian solutions as do their taller European cousins among the Eco Logical architects' buildings. A more fundamental problem with these American "green" buildings is that they do not physically knit together the life of the community. If ecology is not about the individual organism in its community of other living things, what is it about? If ecological buildings are not about their relationship to the rest of their community, what are they about? To date, in the case of these buildings, the answer to the question "Where's the community?" is "The parking lot — that's where you go to connect with it."

In Stockholm and Helsinki it occurred to me that ancient cities, Çatalhöyük in today's Turkey and Jericho in Israel, the Minoan and Mycenean cities on the island of Crete, the pueblos of the American Southwest, the "lost city" of the Tairona in Columbia, and contemporary towns with ancient roots like Shaban, Yemen, and others across the Kasbah belt from Morocco to Afghanistan, have reflected and served a synthesis of ecological and social needs. This old synthesis architecture saw the building primarily as part of a whole community embedded in the natural environment. Enthusiastically praised by Bernard Rudofsky in *Architecture without Architects* [6], it was as much city planning as architecture. The architect Vakhtang Davitaia from Tbilisi, Georgia, has described an art of building design centered on social gathering places, celebrating streets, squares, public monuments, and schools, in which towns, as if growing out of the earth, assume the same skyward-aspiring form as the mountains of his home country: "Nothing could be farther from our philosophy than the English and American statement 'A man's house is his castle.' We are a sociable people." [7]

Five hundred years ago, the European town was a cluster of buildings two to five stories high, many of them built up against or sharing walls with their next-door neighbors, with homes over stores and small handicraft shops, a walled section in the middle, a larger, fancier administrative building, probably flanked by imposing residences or perhaps a castle displaying fear of the outside and control within its bounds. Rising above it all was the church or cathedral spire, spear-like and ready to fling the imagination completely clear of the Earth on its way to heaven. Close in on all sides were farms and forests, the former populated with domestic animals in pens and fenced fields, the latter partially logged and cleared but, connected by a few small roads to the town, an obvious source of sustenance and a pleasant green sparkle to the eye. This picture says something about the people and their relations to one another, to nature, and to the universe as they understood it.

If ecology is not about the individual organism in its community of other living things, what is it about? If ecological buildings are not about their relationship to the rest of their community, what are they about?

In the Southwestern pueblo of the same vintage, the clustered pattern of mixed living-and-working habitat was even more compact, as if the whole town were one building of many separate rooms three, four, or five stories high. Social order, perhaps cooperation or rigid hierarchy, appears to have had very high value in these close quarters and integrally designed community structures. The whole ensemble reveals an Earth-focused and season-conscious cosmology and daily life: ceremonial kivas were dug into the sacred earth. Openings in the architecture faced southeast to gather the sun's warmth in the morning and to cool in the hot afternoon. Gardens were small and variegated, indicating diverse production; the forest was partially cut, providing fuel for a few columns of smoke. Deer, antelope, and rabbits lived in the valley and on the dry plateaus. Here too we can clearly discern a number of signs of the people's beliefs, values, and life orientation.

In contrast, the present-day city has almost no comprehensible form at all. Look out over today's typical city and the tangle of streets and freeways disappears into a gray jumble of low-lying buildings and the yellow-brown-gray haze of auto emissions. Over this miasma rise scattered tall buildings, most of them owned by banks, insurance companies, and giant corporations. Towering even above these symbols of money, security, consumerism, and control, television transmission towers flash sci-fi red lights to ward off giant buzzing aluminum gnats full of jet-powered people. The towers are beaming images directly into the houses sprinkled widely in random patches about the landscape. Therein people worship at the altar of diversion from nature and a deeper confrontation with themselves: television. The sprawl is tethered almost metaphysically to the towers by invisible electronic waves in the air bounced off satellites rolling around in freezing space that, after evening twilight or before morning twilight, trace spooky silent paths between the stars, way up over the city. The whole scene is financed through the tall buildings occupied for a third of the day by thousands of people who spend the rest of their time scattered over hundreds of square miles or moving about that area in metal boxes on thick rubber doughnuts.

Though historically cities and towns may have started with a kind of old synthesis architecture bringing together the pieces of the community into a whole, they diverged and disintegrated outward across the Earth's surface, scattered by ever-increasing mobility. Cities provided the possibility of specialization and hierarchy on a small acreage, but as transport became more dominant, buildings and clusters of buildings became increasingly detached from the whole and from the natural environment. Buildings began serving their own ends, for their individual owners, in competition with rather than in support of the whole. This is not a completely bad idea.

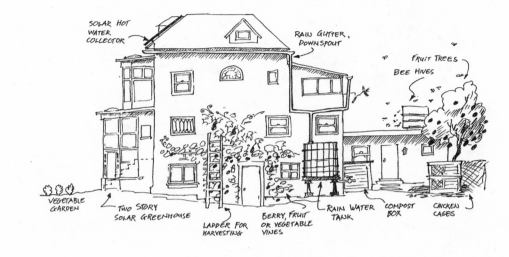

SOLAR HOT WATER COLLECTOR

RAIN GUTTER, DOWNSPOUT

FRUIT TREES
BEE HIVES

VEGETABLE GARDEN

TWO STORY SOLAR GREENHOUSE

LADDER FOR HARVESTING

BERRY, FRUIT OR VEGETABLE VINES

RAIN WATER TANK

COMPOST BOX

CHICKEN CAGES

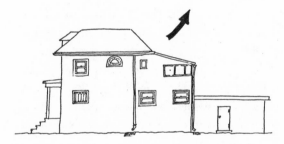

Solar greenhouse remodel-plus.

Turning a common two-story house in the San Francisco Bay Area into a four story house with a two-story greenhouse: Raise the house three feet. Dig out crawl space and turn into occupied basement. Raise roof two feet, making space for room in the attic. Add solar greenhouse, front yard vegetable garden, fruit trees and vines, compost box and water catchment system. Convert side garage into small apartment.

The districts of particular crafts, manufacture, or service intensified the pursuit of particular activities, furthering lively exchange and creativity. Personal expression in the design of buildings also contributed toward the same ends. At a certain scale, though, the spreading of the city meant that it took more and more time to get from one building or district to another. Ancient city builders were almost certainly aware of this and created compact development and narrow streets to address the issue. In a pedestrian world, the notion of access by proximity was undoubtedly obvious and to build in a scattered mode would have seemed strange and impractical. With the use of animals and carts for transportation, and later especially with cars, at some point a threshold

Low-energy vehicle
of the future

of dysfunction was crossed. Problems mounted more quickly than solutions were found, while — and this is worth emphasizing — the increasing distances made it more difficult to see this.

In recent times, architects have often designed their buildings for the clients' egos as well as their own, usually in an attempt to make a unique esthetic statement while creating solutions for buildings' functional problems and clients' needs. Not only does the design tend to disregard the whole, but often it intentionally contrasts with it. Frank Gehry's Guggenheim Museum in Bilbao, Spain, instantly and almost universally recognized as a masterpiece, is an example of contrast creating a more interesting overall environment. When the pieces don't function well together, however, this kind of design is dis-integrative, and this is the case with one-building-at-a-time architecture exacerbated by sprawl.

Looking at the buildings presented at the Eco Logical Architecture Congress, I concluded that it was time to re-unify the disintegrated city and bring the ecological and social strains back together again in a New Synthesis Architecture. In this philosophy, the building is primarily conceived as a part of, not apart from, the rest of the whole built community, and intentionally positively related to the community of natural environments and life forms.

Wilderness and the Wildness of Cities

As much as I like European cities, after traveling to Australia and Brazil and reflecting on my childhood in the American West, I began to realize that the ecocity should have an element of wilderness that is missing among Europe's manicured parks, its highly managed forests, and its working agricultural landscapes. People control nature in Europe — often with harmonious results, but results that nonetheless show the human hand and mind in every detail, every stonewall, every goat-nibbled meadow, every replanted forest. In contrast, people in western America are surrounded by a wilderness that, though by no means unadulterated, still has a pervasive presence. We need much more of it in and adjacent to our future cities, as well as in vast landscapes far from them. Somehow we have to give nature large zones of freedom right up to the urban edge as we design our built environments and give ourselves our own kinds of human freedom.

As the first step, we need to establish the principle of restoration and regeneration. Setting aside existing wilderness is a start, but we need to work toward preserving much larger, contiguous areas quite close to larger human populations. In and adjacent to cities, we should re-establish those living ribbons of blues and greens, making a very serious commitment to restoring waterways, shorelines, ridgelines, and wildlife corridors for continuous habitats of plants and animals.

Wilderness can expand as cities become restructured for pedestrians and shrink back to reasonable limits. Urban centers could coalesce around nuclei of activity in the suburbs, transforming bedroom communities into real towns with their own viable economies and cultural excitement. Establishing greenbelts is that pause in destructive growth, that moment of still reflection that hangs in time before the pattern is reversed and we begin withdrawing from car-dependent sprawl and restoring nature in a serious way. The first greenbelts were established toward the end of the 1800s. Depaving and withdrawal from car-dependent sprawl began toward the end of the 1900s. Now it's time to put both of these strategies on the fast track.

We will do well, first, to use nature as a model for designing our own communities, lifeways, and technologies and, second, to give back to nature large areas for nature's own management. In this scenario, ecocities, ecotowns, and ecovillages will flourish around the world in a wild profusion of forms — from Arctic fishing and hunting villages to sustainable logging and farming cities in the tropics, from oasis research outposts to great centers of manufacturing and commerce. But all of them will maintain enormous biodiversity in their surroundings, employ renewable energy almost exclusively, recycle assiduously, maintain regionally stable or slowly shrinking populations, and produce unique customs and artifacts. Millions of continuous acres of the metropolis will have been broken up into smaller, more compact settlements as asphalt and lawns are plowed under for crops or simply abandoned to the winds and rains and the seeds dropped by birds. Wildlife will move across the long-forgotten suburban wasteland between thriving villages and cities.

Four types of landscape will be recognized: one defined by the ecological city, town, and village, another primarily for human support (mining, forest production, agriculture, fish farming), a third almost natural but inhabited by humans who know it intimately and are part of it, such as the Laplanders and the Inuit on the Arctic coast, and a fourth nature's alone. Together these last two landscapes will cover more than three-quarters of the land surface of the Earth. Immense tracts of desert and agribusiness will return to grassland and forest. The productive landscape will make up most of the rest. The ecocity landscape will be small but packed with activity.

Nature will manage the two larger landscapes. Species will slowly and steadily increase in diversity, metamorphosing at the slow pace of evolution, with human intervention limited and always in defense of diversity. When whole species disappear, it will be because they have been transformed through time into something else or eliminated by some natural competitor or cataclysm.

People will manage the productive landscape. Soils will grow richer over time and the numbers of members of any species still surviving the ravages of the Age of Progress will increase, incidental to the variety of species nurturing and nurtured by people. Logging companies will manage lands with a wide diversity of economically valuable trees, taking care to protect rare plants and animals and rotate harvest areas so that it is difficult to tell a natural forest from one used for production. Where an industry is actually extractive (for example mining), extraction sites will be repaired with esthetic and species diversity as guidelines.

Finally, people will manage the ecocities, once again using nature as a model. Land uses and the built infrastructure of buildings, streets, bridges, rails, wires, vehicles, and so on — the anatomy of the city — will determine the potential for life in it, rather as the genetic heritage of an individual provides for but does not guarantee full realization of that individual's potential. The city is a complex living creation, but if it is arranged and built for a healthy metabolism of the society, the economy, and the exchanges between nature and people, it should be comprehensible, not confusing. As any animal's design is of enormous complexity, it also makes sense in its basic structure. That sense is obvious in the structure of traditional villages, and if the contemporary city is designed with that sort of integral logic — with its parts arranged to function harmoniously, largely on the organic model — it will, because it is comprehensible, be buildable.

Compactness — access by proximity — and diversity will be crucial. Today's car-dominated cities are flat with holes (parks and parking lots) and tangled veins (arterials and freeways). They are enormously larger than, but analogous to, a sheet of snail-eaten paper with a lumpy zone or two representing the taller buildings in the centers. An animal with such a shape would be dysfunctional, and so is a city. In contrast, the anatomy of an ecologically healthy city will be characterized by walkable centers, transit villages, discontinuous boulevards (rising over wildlife corridors or tunneling beneath them between centers), and agricultural areas close to the center. Compact and very diverse town and city cores and neighborhood centers will be universal. The metropolis will have become several pedestrian cities and more pedestrian towns and villages of varying character and size, linked by bicycle, with support for longer distances from transit.

Major downtown and smaller neighborhood centers will be small enough for most people to easily traverse them on foot. Their size will be based on the size and speed of the human body; thus we can be fairly confident that once we figure out the proper scale and form (which will be different for different populations and climates), we will have a human design truth that will last essentially forever. These centers will provide access to a wide variety of land uses and services in close proximity: housing; work spaces; food, hardware, and clothing shops; educational facilities and places to socialize — the whole community in its basic form. Many designers who think about pedestrian environments consider an area of about a quarter-mile radius the maximum for an active pedestrian center. I think this is a good starting point. We can amend it in accordance with the stamina of the people, their willingness to use bicycles, the climate, and the overall functioning and scale of the center.

For those few who must commute there will be transit, almost always rail. Cars will be for specialized and emergency use or for rent; they will be rigorously restricted to limited areas on the city's fringe and travel a network of small country roads that cater mainly to farm and logging vehicles, some electric or alcohol-powered and some drawn by horse, mule, or ox. Leisurely and time-consuming car trips will not be unheard of, but most long-distance trips in this scenario will be done by rail, airplane, or bicycle. Country roads will be narrow and disappear among trees or plunge into tunnels, at times for miles, weaving in and out of the surface like an embroidery thread through cloth. In cities, sky walks and bridges will connect public terraces and rooftops virtually everywhere. Public investment will be lavished on the pedestrian, who will be pampered with arcades, awnings, porches, verandahs, and covered walkways. Mid-block pedestrian passageways at ground level, galleries, and back alleys will become friendly commercial and garden-like lanes that create a fine-grained environment for people on foot.

Tall buildings around public spaces will be spectacular expressions of confidence in the future and a new faith in human creativity guided by responsibility. Large trees visible for miles, seeming to float high over the city, will be enshrined in rooftop arboretums, supported by pillars built into apartment and office buildings. Movable greenhouse glass and windbreaks perched a dozen or two dozen stories above the streets will shelter and bring attention to plant, animal, songbird, and human alike. Fruit trees, flowers, and berry bushes will be everywhere in the streets, in window boxes, and on rooftops, attracting insects, birds, squirrels, and lizards. New and old techniques will combine to store heat in adobe walls and attach solar greenhouse glass on the cascading terraces of the sun-facing sides of buildings. The gardens, cafés, and sports centers high on the shoulders of buildings will be sheltered by glass walls that fold away in pleasant weather or for whole seasons. People will gather there for the magnificent views and daringly move from building to building on bridges high

*More low-energy
vehicles of the
future.*

above the streets. Everywhere there will be water, with brooks, waterfalls, and sculptural fish ladders, both natural and crafted, becoming focal points in the design of public spaces and the arrangement of buildings. Music from street performers will mix with the sounds of flowing waters and the murmuring voices of people sitting in spotlights of sun discussing and pontificating, whispering and laughing, where the café tables and staircases pour out into the streets and plazas. The city will be so small on the land that all its limits will be visible from any central vantage point, and people will enjoy being so close to nature.

A long-term balance with nature is anything but static or culturally predictable. Though resources and biological health need to be sustained into the deep future, "sustainability" in a more general sense is a term strangely out of place in this world that loves real weather and the excitement of wilderness and the unknown, that enjoys surprise and the bracing stimulation of nature's power, calm, and change, that enjoys cultural creativity as well as a fascination with history. A dynamic and healthy balance with nature and between past and future opens the door to infinite explorations in art, design, and science as well as to more authentic human relationships. The ancients of many traditions used the word "harmony" to describe the healthiest, most respectful relationship between humans and nature, and we may never come up with a better one.

The broad outlines of the task ahead of us should now be coming into focus, but one last point needs to be emphasized. Because the truths from which the ecocity idea emerges are based on the human body — its size, speed, and requirements for nourishment, shelter, procreative and creative excitement, and fulfillment — and on the relations of living organisms to each other and their environment, the principles of ecocity building are applicable forever and everywhere. They are not subject to any fundamental change due to changing tastes, styles, trends, or fads. Once they are discovered, their possibilities are limited only by our own imaginations and abilities.

According to an ancient Chinese proverb, the beginning of wisdom is calling things by their right names. Noticing the absence of words to express ecocity ideas, some years ago I began making up my own words for them. I started with "ecocity," picked up words coined by others, such as Bill Mollison's "permaculture" and Paolo Soleri's "arcology," and reduced complex notions to formulas such as "access by proximity" and slogans like "shrink for prosperity." I found ways of more easily visualizing the various kinds of cities by contrasting sprawl and compact development as being two-dimensional vs. three-dimensional and by stressing the analogy with complex organisms. Part of the reason so many people miss the fundamental importance of their largest physical creation — the city — in the problems and solutions that dictate so much of our lives is the simple lack of a language with which to think effectively about city structure, about the functioning of the built community. Deeper than that, because the arrangement of parts of the city has not been well laid out in the dis-integrative phase of architecture and city design, there is little order in the ideas that describe it. Where is the science and art of investigating, describing, designing, and building healthy cities? Not having found it, I have tried to spell out its beginnings here in this book, and propose to call it "ecocitology."

Because the truths from which the ecocity idea emerges are based on the human body — its size, speed, and requirements for nourishment, shelter, procreative and creative excitement, and fulfillment — and on the relations of living organisms to each other and their environment, the principles of ecocity building are applicable forever and everywhere.

Light rail – the best transit fit with ecocities.
Streetcar arrives at Mills Hall, central campus of Mills College, Oakland, California, for which an ecocity green plan has been written.

CHAPTER 2

The City in Evolution

THE CITY EXISTS, like everything else, in an evolving universe. But it appears to have a special role in evolution that may have a great deal to do with how we build and use it. The theologian Pierre Teilhard de Chardin and the landscape architect Ian McHarg have written about this special role, but it's the architect Paolo Soleri who has been the most explicit about it. Perhaps the best way to begin, however, is with Thomas Berry:

> The changes presently taking place in human and earthly affairs are beyond any parallel with historical change or cultural modification as these have occurred in the past. This is not like the transition from the classical period to the medieval period or from the medieval to the modern period. This change reaches far beyond

the civilizational process itself, beyond even the human process, into the biosystems and even the geological structures of the Earth itself.

> There are only two other moments in the history of this planet that offer us some sense of what is happening. These two moments are the end of the Paleozoic era 220 million years ago, when some 90 percent of all species living at the time were extinguished, and the terminal phase of the Mesozoic era sixty-five million years ago, when there was also a very extensive extinction.[1]

Berry describes in rich detail the "lyric period" of life on Earth, the Cenozoic era, in which flowering plants appeared in wild profusion, color and sweet fragrances spread over the land, and mammals expanded from a few obscure rodent-like varieties into the enormous diversity of warm-blooded, quick-witted species that came to fill the lands, waters, and skies of planet Earth. Berry sees us as now entering an "Ecozoic era" in which life in all its ecological interconnections becomes an organism capable for the first time of attaining "reflexive awareness" or deep consciousness of itself. As we become aware, he says, through us the universe consciously awakens to its own existence.

This juncture between the Cenozoic and the Ecozoic era is the point beyond which evolution consciously determines its own fate. By design or default human beings have become the agents of this transition, and in us the universe has acquired physical eyes and hands to see and shape the course of its own evolution. It has gathered and compacted into live physical beings a new creative force — Mind — at once a pilot and a builder of communities, technologies, and landscapes transformed to the scale of planets.

Though most of the materials for this construction were produced by earth, water, air, and other organisms long before humans appeared on the scene, we have arranged new materials in such a way as to constitute something new. Soleri calls it "neomatter," and it ranges from the relatively simple concrete invented by the Romans to the thousands of exotic chemicals introduced into the biosphere and to their accidental combinations and degenerating by-products. Proposing what we should do with these new materials, powers, and responsibilities, and in particular what to build, is the aim of this book.

It seems the universe began with a colossal explosion: the Big Bang. Brian Swimme, co-author with Thomas Berry in *The Universe Story,* [2] more poetically terms it the "primordial Flaring Forth": it sent rushing out into the void sub-atomic particles quickly forming into vast, expanding clouds of lightweight gases, predominantly hydrogen. This happened around 13 billion years ago in a process Soleri identified as miniaturization/complexification /quickening. This process was immediately and everywhere established as one of the primary dynamics of evolution. Out of these vast clouds atoms were pulled together through their own gravitational attraction to create masses of material that, when compacted enough to spark thermonuclear reactions, lit up as stars. Within these stars, a wide range of elements was cooked up over immense time and, in those of appropriate size, exploded again in mini-"flaring forths" called supernovas, throwing a new soup of atomic particles and combinations of atoms called molecules into interstellar space. Relentlessly the process of miniaturization/complexification/quickening continued asserting itself as these particles, too, gathered together matter and energy in relative pinpoints in the vast dusty halls of the vacuum.

With these second-, third- and later-generation stars, much smaller and far more complex accumulations of elements and molecules emerged: the planets. On and in these planets, atomic decay and internal friction produced heat imbalances and convection currents that began mixing and sifting, sorting and distilling physical material in spatial volumes much smaller than the stars or dust clouds from which the planets themselves had condensed. On the surfaces of these warmish newcomers, internal heating met the warming of suns and cooling of space, producing zones of the most complex mixing yet seen in the universe.

Here on the surface of our home planet, in environments of a complexity many orders of magnitude greater than that of interstellar space or the insides of stars or planets, solids, liquids, and gases combined and were zapped by lightning and irradiated by sunshine, cosmic rays, and nuclear decay. In this heretofore most complex and miniaturized of all realms of the universe, in a pattern seen since the beginning, a changing reality characterized in its broadest

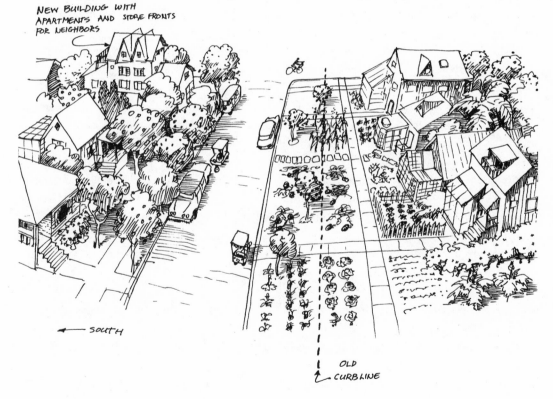

NEW BUILDING WITH APARTMENTS AND STORE FRONTS FOR NEIGHBORS

← SOUTH

OLD CURB LINE

Making streets narrower. Suburbs are redeemable only very close in to town centers. A small step in making these neighborhoods more convivial, water permeable and garden friendly is making streets narrower and depaving.

sense by a process of miniaturization/complexi-fication/quickening was bringing something new into existence: life.

Life gave rise to something so miniaturized and complex and rapidly changing as to be experienced much more in what it does through the living beings it occupies and animates than in itself: consciousness. And I would say, at the same time, the other half of that same phenomenon, namely conscience. In its wholeness it's sometimes called the psyche, spirit, or soul of the physical form whose reality it shares.

It is conceivable that larger coherent systems — organisms by some reasonable definition — could exist with the same degree of complexity covering much larger volumes of space, but such systems don't seem to be "buildable," either by nature or by humans. Proximity — the closeness of parts, on a small scale — is necessary, or the thing cannot hold itself together. Gravitational, chemical, or atomic connections operate over certain distances beyond which they cease to bind. There seem to be no known forces to bind in a diffuse and yet complexly organized manner over great distances. We will be talking later about "access by proximity" as a principle of the ecocity; it seems as good a description of chemical processes as it is a slogan for city building.

Paolo Soleri's *Arcology: The City in the Image of Man* includes one of the most elegant (miniaturized) statements on the subject

I have discovered: "Any higher organism contains more performances than a chunk of the unlimited universe light years thick, and it ticks on a time clock immensely swifter. This miniaturization process may well be one of the fundamental rules of evolution."[3]

That last sentence may well be one of the great understatements too, since there seems to be no evidence to the contrary anywhere, at any time. The miniaturization pattern in evolution is so fundamental, in fact, it seems to me we could think of it as the First Law of Evolutionary Dynamics. Scientists speak of the Second Law of Thermodynamics; the idea that all energy is in a winding-down pattern of entropy that will result in the heat death of the universe. At the same time there are forces of atomic, chemical, and gravitational attraction conspiring to drag it all back together. Buckminster Fuller called this counter-entropic pattern "syntropy." Insofar as the stuff of the universe does get reassembled, it is toward more complexity all over again and in more compacted spaces; the Second Law of Thermodynamics encountering the First Law of Evolutionary Dynamics in a dance of entropy and syntropy that defines and organizes much (or all) of reality. I think of entropy and syntropy as a set of dimensions, each of which, without the other, is nothing. In the matter/energy universe (another dimensional set), matter and energy just don't seem to exist separate one from the other or separate from this dance of entropy and syntropy.

SOLAR
COLLECTORS

ROOFTOP
ARBORETUM

SPIRAL
STAIRCASE
IN SOLAR
GREENHOUSE
EXTENSION

G. FOR
GREENHOUSE:
G

OLD
PARKING
LOT

OLD
BUILDING

SOUTH

Adding fine grain density to a major street. *New structures are shown above the dashed lines.*

I would suggest condensing (miniaturizing) the term "miniaturization/ complexification/ quickening" down to a single word: "miniplexion." The word complex comes from the Latin for weaving or braiding (*plectere*) together or with (*com*). Add "mini" for "small." The process of complexifying, of delicately, subtly weaving together, becomes the almost lyrical miniplexion. It's not just an important process of the universe; it permeates and characterizes everything. Nothing exists separate from it.

In addition, in no way does unpredictability or catastrophe in any given biosphere on this or other planets demonstrate that miniplexion itself is stopped for even the shortest time. Because after the impact of the asteroid, the retreat of the glacier, or the desiccation of the spreading desert, all systems (of those that remain) remain on go: they move toward further miniplexion. If local designs and innovations are lost in catastrophe, the biological and even chemical survivors remain as dedicated as ever to the process, being able, it seems, to do nothing else. In fact, the exotic comet smoke and debris from outer space, the finely ground glacial dust, the radically altered environment of the desert can create new opportunities for evolutionary experimentation as microscopic amounts of rare elements contact living cells, large geographical expanses open up to reinhabitation on enriched soils, and desiccated landscapes provide environments for future species shy of competition but able to endure heat, cold, and dryness.

Gaia, the Self-Regulating Planet

In 1875, the Austrian geologist Eduard Suess proposed that all living creatures on the Earth constitute a sphere of life that he was the first to call the "biosphere." In the vastness of evolutionary time, the biosphere had emerged on the surface of the stony mineral continents and metallic hot core of the Earth's lithosphere, suspended in the watery oceans, rivers, and lakes of the hydrosphere, and bathed in the gasses and vapors of the atmosphere. By the 1920s, the Russian geobiochemist Vladimir Vernadsky had realized that the Earth's biosphere had transformed the non-living material of the Earth into an environment co-evolving with, and managed by, the countless numbers of living organisms on the planet. Life and Earth, he pointed out, were together creating their future collective self.

Look at the chiseled walls of the Grand Canyon — thousands of feet of vertical geological text representing hundreds of millions of years of Earth's history colored by, and largely made up of, once-living organisms. Look at the deep-rooted tall mountain ranges of limestone and marble that once were shellfish, the underground oceans of oil and the vast regions of coal and tar sands that once were great marshy forests and shallow seas depositing layer after layer of organic matter, thus preserving the chemical investment from the energy of the sun. Look at the immense deposits of iron and other metals once, long ago, precipitated out of water by living bacteria to become the ores used in building civilizations from the Bronze Age until today. Look at the blues and the sunset colors and the crystal black night and feel the temperature of the sky — all as they are because of the activity of living organisms. This is the work of life on Earth already accomplished, turning human artifacts like the Great Wall of China and our giant cities into small squiggly lines and smudges on life-built, geologic, and atmospheric layers dozens of miles thick, encompassing a whole planet.

Vernadsky observed that the biosphere seemed to be self-regulating in a way that promoted relative stability on Earth. Oxygen in the atmosphere has been maintained by the biosphere at a level that permits its use by living organisms, but not its abuse. Plants and animals seem to have struck a balance with the inanimate world in maintaining the proper concentration of carbon dioxide in the atmosphere to keep the planet at a temperature that is just right for a large variety of species and to provide oxygen for plants and animals, but not so much as to generate excessive fires.

Refining this idea further, in the 1970s, the microbiologist Lynn Margulis and the atmospheric scientist James Lovelock described the processes by which living organisms transform Earth's atmosphere and geology and regulate the air's composition and chief properties so that life — the living organisms, the regulators themselves — can flourish. They dubbed the Earth/life superorganism "Gaia," after the Greek titan of Earth.

Father Pierre Teilhard de Chardin proposed in 1925 that within the biosphere, through human communications and cultural artifacts such as books, music, art, and radio, people had created an evolving "noosphere," or sphere of knowledge. He argued that cities brought together concentrations of people and their technologies as nodes of consciousness — nodes of intense complexity and enormous leverage for further evolution — in the evolving noosphere. From all these thoughts the notion emerges that the biosphere, or Gaia, is evolving a physical brain composed largely of our human-built infrastructure of cities and their support structures. We begin to see that humanity is part of a physical, living, thinking thing or being composed of parts of the Earth itself and of countless trillions of living creatures. Our specialty seems to be brain building for a collective version of Thomas Berry's "reflexive awareness."

Whether we accept this evolutionary panorama that Berry would call "The Story" and the place of the city within it or adopt a more prosaic view of the city as functioning as a collection of buildings, landscapes, vehicles, tools, people, animals, and plants engaged in a mix of economic, cultural, and natural activities under the partially conscious direction of people, the built human habitat sits squarely in the center of the make-or-break human evolutionary drama. Whether the city is somehow thinking on its own in a way to which we contribute or exists as a neglected, misunderstood tool for advancing evolution or, through misunderstanding, helps derail it, there it is, functioning in the evolutionary context while changing that context in ways we can come to understand and in which we participate.

The Urban Comet

The city's role in evolution that could bring on Thomas Berry's Ecozoic era is far from rosy. We are, on a planetary scale, ripping vast patches out of the fabric of life, rapidly degrading the natural systems, up to and including the weather and climate upon which life depends. There is now, for example, more than a third again as much carbon dioxide in the atmosphere (passing 380 parts per million by volume in 2004) as there was at the beginning of the Industrial Revolution (280 parts per million). At current rates of increase, scientists are predicting that within 50 years the levels of carbon dioxide will be double those of today. This is a major transformation of the atmosphere of the planet. The great majority of experts believes that we can expect global heating, more extreme storms, and massive habitat collapse around the world. The Inuit on the Arctic Ocean coast of Alaska don't predict it, however; they experience it: where they live, it is now six degrees warmer than it was in 1950. They witness the body weight loss in game animals, the failure of winter ice to form, the collapse of populations of certain species including the polar bear, the melting of permafrost, the formerly vertical trees becoming "drunken forests" since they sink at odd angles into mud, and the weird dusty whirlwinds that didn't exist more than ten years ago. Closer to home for most of us, cities are becoming "heat islands" typically seven or eight degrees hotter than the surrounding countryside. As their physical footprints expand and they come to constitute a significant fraction of the Earth's surface, they begin adding to the heating of the Earth directly.

And it only gets worse. Research being conducted in Wisconsin, Michigan, Kansas, South Carolina, and Florida as well as in England and New Zealand is examining what the increasing levels of carbon dioxide hold in store for plants. In various chambers and in semi-sheltered outdoor environments with carbon dioxide levels matching those expected in 2050, plants and animals from insects to sheep are being raised and studied. In a nutshell, most plants grow much faster but are rich in fiber and starch while very poor in nitrogen products and protein. Various caterpillars, for example, eat 20 to 40 percent more plant material but get much less nourishment, take longer to grow and —

Closing the street and adding yet more pedestrian-oriented buildings and activities. Same location as last illustration, but with all-weather arcades and more shops, businesses and housing.

EDIBLE PINE NUTS

BIOREGION FLAG
EARTH FLAG
U.S. FLAG

BECAUSE OF BRIDGES AND ELEVATED PATHWAY, UPPER LEVELS ARE ACCESSIBLE TO WHEELCHAIRS WITH FEW ELEVATORS.

those that survive — end up on average about ten percent smaller. An insect called a leaf miner is twice as likely to die in today's double-the-carbon-dioxide atmosphere. "Although they may eat more per individual, a lot of them die before they've eaten their fill," says Peter Stiling at the University of South Florida in Tampa.[4] "The leaves are not rich enough to support them." As they struggle just to eat, these herbivorous insects and the plants they consume put out visual and chemical signals that make the insects more easily locatable by parasitic wasps and other predators. Add that to temperature changes and major changes in plant species distribution, and the possibility of a world without many of today's species of butterflies and moths becomes a real possibility. Sheep eating the fiber- and starch-enhanced plants in the double carbon dioxide atmosphere of another experiment are having trouble digesting enough nitrogen compounds because the bacteria in their rumens have more work to do to produce the proper amount of protein. They need to pause in their eating to wait for the bacteria to catch up and complete digestion. Farmers might "solve" this problem by adding protein to their diet. Wild animals won't be so lucky.

The heat-island effect comes largely from city streets, parking lots, and roofs absorbing solar energy — asphalt, concrete, and roofing tar baking in the sun. The city built for cars is massively energy-consuming, specifically for mobility. Most people are familiar with the direct results of automobile driving with regard to carbon dioxide production, and we hear over and over that driving less is one of the most important things we can do. Meanwhile scattered small houses consume far more energy per person served than buildings such as apartments and condominiums, since they share heating and cooling energy with no one before it is lost to the atmosphere. Beyond this, the manufacture of automobiles, gasoline, tires, highways, parking lots and structures, gas stations, and so on, the shipping of these items or their building materials, and the disposal or reuse of materials takes up a very large share of the energy use of industry. Sprawl that drives farming off rich soils close to city centers helps transform farming into an energy-intensive industry consuming on the order of ten times as many calories in fossil fuels as calories produced as food. As the farmer/author Wendell Berry reminds us, the average distance that food travels from farm to American mouths is about twelve hundred miles. And we haven't even begun to look at the specific poisons associated with the city: local air pollution, water runoff, ocean oil spills, and so on.

The connection of the city and its particular structure to evolution by way of its immense impact on biology and climate should be obvious. Certainly, the evidence linking our city structures to current extinctions and changes in normal evolutionary patterns is incontrovertible if we simply open our eyes.

"Ecozoic" may sound uplifting, "new era" hopeful. But in any case, the quality of life depends largely on how we build our cities.

Islands and Extinctions

Our cities, farms, highways, and dams are slicing natural habitat into ever smaller pieces. Island biogeography — a branch of biology informing and enriching evolution biology with meticulous study of the species living on islands in varying degrees of isolation — shows us some fascinating patterns. In *The Song of the Dodo*,[5] David Quammen traces the history of the evolution sciences by adventuring to remote islands and following the scientific battles all the way to their conclusions in solid theory. Here are some of the patterns he discusses:

Small islands have smaller numbers of species, and the number roughly doubles as the area of the island increases tenfold. Say there are three species of beetles on a hundred-acre island, then there will be about six on an island of one thousand acres, twelve on an island of ten thousand acres, twenty-four on an island of one hundred thousand acres, and so on.

How frequently a new species arrives on an island from the vast biological supply house of the closest continent is proportional to the distance it has to travel and the vagaries of winds and ocean currents. Rafts of driftwood ripped by floods from the forests of the Andes, for example, did bring reptilian settlers to the Galapagos Islands six hundred miles from Ecuador, but only extremely rarely. For every animal that landed in a million years, many hundreds drifted by, dying of thirst and hunger, some coming close enough to see the grass and shrubs or even hear the clatter of roosting sea birds only to be blown by winds farther into the vast Pacific. A minuscule fraction coughed out to sea by storm-swollen mainland rivers made the journey and survived. The distance factor meant that no mammals could make the trip at all — with their rapid metabolism, they simply couldn't last or fast that long. The delicate skin of the salamanders and frogs could survive the salt spray and sun only very briefly; they are never found on islands that were not once connected by land bridges to continents.

Then, should other species have arrived earlier and settled snugly into all the niches on the island, the newcomer would have had to fully or partially displace one or another of the old timers. In addition, the new immigrant would have had the genetic disadvantage of inbreeding since only a small number, a pair, or even a single pregnant female would have made it to the beachhead to found a new species there. The resident species would have already learned genetically, and sometimes through parental teaching, how to mesh with the local ecology.

In any case, the quality of life depends largely on how we build our cities.

Old islands have relatively stable and large numbers of species, young ones smaller but growing numbers of species. Old ones have a very slow turnover rate as species die out and new ones evolve from the locals or arrive from outside; young islands have a higher turnover rate as species jockey for position in the various niches and compete with a less solid alliance of interrelated species against the immigrants. Early arrivals on remote islands, once firmly established, undergo "adaptive radiation." A single species begins to establish itself in certain niches, and, if barriers are significant — as in Hawaii where the prevailing winds create deep forests on one side of the island and deserts on the other — that species tends to produce varieties, races, and eventually completely different species adapted to different climates and different floral and faunal ecologies.

This process takes a while. When a species is driven to extinction, it will be a very long time before it is functionally replaced in its ecology, if ever. Removing a key species, introducing a foreign predator or disease, changing the climate, hunting a species to extinction, or driving it out with agriculture creates a cascade of extinctions, an unraveling of the fabric of life called "ecosystem decay." A hummingbird dies out, for example, because of a human-introduced avian malaria mosquito, and the flowering plant that evolved with the hummingbird's pollinating beak and a battery of insect species dependent upon the plant for food and shelter follow.

As various forms of pollution increase and the sheer number of people grows almost everywhere, cities, highways, farms, logging areas, and lakes behind big dams are dividing the continents into ever smaller, more isolated islands of life. It is therefore important for us to grasp the relationship between extinctions and evolution, to understand the role of cities and city-based civilizations in evolution, and to explore ways of reversing the process so that cities and their supply zones and supply routes can allow natural areas to expand and reconnect. We can create islands of vibrant culture — ecocities in association with their working landscapes — instead of reducing nature to shrinking islands of isolation in a world characterized by an increasingly uniform culture of clutter. Says David Quammen, "Evolution is best understood in relation to extinction and vice versa. In particular, the evolution of strange species on islands is a process that, once illuminated, casts light onto its dark double: the extinction of species in a world that has been hacked into pieces."[6]

A fundamental pattern for the city's impacts is the relationship between land area and the total amount of land required to provide it with its biological, material, and energy resources — its "ecological footprint." Mathis Wackernagel and William Rees[7] estimate that the consumption level of a typical Canadian, for example, requires approximately 12 acres

Above: Natural drainage bicycle flyway.

In the city designed with high density and ecological principles in mind, elevated bicycle paths are a real possibility and would be a pleasure for many.

of land. If a Canadian city has one hundred thousand citizens, then it will require 1.2 million acres of trees for wood and paper, mines for ore, farms for food and fiber, plants to absorb carbon dioxide generated in energy use, and so on. This island called a city is more ambitious in its exploitation than a colony of birds on an oceanic island sending out its daily flight to harvest fruits, seeds and insects from its limited land area and fish from the ocean offshore. The city's ecological footprint is proportionally much larger than that of the birds' rookery.

The higher the density and diversity of a city, the less dependent on motorized transport; and the fewer resources it requires, the less impact it has on nature. As the city is designed to conserve energy and materials while turning wastes into resources, building soils like a compost box, and restoring nature by revitalizing local natural habitats, the ecological footprint shrinks toward an optimal size for its population. When the city is actually building more soils than it is consuming and helping species to survive — actually giving back acreage to nature and creating new resources for its healthy evolution — the negative ecological footprint is turned on its head and the conversation changes character fundamentally.

The car city not only takes up more land area per person but makes even greater demands on the Earth's resources. The transit city rises somewhat higher from a smaller physical footprint and has a more than proportionally smaller ecological footprint. The bicycle city continues the pattern with its yet smaller physical footprint and far smaller ecological footprint, and the very high-density pedestrian city, probably the truly ecological city, will not only take up much less space on the surface of the planet but require far fewer material resources and might even return a net evolutionary benefit to nature.

The Australian researchers Jeff Kenworthy and Peter Newman, examining cities around the world, have developed extensive data on the subject.[8] In their writings, we see the almost mathematical relationships of density and diversity of cities in relation to pedestrian areas, bicycling, transit, cars, and the health of local and global ecologies. Their work amounts to an urban corollary to the kind of relationships developed in island biogeography.

Perhaps the largest pattern created by human beings on the geographic tapestry is the drowning out of natural life in vast swatches. Though islands exhibit a wide range of unique variation in plants and animals, they are home to a modest number of species. The real reservoirs of biodiversity are the continents. However, human settlement is swamping the coastal areas and the valleys, leaving ever smaller natural environments. As people occupy ever more land, it is as if the oceans were rising. Mountain ranges become isolated, then ridges, and finally, higher peaks turn into separate shrinking islands. At the same time global warming is reinforcing this pattern by driving cooler climate species up and out of existence off the tops of mountains. Civilization would be far wiser to emulate the extraordinary diversity of the unique islands of the world by creating city-islands in a sea of biodiversity rather than leaving islands of "parks" and reserves in a sea of civilization's sprawl. Cities should be semi-isolated and locally highly tuned to their immediate hinterlands and local ecological conditions.

Bikeway into the country. *Pedal power highways will make sense connecting city, town and village when close together, as when carved out of sprawl development, passing through restored natural and agricultural land. Tunnels and high bridges are important for human/wildlife crossovers.*

The very form of the city, by providing access to culture, resource, and nature, has the potential to raise our consciousness of evolution to new heights.

The density, diversity, form, and function of cities and the awareness of their citizens in this regard are now key factors in evolution. On the positive side, the very form of the city, by providing access to culture, resource, and nature, has the potential to raise our consciousness of evolution to new heights. Though the effects of today's enormous sprawled cities on biology and evolution are grim, learning about such cities and about the alternatives to them gives us the tools to solve many urban and evolution-sized problems.

The City as Organism

Just as exploring the patterns of evolution helps us to understand the city, so does examining the analogy with living organisms. Viewed as a whole system that is part of the larger whole system of the biosphere, the city seems to have a number of parts:

- Skeletal system: for providing structural support — architecture, bridges, telephone poles
- Muscular system: for powering locomotion and internal movement — vehicle engines, gas turbines/electric generators, elevator motors, fluid pumps
- Digestive system: for transforming stored energy (food or fuel) into more usable forms — space heating, food-processing plants, gasoline cracking facilities, solar greenhouses.

- Fat, spleen, pancreas: for storing and retrieving energy, enzymes, and chemical reserves — warehouses, garages, cellars, water reservoirs, cisterns
- Nervous system: for enabling internal communications — wires, fiber optics, streets for people taking news back and forth, press services, newspapers, radio and television broadcasting and receiving equipment, computers, post offices
- Brain: for receiving, storing, internally processing, and coordinating information — universities, management offices, government buildings, non-profit organization offices, books, libraries, CDs, tapes, records, hard drives
- Vocal cords: for communicating outward — the above-listed communications devices directed outward toward the countryside and other cities
- Ears, eyes, and other sense organs: for receiving information from the outside — again, the same communications devices used for receiving information
- Heart, veins, arteries, lymph system: for internally conveying materials — streets, gas and water lines, storm and sewage lines
- Skin or hair as an edge/barrier/semi-permeable membrane: for separating

the organism from the outside environment, keeping certain things out while admitting others — city walls or "limits," sometimes a natural feature like a river or a constructed one like a boardwalk by the beach

- Mouth, nose, pores: for allowing selective entry and exit from the whole system, working with the separating membrane mentioned above to regulate entry and exit of materials and living forms for the health of the organism — city gates, legal city limits, entry and exit points of waterways, highways
- Liver, kidneys, spleen: for filtering and recycling — compost boxes, chemical or biological waste-water treatment ponds, recycling systems
- Bladder, rectum, anus: for excreting that which can't be recycled (or can only be recycled by being contributed to the larger system outside where other participants in the system can recycle) — sewage system and outfall, waste incinerator, crematorium, freeway to the landfill
- Sex organs of both sexes: for reproducing the system — colleges, design offices, environmental advocates, general voters, construction companies

Bridge building with monumental tree. *The architecture of larger buildings is largely the salvation of cities — if energy conserving and functionally and culturally diverse, vital and friendly. Here, a new bridge building spans a street, angles toward the sun (is passive solar), and features a large planter box on strong columns supporting a large native tree, attracting native birds and insects; a kind of eco-art piece celebrating nature.*

preparing to build more of the same, or perhaps ecocities

According to James Miller,[9] author of an enormous textbook called *Living Systems*, every living system has nineteen irreducible subsystems. Though I think more in terms of my list above, he breaks down his nineteen systems in the following way:

- First, those that process both matter-energy and information: (1) Reproducer and (2) Boundary.
- Second, those that process matter-energy: (3) Ingestor, (4) Distributor, (5) Converter, (6) Producer, (7) Matter-energy storage, (8) Extruder, (9) Motor, and (10) Supporter.
- Third, those that process information: (11) Input transducer, (12) Internal transducer, (13) Channel and net, (14) Decoder, (15) Association, (16) Memory, (17) Decider, (18) Encoder, and (19) Output transducer.

Though this list almost requires that you read a book that long, a little mulling over gives you the idea that, indeed, there really is a pattern of order at the core of living systems. I'd add that at least a rudimentary knowledge of the dynamics of such systems is indispensable for designing better towns and cities. Miller includes among his living systems with their subsystems, in both physical form and functions to fit the form, the cell, the organ, the organism, the group (using the example of a business office), the organization (a ship), the society (a national government), and the supranational system (for example, the European Economic Union). He could easily have used cities and bioregions as examples and even complex buildings, if housing an architect's office could qualify for the reproductive function. All have these analogous subsystems and functions and are related in analogous ways to their environments.

The city, then, is a system somewhere between the human animal and the bioregion and biosphere in size, a living system in which we reside. An organism mutates and physically evolves over thousands or millions of years and countless generations directed by ever-so-slowly changing DNA. The traditional village also mutates into something different very slowly, its subsystems having functioned dependably for thousands of years, directed by relatively stable human traditions. But the city has mutated rapidly, directed by commerce, creativity, greed, and colliding concepts of the good and beneficial — and the bad and detrimental. All are living systems manifesting the subsystems and functions Miller identifies. If we understand this, we realize that the city needs to work at the extraordinary efficiencies that only the miniaturized, three-dimensional pedestrian city can deliver.

The Three-Dimensional City

In the late 1950s, Paolo Soleri lived in the desert outside of Phoenix, Arizona, in a place called Paradise Valley. In those days before sprawl, the days of fresh desert air, blue sky, and stunning tall rocks placed in stately clusters in the immense landscape, the place deserved its name. He was drawing, building models, and writing about better ways of building cities in that climate and topography. The city he proposed, called Mesa City, having a large population, covered a great expanse, but he imagined that there would be considerable concentrations of people in

Bridge building with generous "hallway." *Similar in design to the previous illustration, this structure spans a covered street illuminated partly by beams of sunshine streaming through skylights.*

much of it. He recalled the kind of vitality he had experienced growing up in European towns and cities, and he knew that a certain density of population and intensity of activity were required for a vital economy, a healthy ecology, and an involved community. Thus the wide-open spaces of Arizona played off against the compact hill towns of northern Italy. Expansive land uses and compact vital centers stretched out and pulled together in his imagination. Following the planning *modus operandi* of the time, Soleri labeled various parts of his Mesa City "housing," "industry," "civic center," and "education," like organs of a living system, in generous landscapes of specialization.

Then, rather suddenly, his thinking shifted, as if his stretched-out city had sprung together into a single springy three-dimensional form, with all the various parts lying beside, over, and under, rather than simply beside one another. Now the drawings included "sections" (vertical slices viewed from the side) through the town with "housing" above or around "education," and "industry" below that, and so on. Soleri realized immediately that such a three-dimensional city, though large for a single structure, would be very small for a city. Such a city would bring enormous complexity into very close proximity. It could cover one-tenth the land and consume one-twentieth the energy, making renewable energy systems immediately practical and massive, polluting energy systems

anachronistic. It could recycle at extraordinary efficiencies and produce virtually no pollution. It could completely liberate its citizens from automobile dependence and its costs and hazards and do all this while creating a far superior container for a creative culture.

Interviewing Soleri in 1970, I asked him when he first made the urban/evolution connection. He told me that shortly after the dawning of his three-dimensional consciousness he had been reading Teilhard de Chardin's writings on evolution, which had been suppressed by the Catholic Church but became available shortly after his death in 1955. Lying near the Cosanti swimming pool taking his afternoon siesta, thinking about evolution and the city forming in his imagination, Soleri suddenly realized that the three-dimensional city was an instance of evolutionary miniaturization/complexification/quickening, and that, if built, it could be the instrument of the next quantum step in evolution. The three-dimensional city was the new physical form, the mutation, to take evolution to its next plateau. The formula for gravity came to Newton, the story goes, while he was sitting under an apple tree contemplating falling apples. Charles' law relating gas volume, temperature, and pressure came to him in a dream that he recalled and wrote down with a pencil and paper next to his bed. Similarly, after rigorous preparation, seeing the pattern came almost effortlessly for Soleri. I said, "That must have made you rather happy." He answered, "Happy? I was ecstatic." De Chardin had seen the broad outline of the city's role in evolution but not the physical form it would have to take. Soleri identified that form and recognized that such a city would have to emerge from creative intent. Mindful creativity would supersede mutation in this step in evolution.

The new outpouring of city drawings, models, and writings from his workshop had a radically altered form: they were very three-dimensional. Soleri called his concept "arcology" for the fusion of architecture and ecology — architecture that becomes a new ecological reality and architecture that is part of, and participant in, evolving ecosystems. He adopted terminology from de Chardin and made up some of his own, arguing that humanity was developing a "technosphere" producing "neomatter" that must now come into balance with the biosphere through conscious direction of the evolving "noosphere," the sphere of knowledge.

Each Soleri town was designed as a single structure — a single complex building or a condensed and consolidated whole community. A village might rise to five stories, a small town might be twelve stories, a large city 150 stories high. Not everyone was enchanted by his schemes. Critics called them file boxes — never mind that his drawings and models were radically different from any cubes or rectangular solids you'd ever seen before and showed generous semi-interior spaces,

panoramic vistas opening up to nature, and expansive cathedral-like pedestrian environments. Rather than claustrophobia, to many including myself, they conveyed new freedom in their vistas and in the possibility of simply walking out of the city and into nature.

Most detractors made assumptions that did not cover the range of possibilities, stating their own assumptions about his work as if they were Soleri's proposals. It was Buckminster Fuller, not Soleri, who suggested putting a giant climate-controlling dome over Manhattan. Soleri built one small residence set into the cool earth of the hot desert and drew two or three cities out of dozens set partially into the landscape in a similar way, yet half the people I talked to about his work over the next twenty years said, "Oh, you mean the guy that wants to build underground cities." On his identification of the role of cities in evolution his critics were almost completely silent.

Where are the environmentalists who care so much about the fate of the Earth now that the idea of building for a radically reduced impact on nature is circulating widely? Soleri's experimental town in Arizona —Arcosanti — could be built for a fraction of the cost of an aircraft carrier or submarine or a major freeway expansion, and what we would learn from it would be among the most important things humanity ever learned. Soleri discovered early that support for the ecologically healthy city is very limited relative to the

Soleri's experimental town in Arizona – Arcosanti – could be built for a fraction of the cost of an aircraft carrier or submarine or a major freeway expansion, and what we would learn from it would be among the most important things humanity ever learned.

needs of our times. "There's an enormous gap between interest and commitment," he has said. Build a better geegaw, doodad, or thingamajig and the world will beat a path to your door; design a better future and it will suspect and marginalize you.

Soleri has received virtually no work contracts or assistance from government, business, or foundations. The sale of wind bells made at Arcosanti, his books and talks, and the steady flow of students earning college work/study credit for helping to build Arcosanti have provided almost all the funds and labor for his work. But the lessons that might be learned from his relatively purist approach to ecocity-building can be applied to changing existing cities in many different ways. The support is so low and the importance so high it's a national disgrace.

Lessons

Evolution biologists tell us that in all the history of life on Earth there has never been a species in the general size range of the human being with a population even close to ours. In fact, says Edward O. Wilson, since about October 12, 1999, when the human population passed six billion, "we had already exceeded by as much as a hundred times the biomass of any large animal species that ever existed on land."[10] In other words, human numbers are now more than one hundred times greater than the runner up — ever. At 6.5 billion individuals, our species constitutes one hundred times the biomass of any small antelope or deer, large wolf or pig species that ever existed. With our food animals, tree and agricultural crops, and human body biomass, humans appropriate about 40 percent of all solar energy accumulated by life on the land surface of the Earth and 25 percent of that accumulated by all life on the planet, including life in the waters. That does not count stored energy in fossil fuels that we are burning up at the rate of approximately one million years of deposition per annum.

The lesson from the immensity of evolution, ironically, is to think small. The concept of the small, tall city will come back over and over to help us think through the means to a better future. Complexity in those smaller, taller spaces will translate more gracefully into "diversity" — diversity of human uses, economic activity, cultural systems, ecological building features, and species. Whether it's small and tall in Soleri's single-structure city or in the vital neighborhood center going from one- and two-story buildings to three- to five-story buildings, this city will be a model for reshaping civilization.

Earlier I mentioned the forces that bind things together in the evolving universe, and the distances over which they operate. Similarly, people are united by certain forces of attraction for personal and mutual benefit, whether it is for economic transactions and education or for delight and reproduction of the species. Physical nourishment, shelter, security, curiosity, affection, sex, personal and social fulfillment — for all these forces there are distances in space and time beyond which they fail to function. The community simply can't be scattered too thinly. Gasoline is the most typical substitute for the social forces that bind community when the distances get too large, and it doesn't work very well. In fact, it's a disaster measurable in social segregation by distance, human alienation, atmospheric pollution, death on the highways, habitat destruction, species extinction, and climate change — that same old list of victims of myopic collective design and planning. From the evolutionary perspective, sprawling suburbs have to go. The relentless spread of humanity, both in sheer numbers and in the two-dimensional space taken up across the landscape, has to be reversed, and the restoration of nature has to proceed on a grand scale. We need to roll back sprawl and rebuild civilization.

The lesson of the end of the Mesozoic era and the beginning of the Cenozoic era 65 million years ago, when the dinosaurs disappeared, is the same one, with its own twist: the large and violent are not necessarily well treated by evolution. Now, depending on how well we understand our place in the noo-sphere and the biosphere and are able to transform the means by which we evolve, we humans have a once-in-the-lifetime-of-the-planet opportunity to transform a grand mistake of consciousness into one of its finest achievements.

The relentless spread of humanity, both in sheer numbers and in the two-dimensional space taken up across the landscape, has to be reversed, and the restoration of nature has to proceed on a grand scale. We need to roll back sprawl and rebuild civilization.

Restored creek flows through
"downtown" Mills College campus in
proposed ecological general plan.

CHAPTER 3

The City in Nature

ALTHOUGH THE FORM of most cities today is destructive of nature, it has not always been and need not continue to be so. Parts of many cities today function far better than other parts. If we were to put the best parts together, such a city would be vibrantly healthy. A compost box, a simple device built and managed by people, is a modification of the forest floor where organic matter is transformed from waste to a great new resource. Similarly, a city could build soil and support biodiversity if it were designed, built, and maintained for the task.

Many of our city dwellers show a deep appreciation for nature. This appreciation may have come from a nostalgia for nature lost, but it may also be caused largely by the city itself — its sociability and its gathering together of knowledge. Possibly the urbanite's appreciation of nature is taught by the physical form of the city — the fact that it does function something like a living, natural organism, at least in its potential efficiencies, even if for almost everyone this lesson will be subliminal.

Appreciation for nature takes many forms among city folk. The average city is a complex botanical garden of trees, bushes, and flowers from all over the world: on streets, in parks and yards, in containers on porches, in lobbies and living rooms, hanging from macramé, attached to chunks of dried peat moss — from tiny lobelia and Johnny-jump-ups to giant sunflowers and towering redwoods, from showy palm trees to bushy ferns loved passionately by their owners. Aquariums, terrariums, and birdcages are perennially popular. Biodiversity in cities, despite the acreage given over to buildings, barren rooftops, and asphalt streets and park-

Biodiversity in cities, despite the acreage given over to buildings, barren rooftops, and asphalt streets and parking lots, is very high, surpassing that of many environments where few or no people live.

47

ing lots, is very high, surpassing that of many environments where few or no people live. Dogs, cats, guinea pigs, hamsters, snakes, geckos, chameleons, tarantulas, ants, and worms are all living in our homes and schools.

Cities are perhaps most exciting when storms hit, windows shiver, thunder and lightning crack the night, lights go out, candles are lit, snow silences the streets, stars come out sparkling — and birthrates are statistically higher nine months after a power failure and blackout. Even moderate earthquakes and volcanoes at a just slightly dangerous distance, like Mount St. Helens erupting within sight of Portland, are exciting, breathtaking experiences that create respect for nature in most of us. As Ian McHarg says, "We need nature as much in the city as in the countryside."[1] Most urbanites and, in fact, suburbanites too, spend a good deal of time and energy dreaming, planning, and doing something about getting out of town in quest of more nature. Suburban life in itself is an expression of the desire to have both an urban and a rural life at the same time.

The Nature of Cities

If we want to understand the nature of cities, distance is a key factor. The principle of access by proximity applies to living organisms and cities alike. Gathering people together reduces distances, which in turn reduces the need for travel and the expenditure of transport energy, the level of pollution produced, and the quantity of land paved. If cities are to fulfill an evolutionary and environmentally healthy purpose of some sort, it is necessary that they be as trim and energy-efficient in their activities as natural ecological systems invariably are.

Country people who actually live off the land and do no harm in the process, don't commute, travel only rarely, compost, recycle, and use renewable energy technologies that have little impact on nature. But the typical energy requirements and pollution production per person in the country can be very high. The rural or exurban lifestyle, where a house stands alone, sharing no walls, common spaces, or tools with others, and where long-distance travel is required for sustenance and socializing may appear to co-exist gently with nature but in fact is a major cause of environmental degradation. Extra energy is required for many reasons, and though pollution per person is diluted in the larger landscape, the total CO_2 output and its global effects can be very high.

When we build a city, as compared with the rural or suburban infrastructure, the investment per person is far smaller — another efficiency in the nature of the city that benefits nature. The building materials, streets, rails, transportation vehicles, pipes for gas, water, and sewage, electric and telephone lines, and energy required are far less per person than in sprawled suburbs and rural areas.

If cities are to fulfill an evolutionary and environmentally healthy purpose of some sort, it is necessary that they be as trim and energy-efficient in their activities as natural ecological systems invariably are.

Services are more compacted spatially, too, allowing speedier delivery and less expenditure of money and energy. In the city, the postal delivery person serves many more people in a day's work than in the country, delivering mail to dozens of people at an apartment building or two when those dozens might take five, ten, or twenty times as long to serve in the suburbs or country, requiring the post office to buy more trucks and gasoline and to hire more delivery people for any given number of deliveries. Fire trucks and ambulances arrive more quickly in the city. More people with serious injuries die in the country than in the city because of the time it takes to get to hospitals.

Cities are blamed for the sheer quantity of their population and pollution and for their demand for energy and other resources, but if the same number of people with the same level of consumption were more dispersed, their impacts would be far greater. When people blame the city in general for environmental damage, they fail to realize that the same number of people in a suburban context would cause even more damage. They fail to understand the connection between the all-important arrangement of services and the density of population, lumping dense, diverse communities in with sprawling suburbs. Compact cities can actually constitute a long step toward reducing a population's assault on nature. Reducing overall population, reordering the built community, cutting demand,

changing the technology, or any combination of the above will help solve the problem. When city building becomes ecocity building, all approaches will be taken at once (Impact = Population times Land use/infrastructure times Affluence times Technology). When all these parts are brought together in the logic of a well-tuned organism, then we will have a synergistic combination with far greater benefit than otherwise conceivable.

The well-formed city is a kind of economic/social machine of very high efficiency. People live in cities partly because the very structure of the city means that they can get more done there with less energy, effort, time, and money. The city is a natural pattern of organization for cultural living; it is as natural as anything else about us might be natural. When built and functioning well, the city can be an excellent tool for bringing culture into harmony with nature. It's just that nobody has bothered to pursue its design and construction as such.

Elephants in Berkeley

Berkeley's natural historian and expert on Native American history, Malcolm Margolin, has described the rich tapestry of living creatures, landscapes, climates, and micro-climates that characterized the Berkeley-Oakland area before the Europeans arrived, at a time when the local Native Americans managed the landscape with fire. Grassy hills on the east rose over a gently sloping savanna covered with

When built and functioning well, the city can be an excellent tool for bringing culture into harmony with nature. It's just that nobody has bothered to pursue its design and construction as such.

Proposed bridge from the University of California campus to downtown Berkeley

more grasses and dotted with occasional oaks. These sloping flatlands terminated in the west at the marshes and sandy beaches edging San Francisco Bay. The canyons in the hills were wooded with gigantic redwoods and fragrant bay laurels. Many creeks, emerging from the hills, took roughly parallel courses across the flatlands to the bay, connecting the ridgeline to the shoreline with green strips of mixed bushes and willows and occasional larger trees. These colonnades of trees seemed to pour out of the canyons through the year-round grasses, past the scattering of seasonal ponds and freshwater marshes, toward the salty realm of crabs, shrimp, salmon, sturgeon, seals, otters, and whales.

This was the most popular avian resort on the migratory flyway of the west coast of North America. Seabirds and inland birds by the millions blotted out the light in dark, swirling clouds, raising the sound of thunder with their countless wings. Hawks cut their way through the confusion to rodents and rabbits while enormous California condors hovered and wheeled overhead. Elk, antelope, and bear moved across the landscape with other large mammals, not the least of which were the people.

The Oakland-Berkeley Hills Firestorm of October 20, 1991, swept away whole neighborhoods — 3,375 homes — with such an intensity that concrete foundations crumbled into softly rounded forms. Despite my having known the landscape so well, so complete was its transformation that I found it hard to remember what had been there before. When a few days later I asked Malcolm, "What was the landscape like here before the Indians arrived and began managing it with fire?" he paused for a long moment.

"Stunning question," he replied. "I don't

know. I don't know if I've ever thought about it."

In the next few days I kept asking the question of friends until one of them, a remarkable naturalist named Sterling Bunnell, said, "Well, it probably looked a lot like East Africa does now, because it was managed by elephants."

Elephants in Berkeley? That was a mind-expanding idea. Probably, Sterling suggested, the landscape here before the Indians, before about fifteen thousand to ten thousand years ago, was very complex and richer in biodiversity than the same land the Indians inhabited later. There were two species of elephant, a giant camel, a giant ground sloth, a giant bear, a very large dire wolf, the American lion (slightly larger than the African lion today), and horses. The Indians, he said, seem to have eaten them all — maybe not the wolf and lion — and were probably the main cause of their extinction. In addition there would have been most of the animals the Indians were living among when the Europeans arrived, such as elk, deer, antelope, and grizzly and black bear.

The elephants' role? They rummaged about tearing up trees, creating large areas of open space, and trampling paths in wild but logical patterns across the landscape, clearing portions of forests in some areas, slicing through brush lands, leaving trails that other animals could use. Elephants have a knack for knowing where water is located just under the surface. They rip into the soil with their tusks, throwing the earth to the side, and dig down to the shallow water table. Thus they create ponds that they and many other species enjoy for drinking and bathing. In other words, they manage the land like a special kind of gardener who somehow, without a stated plan, ends up creating tremendous biodiversity. Furthermore, in Africa they are known to deforest very large areas of land, giving rise to a whole new sequence of ecological zones, with the forests growing back over many decades. They act something like great fires, rotating around immense landscapes and creating variety over long periods of time as well as large geographic areas. It seems they are bent upon creating diversity in almost every way they can.

What the elephants do on their canvas of millions of acres over hundreds of years, said Sterling, is essentially to create convoluted edges between various habitats and, for that matter, time periods. Edges are far richer in

Below:
Plastic bridge
installation

life than large habitats of a single type or ones existing in the same condition without change over time. He added that the appreciation people have for landscapes with edges (a forest edge on a meadow, a moist oasis up against cactus desert), which are called "ecotones," is probably the source of some of our esthetic sensitivity and comes from the fact that they are full of possibilities for things to eat, building materials for nests, and places to escape from predators. I asked Sterling, "If elephants are gardeners of this sort, creating such rich environments, could people do the same?" "Undoubtedly," he answered.

Margolin has described how people can insert themselves into a natural environment and create subtle changes in that environment that make it richer in wildlife. In *Earth Manual: Working Wild Land without Taming It*,[2] he writes of collecting seeds from native plants, carefully labeling them, then planting them strategically at proper times of the year. In the same breath, however, he may propose chopping down a tree or even cutting half way through the trunk and toppling it over so that, in the late winter and early spring when food can be desperately scarce for deer and rabbits, the animals can find browse in its twigs, buds, and bark. This tree cutting is based on the assumption that there is no scarcity of trees in the area, but it is still startling — except for the fact that we can see in it the elephant gardener at work. Similarly, he suggests gathering broken limbs and arrang-

ing logs and rocks to create cover for small animals so that they can more easily avoid predators. He also advocates damming watercourses far more enthusiastically than I had anticipated, saying that it is hard to go wrong creating more bodies of water and land/water edges provided that proper fish ladders and special diversity-maintaining strategies are employed. Cataracts should be saved. Flooding rare habitats and beautiful features should be avoided. As Sterling Bunnell points out, an esthetic sense helps us identify the areas of most diversity — again, the elephant at work, or maybe in this case the beaver.

Permaculturists, with their subtle design approach to a "permanent agriculture," agroforesters, and organic farmers all use design principles to produce rich harvests, although the biodiversity here is likely to be mostly among non-indigenous food plants and animals. Crop rotations, companion planting, the use of manure, and composting have the potential, taken together, to get enormous production out of the soil while leaving it as rich as or richer in nutrients than before.

Native Americans and Invaders

Sterling Bunnell theorizes that in their ancient oral traditions the Native Americans remember with regret the earlier days when their ancestors hunted and exterminated the great Pleistocene mammals of North America. He thinks that it is likely that the traditions that focus such reverence on nature come from such an understanding and that a determination to avoid any such future catastrophe is partially responsible for their ecologically informed philosophy and lifeways. By this theory, the Bay Area Native Americans managed their landscape with fire specifically for maximum species diversity; that is, for the preservation of all species. Other students of the subject say that the warming and drying of the continent in the waning centuries of the last Ice Age, which coincided with the appearance of humans with relatively advanced hunting weapons, was the more important cause of extinction and that, in any case, there is no way an oral tradition could preserve memory over such an enormous time span. In either case, some people knew a lot more about their environments than the European immigrants and their descendants and used that knowledge in arranging their lives and communities. We can learn from them.

Kirkpatrick Sale and David E. Stannard both speak of the relationship between the white conquerors of the Americas and those they conquered, and contrast their perspectives on nature.[3] For those who think that recent industrial society has been the first society to abuse nature and that the science of Newton and the philosophy of Descartes represented the almost biblical fall from grace, these scholars' work opens up new territory. The poison had been alive and well in a form as damaging as today's and goes back far beyond Newton and Descartes; in fact, it *is* biblical and goes back even further than that. In many ways the pre-industrial conquest of peoples and nature was even more intentional than today's environmental destruction, which is often a by-product of preoccupations so far removed from the site of exploitation that the beneficiaries of the exploitation scarcely know it is happening and, furthermore, don't want to know. The older style of exploitation was far more personal and intentional. People just went out and killed, stole, and extracted with a righteous rationale, clad in armor, by the strength of the flailing arm, with blood and wood chips flying.

"When Colon [Columbus] set foot on his landfall island he brought this ecological heritage with him," says Sale:

> We must begin, alas, with Europe's fear of most of the elements of the natural world — a fear based, as it always is, on simple ignorance .... The church offered no encouragement for any investigation .... Common lore ... was not much better, filled with either mundane and stereotypical views (lambs are meek, lions brave, wolves crafty) or fanciful and erroneous ones (toads suck cow's milk at night, woodpeckers are dangerous predators, beech trees deflect lightning...) ... All this platitude and misinformation about the real world was glued together with non-

sense about the monstrous and fantastic world, and held to with the same level of credulity by even the most inquiring minds of the day. Laurence Andrew's very popular bestiary *The Nobel Life & Natures of Man, of Beasts, Serpents, Fowls & Fishes That Be Most Known,* for example, the first printed work on animals in the English language, lists with equal credulence 144 known animals, 8 entirely unknown, and 21 strictly mythological.[4]

The Alps that we tend to think of as beautiful today, draped in snow, wreathed in clouds, and looking up at us from coffee-table books and down from wall calendars, were "distorted," "chaotic," and "hideous" to numerous commentators at the time of the early European colonial conquests. The forests were full of monsters, criminals, and diabolic forces enough to terrify any child in hearing fairy tales or adult in contemplating long travels — the sooner they were cut down and put to use, the better.

And that they were. Sale estimates that by 1500 Europeans were using up a ton of wood per person per year.[5] John Perlin in *A Forest Journey*[6] documents forest after forest in Europe falling faster than they could regrow, from Roman times on. This followed a legacy two or three thousand years older in the Middle East, where deforestation upstream in the Mesopotamian Valley watershed helped drive one civilization after another over the edge and into oblivion. By Columbus's day, many European countries were searching outside their borders for the natural resources that had become scarce or economically or technologically unavailable because of over-utilization. Just in time, a whole new rich and relatively defenseless world was discovered.

Just as the state and church controlled life and afterlife in the Europe of the time, so the Europeans saw the wild areas of the world and their people as dangerously out of control and needing to be disciplined, civilized. "Wild," Sale points out, comes from "willed," that is, "self-willed, self-determining, independent." People and living things that functioned on their own without adherence to a clearly hierarchical authority were somehow out of line in a proper cosmology, and therefore their exploitation and even brutal destruction was of no particular concern — unless the concern was to improve things:

> This separation from the natural world, this estrangement from the realm of the wild, I think, exists in no other complex culture on earth. In its attitude to the wilderness, a heightening of its deep-seated antipathy to nature in general, European culture created a frightening distance between the human and the natural .... To have regarded the world as sacred, as do many other cultures around the world,

would have been almost inconceivable in medieval Europe — and, if conceived ... punishable by the Inquisition.[7]

In addition to ships, guns, and a willingness to kill, steal, and rationalize, says Sale, "it has been estimated that because of its animals of transport and burden fifteenth-century Europe had a source of power five times as great as that of China. If one considers the almost total absence of large domesticated animals in the New World, it might be said to have had as much as twenty times that of the Americas."[8] If we consider the fact that no wheels were used in the New World except, oddly, on toy animals for children, we can see another vast advantage of the Europeans in their conflict with Native Americans.

But the conquest and subsequent colonization of most of the world by Europeans was perhaps based even more on cities. First, the city itself was an invention that made possible an extremely rapid exchange of ideas, resources, and tools and provided an immediately available workforce. We seldom hear that the city itself conquered nature and the more nature-based cultures of the world, but the suggestion was not lost on the conquerors themselves. Their entire enterprise was named for the city and was very consciously called "civilization." It was held in such high moral, mythological, even cosmological esteem that it became the ultimate rationale for exterminating or expropriating all cultures, species,

The city itself was an invention that made possible an extremely rapid exchange of ideas, resources, and tools and provided immediately available workforces.

properties, and habitats in its path. The objective was to bring civilization to all people and to tame the wild for industrial and agricultural production, with the city at its center. Heaven itself was the City of God, and it was its streets, not the rural paths in the woods, that were paved with gold.

In contrast, the indigenous peoples lived more harmoniously with nature. One such tribe still exists and was recorded in 1988 in a remarkable BBC film by Alain Ereira called *From the Heart of the World: The Elder Brothers' Warning.* For over 1,000 years the Kogis of Columbia have lived high on the forested mountain slopes twenty-five miles from Santa Marta on the Caribbean coast. Since they melted into the forest to escape the Spanish in the 1500s, they have been practically unknown to the outside world. A nineteen thousand foot mountain massif eighty miles on a side called Pico Cristobal Colon or the Sierra Nevada de Santa Marta is the glacier-draped center of their world and, they say, ours as well.

In the 1970s, with their glaciers receding, waterfalls drying up, and species disappearing, the Kogi elders (the Mamas) trained one of their own to become their spokesperson to plead with the outside world for restoring balance to human/natural affairs. Coincidentally, at about the same time, a grave robber digging for the gold of the Tairona, the Kogi's ancestors, wandered into a particularly forbidding and wet sec-

Wind screen and glass
elevators.

*These create a pleasant
environment on roofs,
which are generally
windier than locations on
the ground. Glass elevators
are just plain fun, plus, if
there are concerns about
security, you can see who
you may or may not want
to ride with.*

tion of the lower mountains, and there, just
fifteen miles outside Santa Marta, stumbled
upon a stunning "lost city." The anthropolo-
gist who directed the early excavations of the
city, Alvaro Soto, said, in Ereira's film,
"What the Indian mind wanted to show was
that it is possible to have a good density of
population in a very beautiful environment

without destroying it. They adapted the city to fit this environment perfectly."

Despite over four hundred years of erosion by water and disruption by plant roots in this dense, wet jungle, the stonework is only slightly damaged. The Colombian Director of Indian Affairs, Martin von Hildebrand, said of the city, "It was a whole integrated organism. It is the interrelation of these sites that keeps the world in harmony, and it is the duty of the Mamas to see that the world remains in harmony. If you excavate ... you take out part of the system. Each part is integrated into the whole. They consider that fundamental."

Said the Mamas, "People were made to care for the plants and the animals. If we act well, the world can go on." Their directions for restoring the well-being of the world included the request that anthropologists, gold seekers, and others leave their lost city so that they could return to it and restore it to well-being and, in so doing, increase the well-being of the world. Their request was honored, and they have begun to reestablish themselves there.

Close-ups in the film about the Kogi provide some notion of how building in that location could be substantial and long lasting. Rocks in walls and walkways on the steep slopes and ridges lap in such a way that dripping water from the frequent rains falls free of the walls rather than running down cracks where stones meet. The gardens of the Tairona and the Kogi appear to be a cross between complex companion planting and a kind of forest management. They plant in the forest while caring for the self-propagating plants that they value for their many different uses. What they are doing amounts to cutting trees selectively, choosing carefully to maintain species diversity, and enhance their usefulness to people; introducing special food plants and carefully managing the ones already present; protecting favored animals from predators; feeding individual animals and plants in need; clearing small areas to get a little extra sun to a flower considered especially beautiful, planting to give a little more shade to another; adjusting a rock to direct a rivulet in times of rain to a dry plant, removing a big rock in a nearly dry stream bed to create a small frog pond. All this and more than we can imagine is the reality there in this landscape that ranges from rain forest to desert, foggy forest to icy tundra, palm-fringed Caribbean beach and tidal pool to blue glaciers and rainbow frosted waterfall — all separated by less than twenty-five miles.

The lost city of the Tairona, center of this culture, was of course strictly pedestrian and so were the other ancient New World cities and towns. Their citizens probably didn't think much in terms of "compact development" because, given a foot-transportation reality, they simply built so that people could get to one another and out into the countryside. It must have seemed natural — as natural as clusters of cliff-swallow nests, beaver dams, honeycombs, and caterpillar tents.

Not only the environments and cultures in the Americas but also the types of community were extraordinarily diverse. Many of the works of Native Americans were formidable, reports David E. Stannard in *American Holocaust*. The Adena culture, beginning 1,000 years before Greece, came to cover an area from Vermont to Indiana and from New York to Virginia. Its people built "towns with houses that were circular in design and that ranged from single-family dwellings as small as twenty feet in diameter to multi-family units up to eighty feet across ... in close proximity to large public enclosures of 300 feet and more in diameter" that were called sacred circles by archeologists "because of their presumed use for religious ceremonial purposes."[9]

The Hopewell culture that followed it covered a territory from New York to Kansas and from the Northern Great Lakes to the Gulf of Mexico. Like the Adena culture, it had intensive horticulture and large monuments to the dead. Says Stannard,

Literally tens of thousands of these towering earthen mounds once covered the American landscape from the Great Plains to the eastern woodlands, many of them precise, geometrically shaped, massive structures of 1,000 feet in diameter and several stories high: others — such as the famous quarter-mile-long coiled snake at Serpent Mound, Ohio — were imaginatively designed symbolic temples."[10]

The fantastic carvings and paintings of the Northwest Coast Indians, adorning totem poles, large communal buildings, and seafaring canoes, are relatively well-known, but even in the supposedly barren lands of the Inuit the seas, shoreline, and skies often seethed with life, and the people built not only the well-known igloo but also, Stannard points out, "the huge semi-subterranean *barabara* structures of the Aleutian Islands, each of them up to 200 feet long and 50 feet wide, and housing more than 100 people."[11]

Then there were the Anasazi of Arizona, New Mexico, Colorado, and Utah. In Chaco Canyon, New Mexico,

...more than 1,000 years ago, there existed the metropolitan hub of hundreds of villages and at least nine large towns constructed around enormous multi-storied building complexes. Pueblo Bonito is an example of one of these: a single, four story building with large high-ceilinged rooms and balconies, it contained 800 rooms, including private residence for more than 1,200 people and dozens of circular common rooms up to 60 feet in diameter. No single structure in what later became the United States housed this many people until the largest apartment buildings of New York City were constructed in the nineteenth century.[12]

Today's Hopi people, who claim the Anasazi as their ancestors, live in much the same manner. Vernon Masayesva, chairman of the Hopi tribe, describes their cities this way:

Some of our pueblos date back to 900 A.D. and some of our oldest villages pre-date that period. Oraibi is the oldest continually inhabited community on the North American continent. The only message I can give you is what Hopis have accepted as their responsibility for the privilege of living on this Earth. We are concerned for all living things; our architecture reflects this same spirit of reverence for what we call Mother Earth. Hopi villages have pueblo style housing. We were the first known apartment builders constructing houses about four stories high.

The Hopi village always has a plaza, which is the heart. A village to Hopis is a living entity, and so the center, the plaza, is the heart where all sorts of public ceremonies are performed, where children are entertained. Many religious ceremonies were reserved for special places called kivas, built in outlying areas around a village for special ceremonial purposes. The old villages, old houses, were not necessarily built by men; many were built by women, because it was like giving birth to a house. The Hopi architecture was built by the people, for the people. This is an important point to make because I've been through several cities, and the new architecture seems to me to be all science. It's all brains and no heart. They aren't cities where we can grow and be

Snap-on umbrella cables
These can keep umbrellas from flying off windy roofs.

sensitive to our environment, where we nurture our values, where we can teach our kids the important values we never want to forget. When the Americans came they wanted to make us in their image and they brought with them schools and their style of architecture, which was pitched roof, tin-covered, square box houses. And they built these houses primarily to entice the Hopi leadership to move down off the mesas and into the houses, the new houses, the white man houses. It was a way to break up the community in the villages. The buildings that they built had no relation to what the Hopi values were .... So eventually we forgot that communities can be built in such a way that they're environmentally sensitive. If the Hopi has any message for future cities, it is to keep these things in mind. Building should respond to the needs of mankind, and the need of mankind today is to be stewards. So the ecocity is not just the city of the future, but is also the city of the past. Because they all have to now be tied together.[13]

How can the Kogi and the Hopi, with such modest technologies, consider themselves the keepers of the Earth when they lack the powers of industrial society to mine, techno-farm, alter enormous landscapes, and build gigantic edifices? In the same way that the bacteria in the oceans three billion years ago managed to change the atmosphere of the entire planet and to transform its stony shell to the depths of many miles — by small acts of commitment sustained over a long time all going in the right direction. Small things can replicate. Small steps toward ecocities can add up to a healthy planet.

The Kogi and the Hopi actively seek the ear of the outside world because they identify with the patience of nature, knowing it well, and understand the power of the deceptively small thing, the idea, the thinking itself. Oddly, we in the technological world around them, steeped in massive flows of information, give very little credit to thinking itself. What really counts in our high-consumption world is the bottom line: how much wealth, power, and supposed resulting pleasure we have accumulated. We collect acquisitions and accomplishments as if making a list of our lives. "Been there, done that," and then we hurry on. Those who know the power of a true idea are different, and they have included many Native Americans. They have what I call "conceptual integrity," which means they believe in something not because it has high ratings, big sales, voter support, multiple endorsements, and popularity but because it seems to be true in its own right. A few good words at the right time are like a minute cosmic ray on its trajectory toward a strand of DNA about to bring on a positive mutation, to create a new species and thus a new world, not out of arrogance but by simply being there, heading in the right direction, at home in the universe,

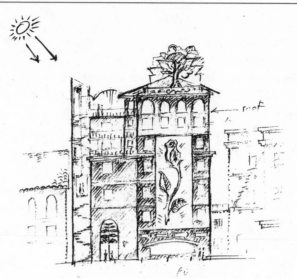

The Deja Vu building. *A proposal for a building in Berkeley for the Berkeley Psychic Institute, exhibiting a glass elevator sharing the sound side of the building with a glass atrium entry. There are wind screens, street-side window gallery, rooftop tree, great views and the rose symbol of the organization.*

doing the right thing. That, which the Indians understand and we don't quite yet, is the big secret for the healthy future.

Bioregions, Hinterlands, and Cities

The natural system that expresses a kind of unity is known as a bioregion — a landscape within natural bounds, often a watershed, made up of a distinctive set of species in inter-relationships specific to that region. The demarcations can be subtle, as a rich grassland fades into a desert, or abrupt, as a mountain ridge cuts a sharp line between dense upwind forest and downwind desert. Within a bioregion there may be extremely complex and highly differentiated smaller environments, such as a river with all that it supports in its waters, a marsh system, and a grassland valley, each with fish, birds, insects, and large mammals, some of whose range overlaps into the adjacent environments.

Peter Berg of San Francisco's Planet Drum Foundation has added the human dimension to earlier notions of the bioregion, acknowledging that cities can fit comfortably into a bioregion. Just how has not yet been thoroughly worked out, partly because detail is of the essence, and it takes a long time for the complex detail of a bioregion to be organized into clear, large patterns of order in the mind of any person living there. But the fine threads are weaving together (miniplexing) right now. We are learning in bioregions by long sensitive study, observation, and careful participation in trial, error, and course correction. Traditional indigenous cultures have had a strong tendency to respect their elders, in part, because the nature of much of their knowledge is complex and detailed and takes more time and patience to learn.

Jane Jacobs, bringing her keen observations of urban life to bear on the bioregional perspective, links cities intimately with what she calls "the hinterlands;" the basis in nature providing resources for the city while the city provides the vital economy for both itself and the hinterlands. Summarizing economic theories of recent history, she points out how difficult it has been to apply these theories to reality. It seems that all of them have failed to function as predicted. In fact, economic surges and disasters have come and gone in a pattern so much their own, so independent of control or clear understanding, that theory is little improvement over muddling through by habit or tossing dice. "In the face of so many nasty surprises," says Jacobs,

> ... we must be suspicious that some basic assumption or other is in error, most likely an assumption so much taken for granted that it escapes identification and skepticism. Macroeconomic theory does contain such an assumption. It is the idea that national economies are useful and salient entities for understanding how economic life works and what its structure may be; that national economies and not some other entity provide the fundamental data for macro-economic analysis.[14]

What Jacobs offers instead is the concept that cities, generally the relatively large and economically diverse ones, are the key organizers of economies, the basic units of economic prosperity. It is in the nature of vital cities to produce a great variety and number of products for their own people and to trade actively with suppliers in the hinterlands and with other cities. Cities can do this because they have assembled many productive, creative people and their tools, facilities, and resources close enough together to allow frequent exchange with each other, to allow them to share ideas and provide raw materials or parts or services for one another. Without all the parts, including the people, linked and actively participating together in a vital city, the economic processes cannot thrive. Some people — the Shah of Iran and Peter the Great of Russia, for example — have tried to buy "development" without assembling the parts of a city, including the traditions of innovation, but it has never worked. Lewis Mumford, advancing the same thesis twenty-three years before Jacobs, argued that "the industrialization and commercialization we now associate with urban growth was for centuries a subordinate phenomenon, probably even emerging later in time."[15]

Jacobs also points out that the city depends on the hinterlands from which food and resources come. The people in the hinterlands in turn receive from the city tools, cultural items, and access to markets. If the city and its hinterlands can thus be mutually supportive, ecological diversity in the country can be encouraged via an increased variety of markets, which, paralleling biodiversity, encourages diversification of products from the hinterlands. In one case she mentions, in Japan, rare mushrooms attracted the attention of farmers and began to be harvested for sale in a nearby city. After that, the mushrooms and their associated environment thrived under the stewardship of people selling to the markets in the city. With the mushrooms as a new source of prosperity, the people reduced

Vancouver, Ken Yean, and putting pedestals and towers together.

Four images follow in sequence, starting here with towers in Vancouver, British Columbia. Trees are planted on pedestals and sidewalk edges, creating an interior streetscape of urban forest. Towers are relatively far apart to preserve views of ocean and mountains, yet density is high on average.

forest cutting and invested some of their new income and liberated time in imaginatively expanding their range of products even more, while obtaining new tools and information from the city. They discovered a vested interest in natural biodiversity and launched initiatives to preserve it. In this way complex cities can assist in restoring and maintaining complex ecologies.

At the core of the dynamic of cities are the various business and production activities Jacobs calls "import replacement": producing varieties of items that make the community thrive, instead of importing and buying from others. She considers import replacement very basic, going so far as to say that, to exist at all, a vital city needs a pre-existing vital city producing what it needs but is not yet able to produce. She traces cities slowly emerging out of the early unknown in ancient times and overlapping one another temporally, more recent ones firing up on the basis of responses to their earlier trading partners.

Jacobs speaks of lonely Venice, established in the seventh century on a cluster of islands off the Adriatic coast of what is now northern Italy. Retaining knowledge of boat building and other fading memories of Rome and Greece, its citizens found refuge from the barbarians on the European shore who had failed to learn from the empire they had helped destroy, had no ships, and simply couldn't get at them. But there was a city that Venice, struggling up from the mud and sandbars of the Great Lagoon, could trade with: Constantinople, fourteen hundred nautical miles away, maintaining the functions of a vital city. Venetians, says Jacobs, were clever enough to begin replacing their Constantinople imports by making many of them for themselves. Virtually all they had to trade was salt evaporated from the shallow waters around Venice, but the people in Constantinople wanted it. As they got good at replacing their imports by making the products themselves, the Venetians began trading their new products with later European cities, playing the same parenting role that Constantinople had played for Venice.

Later European cities, launched largely by Venice centuries earlier, played the same parenting role for American cities. The Native American cities parenting one another in this manner couldn't influence the new wave of cities after Columbus because, simply put, they were summarily annihilated. Jacobs writes:

> City import-replacing is not all that economically glamorous. The replacements are usually small initially, frequently involve items that in themselves are frivolous, and in many cases are absolutely imitative — but nevertheless, in the aggregate, they add up to momentous economic forces .... Indeed, as far as I can see, city import-

Complex cities can assist in restoring and maintaining complex ecologies.

Ken Yeang's "bioclimatic skyscraper."
Architect Ken Yeang designs buildings for hot climates that use incised "skycourts" to create shade and capture breezes for cooling, and feature planting of large, elevated interior trees in the skycourts.

replacing is in this way at the root of all economic expansion.[16]

This may be something of an explanation of the nature of cities as they go about exploiting nature and other cities for their own development. But another question immediately comes to mind: Where do the imports that get replaced come from originally? Some cities may be replacing imported computer keyboards in the early 2000s and supplying themselves and their region with them, for example, but cities weren't doing this in 1940 because these items didn't exist yet. Who came up with the original? On this even more basic level of creativity it seems that the city is

at least as important as it is in the import-replacing business, serving to gather resources not just for the cleverness of fabrication but for the genius of creation itself.

At yet one level deeper we can ask how the city itself was created. Thomas Berry, the theologian of the Ecozoic, says that the most creative period of history was the Neolithic village-making phase. Perhaps the basic building blocks of today's economics, the germ of the first city, began then with the simple gathering of homes together into the first village that added a few community buildings with special designs for special functions. The Agricultural Revolution, with its domestication of animals and plants and slow buildup of items — stone tools to copper to bronze to iron, baskets to clay pots to metal pans, sun-dried brick to fired brick, dry-stacked stone to stone and mortar, and so on — initiated much of the basic village creativity elaborated and refined later in the cities. The logic of the whole physical community-building venture, including the social and economic dynamics of it, was probably as unconscious then as it is now for almost everybody. The creativity of the myth-building and spiritual awakening that concerns Berry predates cities and developed as language — another immense cultural creative product — complexified. All this was sheltered and facilitated in the early organizing unit that was the village in its natural hinterlands.

All these elements and the village itself grew, and grew very slowly together. Mumford believes that "the city of the dead antedates the city of the living"[17]: nomadic people came back to burial grounds to remember their loved ones and their past, set up temporary camps nearby, probably noticed discarded seeds from earlier visits sprouting into berry plants and fruit trees, set up permanent camps after that, and, *voila*! There is evidence, too, that the complexifying village was a creation of hunting and gathering societies predating advanced agriculture instead of vice versa. In this view, the magic of "access by proximity" and per-haps an organized sort of leisure facilitated by the village led not only to the culture of complex cities but to agriculture, too.

Buffalo Commons

Today many towns and rural areas do not flourish without massive subsidy from outside, usually provided by a national economic policy establishing an extractive economy. In a sense they are "unnatural" and are forced into a non-reciprocal relationship with the land, climate, and biota, that is the flora and fauna of a region. Usually they are located in economically dependent mining, logging, or cattle-raising regions that Jane Jacobs calls "supply regions," often far from the cities to which they deliver resources or products. Without a national policy to support it, habitation of such supply regions would barely exist since nature can't support it and distances from a vital city economy are too great to be conveniently bridged. But nation states are another thing. They look for defense resources, cultivate national pride, want to extend the common language and customs, are fixated on grand economic strategy, believe in the conquest of nature and savages by civilization, and want more wealth for their rich people, power for their politicians, dreams for their poor, and so on.

Looking at the Great Plains of the United States gives us an example of what a nation state, as contrasted with a city/hinterlands system, wanted to extract from the countryside: in this case wheat, and Indians. Initially, grain agriculture was the economic idea, and an Indianless frontier the — racist — cultural one.

In 1937 the painter Thomas Hart Benton wrote, "Cozy-minded people hate the brute magnitude of the plains country. For me the great plains have a releasing effect. I like the way they make humans beings appear as the little bugs they really are. Human effort is seen there in all its painful futility. The universe is stripped to dirt and air, to wind, dust, clouds, and the white sun."[18]

Frank and Deborah Popper of Rutgers University's Urban Studies Department think it's time to face up to the reality that the part of the United States called the Great Desert on early nineteenth-century maps is simply not economically substantial enough — and far too far away from places that are — to be anybody's hinterlands but its own:

At the center of the United States, between the Rockies and the tallgrass prairies of the Midwest and South, lies the shortgrass expanse of the Great Plains. The region extends over large parts of 10 states and produces cattle, corn, wheat, sheep, cotton, coal, oil, natural gas, and metals. The Plains are endlessly windswept and nearly treeless; the climate is semi-arid, with typically less than 20 inches of rain a year.

... A dusty town with a single gas station, store, and house is sometimes 50 unpaved miles from its nearest neighbor, another three-building settlement amid the sagebrush .... Although the Plains occupy one-fifth of the Nation's land area, the region's overall population, approximately 5.5 million, is less than that of Georgia or Indiana.

The Great Plains are America's steppes. They have the nation's ... coldest winters, greatest temperature swings, worst hail and locusts and range fires, fiercest droughts and blizzards, and therefore its shortest growing season.[19]

This starkly beautiful big-sky country, with its ferociously unforgiving weather, has absorbed and flung back wave after wave of government-subsidized settlement. First the 1862 Homestead Act sent a flood of farmers and ranchers into the Plains with large subsidies, not the least of which was the clearance of the Plains Indians by massacre and starvation via destruction of their food supply, the buffalo. Exterminating the buffalo also helped the future homesteaders because, as a friend of mine once said of the vast stampeding herds, "Those guys don't stop for fences, you know." The intense blizzards of the 1880s and the drought and financial panic of the 1890s drove out most of the first wave of homesteaders.

During World War I, with competition reduced and demand increased because of Europe's collapsed agricultural production, there was another subsidized boom. But even before the depression of the 1930s, the economy of the Plains was collapsing again. Then came the Dust Bowl.

From the 1950s through the 1970s dams were built with federal funds, lands purchased or reclaimed from the second departing wave of homesteaders and rented below market to new farmers and ranchers: the third wave of subsidy. In the 1970s oil and natural gas were extracted, coinciding with a quadrupling of the value of these fuels due to OPEC's increased oil prices, creating two hundred energy boomtowns. Today we are in another bust cycle and the Poppers are asking why we should throw good money after bad. Water supplies are diminishing in the area. The enormous Ogalalla Aquifer, a veritable underground sea supplying eleven million acres of farmland in Colorado, Kansas, Nebraska, New Mexico, and Texas, is being depleted far more rapidly than it can be replaced by rain soaking into the terrain. Many counties in this region are in poverty conditions, with virtually all their young people

Integrating pedestals with towers and planting.

departed. The median age is more than fifty. Meanwhile, buffalo ranching is becoming firmly established. When cattle freeze and stiffly topple over dead in the harsh northern winds, the buffalo just hunker down in little clusters of deeply matted furry insulation many inches thick while land values fall and schools, banks, and whole towns close down.

Restoration of some of the wild lands has worked already. "Beginning in 1937," say the Poppers, "the federal government bought up 7.3 million acres of largely abandoned farm holdings of the Plains (an area bigger than Maryland), replanted them, and designated them national grasslands. Today the national grasslands are used primarily for low-intensity grazing and recreation. Often thick with shortgrasses, they rank among the most successful types of federal land holdings."[20]

Why not support what works? ask the Poppers. Why not take nature's hint and reintroduce what they call the Buffalo Commons: a gigantic swath of the United States from Texas to Canada where wire fences snap, curl, and rust into the ground; where wooden posts dry, crack, fall, and disappear beneath wildflowers from horizon to distant horizon; where the plodding of cattle hooves recedes into oblivion and the thunder of bison returns? In this wild landscape, a few oasis-like small towns could be cultural watering holes for hearty Indians, buffaloboys, and occasional ecotourists from distant cities. The Poppers predict that we will, in fact, learn from our past mistakes:

During the 21st century, the American frontier will expand and become more visible. Large chunks of the rural West will be privately preserved — for instance, by ranchers who discover they can do far better by renting their land to hunters for part of the year than by laboriously running cattle on it year-round. The Nature Conservancy and similar preservation organizations will make extensive land purchases. The ecological restoration of land damaged by previous extractive use will be big business; so will ecological tourism .... The combined rise of preservation and decline of extraction will present a remarkable chance to undo the nation's past mistakes .... We are no longer a frontier nation, but we are still a nation with a frontier. And it will be a frontier that will expand far into the next century.[21]

So said the Poppers in 1991. Today, they are seeing their Buffalo Commons "metaphor" moving toward reality. It makes sense in this place. In a new flurry of scholarly papers they are now touting the use of regional metaphor to create themes that will shape sensible relations between people and their natural and cultural foundations. A metaphor like the Buffalo Commons, they say, is "soft-edged planning."[22] Its objective is not to write laws and ordinances, General Plans, and zoning codes but to create the almost poetic images and understandings that lead such political agreements into a world where our imaginations do right by the place where we live.

Learning from Nature, Learning from Cities

The cultures that lived close to nature, rather than behind the (deceptively) secure walls of cities and suburbs and layer upon layer of technology, of course learned something from nature. They had to in order to survive. But with our whole-systems sciences, like ecology and evolutionary biology — and let's not forget "ecocitology" — we can break through the barriers, the concrete and asphalt and automobile habits, and learn even more.

Perhaps the nature of cities was more accessible to those living in the first of them, the logic of "access by proximity" arising "naturally" before humans had the technologies of transportation that made ignoring spatial relationships possible. In the earlier indigenous ways of learning, the logic of nature and the nature of cities, important relationships with wild animals and native plants were obvious, respect for such large-scale concepts as conservation and recycling was basic. As cities increased in size, however, their citizens forgot many of those lessons, exploiting to the point of expunging the very ethic of conservation and recycling, sweeping it away with the exhaustion of numerous

Further integration of pedestals and towers. *An essentially "single-structure" community can be created and linked to surrounding buildings with bridges*

resources, thoughtlessly dumping waste, and even laying waste entire rival cities. Now we can see a whole list of second-generation lessons from nature, from the subtlety of ecological interconnections reinvigorating the old lessons of conservation and recycling to the recognition of the place of our civilization in evolution.

Hints are available from the deeper layers of biology, too. Nature shows us a sequence of events in healthy biological relationships that we ignore only at great risk. Here the idea of the "builder's sequence," which I will discuss in more detail later on, can be useful. When building a house, we have to start from the foundation. Likewise, when building a city, we have to start from the foundation of good land uses and build architecture, urban functions, and people's lifeways on that layout. The model not only applies to the sequence of evolution, from which we can learn much, but also mirrors the fact that "ontogeny recapitulates phylogeny" — that the whole evolutionary life of a species (phylogeny) is played back all over again in a kind of shorthand as the individual embryo of a living creature of that species develops (ontogeny). We don't just start gestating into full-blown tiny human beings immediately after conception, getting larger for nine months in the womb. Instead we look rather like a jellyfish at first, then develop shark-like gills, then amphibian flipper lobes, then a reptilian tail, and so on, step by step turning into humans, building physically upon a pyramid of life going back more than 3 billion years. Skipping any of our earlier evolutionary stages in the recapitulation simply doesn't happen.

Building upon a 3.5-billion-year biological foundation as well as 15 billion years of cosmic evolution, we are all made of twenty-nine basic molecular building blocks: various amino acids, sugars, nitrogenous molecules, and fats.[23] Later evolution does not create new building blocks of exotic materials from somewhere off the band of organic chemistry or even a different set of organic chemicals but uses the same old ones that are in all of us, from paramecium to redwood tree, from starnose mole to bird of paradise, from deep-sea squid to Albert Einstein. Unless those building blocks get stacked up in a particular sequence and order, the organism doesn't even get started, much less yet attain any kind of health and success among the living. And all or almost all of James Miller's nineteen subsystems of "living systems" have to be present, too.

And so, within the realm of our immense freedoms and rights, the lessons of nature suggest that we need profound grounding in very specific sequences of development and have a duty to honor that biological history. If we neglect it, we will very likely forfeit our freedoms and rights while inflicting immense damage upon the rest of our biosphere. Regarding cities, this will mean respecting the builder's sequence and carefully watching it play out.

A final lesson from nature is that all organisms — from the bacteria that laid down the bands of iron ore on the ocean bottoms from dissolved minerals in the sea water to human beings building civilizations — intervene in the rest of nature, no matter what we do. Julian Huxley once said that we've got the job of managing evolution now, whether by design or default, so we might as well get good at it. We might learn from nature about fires, floods, and the disruptions of elephants as they go about creating high biodiversity that at first glance looks like destruction. The lesson here is that deeper knowledge gained from observation is very important to long-term health. We may thus learn that we need not be especially timid in altering certain particular environments, but must understand life systems with some real sensitivity before taking action. Can we, with our growing planetary, even evolutionary consciousness guided by the timeless conscience of contemporary ancients and aided by modern scientific inquiry, build cities in balance with nature? If we learn from nature, assuredly we can.

Structures at Paolo Soleri's
experimental city, Arcosanti, Arizona

The City in History

THE QUEST FOR THE CITY in balance with nature has almost certainly been part of a dream of human fulfillment since cities first emerged. But cities worked so well, so efficiently, that production became over-production and then surfeit and treasure enough to distort the saintly. Once established among its more nature-rooted kin, the villages, the city swept its citizens away with its confused passions, love of arts, materialist greed, creative brilliance, power hunger, service to humanity, and war. Perhaps, in looking at that history, we will discover that now is the time to pull those two historic lines of development — city and village — together again.

Village Foundations

Brian Swimme and Thomas Berry[1] say that the Neolithic village of some eleven thousand years ago represented humanity's greatest flowering of creativity. As the container/artifact called the village was invented, so too were the first basic small containers: baskets, pots, granaries, and cisterns. Weaving and the beginnings of metal tools and weapons originated about then, probably in early villages. The use of animals for transportation and work as well as animal husbandry and horticulture began in this period. Lewis Mumford writes: "Historic man has not added a plant or animal of major importance to those domesticated or cultivated by Neolithic communities."[2]

The Neolithic period appears to coincide with the first flowering of complex and subtle language — an awakening from unexpressed memories and dreams and a launching into the manipulative arts that coevolved with language. The universe, through humans, suddenly awoke and began talking about what it saw, weaving myths and tales, trying to figure everything out for the first time. The first simple stutterings toward solving the great mystery of life and the universe, which took place in the hunter/gatherer Paleolithic, was a crack in the door that in the Neolithic village was flung open to the light of discovery and invention. And the invention of the village had the potential to allow people to establish materially productive and satisfying human communities in a relatively healthy relationship with nature. With the considerate use of agriculture, human needs could be met with reduced pressure on the natural flora and fauna. The early village was very close to, or actually was, a real ecovillage.

More complex towns and cities appeared among the villages 8,500 years ago. Çatalhöyük in present-day Turkey appears to be the oldest. (Jericho in today's Israel, a substantial village of around two thousand people, had preceded Çatalhöyük by about five hundred years but was far simpler.) Thirty-two-acre Çatalhöyük, which sheltered an estimated 5,000 – 10,000 people, was not just a large village since it featured many specialized structures and revealed the existence of a far more complex material culture. Here we find the earliest known fired pottery, woven cloth, and earthen building blocks, sun-dried "bricks," and some of the oldest copper, lead, and gold implements. The first mirror, made of polished obsidian, was there along with murals of people, animals, and birds. Vultures and voluptuous female human figures. Erupting volcanoes on their contemporary horizon only twenty miles away, were featured subject matter. Sculptures of bull's heads built around actual skulls, with horns intact, protruded from walls into what appear to have been ceremonial rooms.

In many ways Çatalhöyük was the first expression of the urban age, and it lasted almost 800 years. But cities didn't really erupt upon the scene until almost two thousand years after Çatalhöyük was abandoned. Then, around five thousand five hundred years ago, the cities of the Mesopotamian Valley appeared — with populations of several tens of thousands each, with kings, slaves, high arts, wars, written language, legal codes, epic forest cutting, abstract science, exhaustion of soils by the hundreds of thousands of acres, and fantastic architecture.

Though the Neolithic village represented an unprecedented flowering of creativity, the inhabitants of such a village would have seen few novelties in a lifetime. If the ancient villages were similar to the ones anthropologists studied in the late 1800s and early 1900s, when such villages were far less tainted by the outside world than our remaining "primitive" villages today and when anthropology was in its early thriving, then not much that was novel was readily accepted. Having lived in a remote mountain village in New Mexico, I know from experience as well as legend that it takes three or four generations to become accepted, and that you are not really one of the locals until your family history is lost in antiquity, more or less equally with all the others'. It helps speed things up for your children's children if you marry into old-time families, though resentments and clan feuds can last generations, too. The village perspective can be as narrow spatially as it is enduring temporally. Says Lao-tzu of villagers, "to delight in their food, to be proud of their clothes, to be content with their home, to rejoice in their customs .... They might be within sight of a neighboring village, within hearing of the cocks and dogs, yet grow old and die before they visited one another."[3]

Village culture has always cherished long continuity and resisted change, and in this it has been extraordinarily successful. As Mumford puts it,

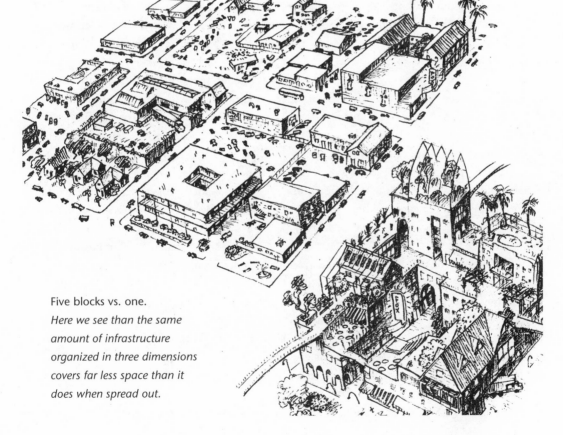

Five blocks vs. one.
Here we see than the same amount of infrastructure organized in three dimensions covers far less space than it does when spread out.

The village multiplied and spread over the entire earth more rapidly and more effectively than the city; and though it is now on the verge of being overwhelmed by urbanization, it maintained the ancient folkways for thousands of years and survived the continued rise and destruction of its bigger, richer, and more alluring rivals. There was sound historic justification, Patrick Geodes pointed out, for the boast of the village of Musselburgh: "Musselburgh was a burgh when Edinburgh was nane, and Musselburgh will be a burgh when Edinburgh is gane."[4]

A major reason for this is evident: villages didn't have much that cities craved. Other than humble food, emergency shelter on a stormy night, and a small, undependable tax base, villages provided little to make them worth conquering. When damaged, they also had much less to recreate in their recovery. Happily for many, the urban kind of history marched right by.

The creativity of the village was collective not only in that many people acting together were doing something new but in that the process, undoubtedly taking generations in many cases, involved more people than were even alive at any one time. In urban society today we are used to amassing considerable resources to launch any venture. Prototypes are expensive. In the village, a very different dynamic prevailed. Creative innovation in a short period of time would have been difficult partially because amassing almost anything was contrary to village values. Jules Henry writes:

In primitive culture as a rule, one does not produce what is not needed .... This helps to give primitive culture remarkable stability. The primitive workman produces for a known market, and he does not try to expand it or to create new wants by advertising .... Related to our contemporary dynamics is the lack of a *property ceiling*. Most, though by no means all, primitive societies are provided with intuitive limits on how much property may be accumulated by one person, and the variety of ways in which primitive society compels people to rid themselves of accumulated property is almost beyond belief. Distributing it to relatives, burning it at funerals, using it to finance ceremonies, making it impossible to collect debts in any systematic way — these and many other devices have been used by primitive cultures, in a veritable terror of property accumulation, to get rid of it .... The fact that our society places no ceiling on wealth while making it accessible to all helps account for the feverish quality Tocqueville sensed in American Civilization.[5]

Emerging Cities

Though things moved slowly in the village, what it created, it created very well. Perhaps that collective, careful, and slow-moving village creativity in itself a dimension of evolving consciousness, was exactly what made the next kind of creativity possible. With the cities emerged that other kind of creativity, one so flamboyant that it has blinded us to the earlier kind. This original Renaissance — village begetting city — probably saw its first glimmerings in Çatalhöyük. By the time of the Mesopotamian cities, it was driven by the creativity of individuals thoroughly aware of their own powers — the creativity of personality identifying with the powers of the universe, the gods of the city itself, or, in later days, a single individualistic creative God. This was a creativity discovering and expressing itself through art, science, manufacture, political and institutionalized religious power, grand waterworks for agriculture, and, of course, city building.

Villages satisfied the needs of the body and celebrated the mysteries of the universe. But the cities ran away with the imagination and sought to solve the mysteries while creating new miracles. They were genius and madness, creativity and destruction, generosity and greed, love and cruelty wildly intensified. From their beginnings they symbolized — they *were* — the physical manifestation of human passions, fruits of the Faustian bargain. Sibyl

Moholy-Nagy writes:

> Man has built and loved cities because in the urban form he constructs the superimage of his ideal self. The common denominator of cities, from Nineveh to New York, is a collective idol worship, praying for power over nature, destiny, knowledge, and wealth. The gods of cities are supermen, of whom Don Marquis wrote, "And he clothes them with thunder and beauty, / He clothes them with music and fire, / Seeing not as he bows by their alters, / That he worships his own desire."[6]

As the cities made possible the highly productive economy of which Jane Jacobs speaks, they generated a new, very dangerous, and often exquisitely beautiful kind of creativity: specialized, narrow, blind to its long-range effects, and powerful in its new integration of ideas, tools, and products. It quickly produced magnificent architecture and sculpture, writing, math and astronomy, dangerous concentrations of power, the sweet smells and moist airs of gardens, shady luxury with occasional fine things to eat and drink, pungent marketplaces, and hot dusty streets filled with strangers and acquaintances in a pact to accept one another. The city was like no other environment on Earth — imagine walking into one when there was only one or a few and the whole rest of the world was the hot valley bottoms and rolling mountains of Asia Minor. City creativity was part and parcel of specialized groups playing different roles in a hierarchical economic order, the emergence of systems of privilege, the centralization of bureaucratic power, the development of military weapons, the emergence of social and economic classes, slavery, greed, and rampant egotism to the point that high leaders regularly proclaimed themselves gods.

It's doubtful this kind of creativity could have crept upon the scene. Instead, it strode into the room and kicked over the table. In all likelihood it was incarnated as men with weapons wanting to take over, allied with creative individuals tired of being told they couldn't create. Patron and innovator, civic despot and artist, proto-industrialist and intellectual, power-wielding priest and earliest astronomer — together they established a new order to lay claim to the powers and secrets of the universe. They were not necessarily very nice, but they did inspire a good deal of imagination and admiration, and not just in themselves, but in individuals everywhere who could avoid severe subjugation.

Bridging History, Nature, and Evolution

At some point as the city was beginning to emerge from the village, large enough numbers of people would have been gathered together with appropriate tools to cause the pace of change to quicken. Close proximity ("access by proximity") and cross-fertilization of ideas, tools, and products would have stimulated creativity to an unprecedented degree, bringing the novelty of conspicuous change into the short span of a single lifetime and giving people the experience of witnessing personal creativity right before their eyes. This was a form of creativity rejected as dangerous in the village but recognized, utilized, and honored in the city. It represented consciousness evolving with little conscience — a kind of creativity more than willing to exploit, in some cases even destroy, the village and much of nature in its service. In the village, the idea of cutting down a tree might have caused the elders to point out the value of the tree to useful and beautiful birds, to children who liked to climb in its branches, to villagers who enjoyed its fruit in its summer shade. But the city's style of creativity would have seen the potential for wood to produce more housing, tools, and furniture, and as fuel for firing pottery and melting gold for art and bronze for tools and weapons.

With this new kind of creativity came a new kind of time. The cyclical sense of time, endlessly returning in harmony with sun, moon, and Earth, was transformed into a vector later thought of as Progress. In the brief cycle of one life, the personality could now see oneself as participating in the process of change in society and nature, experiencing a tiny shred of earthly immortality and god-like creativity,

Zoom in on a tunnel.
A larger melding of pedestal and towers beside the Radio Building with spire in Shenzhen, China, which is about 40 stories. Above right is a zoom in on a six-story high pedestrian hallway, a new kind of architectural feature that would appear with larger ecocities.

causing something to actually happen and be part of a permanent change. It must have been fun, the dawning of the idea of personal creativity. Its products, which might have been regarded as deviant and dangerous in village cultures, would have become adopted in the early city, even celebrated. Feelings of personal mastery and joy in creation and the discovery of math, writing, astronomy, and engineering must have been a heady brew. It is understandable that early urbanites tended to ignore the restraints of village culture wagging its finger at novelty, calling it dangerous to nature and the society of people. "I don't know about this newfangled idea ..." in the village gave way in the city to "Why not just try it out?" — a

rhetorical question rejecting qualms as far less interesting.

What happened as the city emerged can be viewed in evolutionary terms as the emergence of conscious personal creativity out of the conscience that preceded it. Conscience is intuitive, synthesizing, integrative, emotional, limitless, and complex, concerned with endless chains of subtle causes and effects in a very personal way that feels care, guilt, love, and affection, and so poignantly that consciousness is a bit embarrassed. It is poetic, willing to entertain far-flung, personally enriching inferences and analogies. It is often aware of how everyone is feeling and sees itself as essentially part of a community of other beings. It has an ego but not much need for material props, and it cooperates well. It is not just self-conscious but often painfully, responsibly and sensitively so. Consciousness is more logical, unemotional, limited, purist, and scientific, concerned with particular affect, applications, testable results, expediency. Aware that it is aware, it has a strong ego and competes well.

What happened as the city emerged can be viewed in evolutionary terms as the emergence of conscious personal creativity out of the conscience that preceded it.

These two patterns of thinking, as I am contrasting them here, are obviously not mutually exclusive, as many people of conscience, such as the Gandhis and peace activists of the world, are also highly conscious, and many of the most conscious, such as the Einsteins, are among the most powerfully motivated by conscience. But some intriguing distinctions can be drawn that parallel the distinctions between the thinking in villages rooted in nature and the thinking going on in nature-exploiting cities. Conscience gave us the Neolithic village, with its reluctance to rock the holy boat of nature and the sacred home of community. Consciousness, the second to evolve, set up marvelous cultural structures rationalized for the sake of enjoying its new glories and avoiding criticism that might slow Progress. Conscience feels, then consciousness builds and rationalizes. Consciousness invented paradigms to make its work more manageable, constructing not only coherent frames of reference for knowledge and experience but firm preconceptions and conclusions to screen out and limit disturbing incoming information. Consciousness was highly educated but in some areas intentionally ignorant. Conscience and consciousness should have evolved together as the emergence of the sum and unity of both, the emergence of a kind of "conscienceness" in the universe, in both cities and villages in a rich exchange with each other and in balance with the Earth. Perhaps in at least one case they did.

Minoan Crete seems to have evolved from village to city and thoroughly integrated both conscience and consciousness. The ancient culture on the island of Crete lasted about fourteen hundred years contemporaneously with the ancient cultures of Mesopotamia and Egypt and ending about seven hundred years before the Greek Golden Age. Riane Eisler calls the Minoan culture a "partnership" culture, as contrasted with the "dominator" model followed by the other ancient city cultures in the region surrounding the eastern Mediterranean Sea.[7] The cities of Crete had no defensive walls, the art no depictions of war and weapons but many of exuberant sensuality and sport, including acrobatic gymnastics with dangerous bulls being pursued by young men and women alike.

Writes Nicolas Platon, who began modern excavations in Crete in the late 1920s, "All the urban centers had perfect drainage systems, sanitary installations, and domestic conveniences. There is evidence of large-scale irrigation works with canals to carry and distribute the water ... viaducts, paved roads, look-out posts, roadside shelters, water pipes, fountains, reservoirs, etc."[8] Eisler argues that "Cretan palace architecture is also unique in civilization":

> These palaces are a superb blend of life-enhancing and eye-pleasing features, rather than the monuments to authority and power characteristic of Sumer, Egypt, Rome and other ancient war-like and male-dominated societies .... There were vast courtyards, majestic facades and hundreds of rooms laid out in the organized "labyrinths" that became a catchword for Crete in later Greek legend ... laid out over several stories, at different heights, arranged asymmetrically around a central courtyard.[9]

The irregular form of this architecture contrasted sharply with the rigid order of the walled fortresses on the mainland.

The Creator was seen as female, says Eisler, rather than male, and depictions of ceremonies and celebrations on walls and ceramics indicated that women held high and often the highest positions of status and decision making. Despite female leadership, there appear to be no signs of oppression of the male in Minoan Crete — hence Eisler's classification of the society as a "partnership" society. According to Nicolas Platon, on Crete "the fear of death was almost obliterated by the ubiquitous joy of living."[10] In any case, much about peace and vivacious ways can be surmised from Minoan artistic history. Meanwhile, in the "dominator" societies surrounding Crete geographically and following it temporally, depictions of exploitation and oppressions of the female, along with numerous images of war, are in evidence everywhere in the archeological record.

Ultimately Minoan civilization was destroyed, apparently by a mix of colossal geologic convulsions — volcanic explosions and earthquakes — and invasions by warlike neighbors with iron weapons and little regard for a "partnership" culture.

Looking back at these two divergent streams of village and city, and at the inspiring exception that seems to have held the best of the two streams together for several hundred years, we could ask ourselves whether it might not now be possible to forge a partnership between conscience and consciousness, village tradition and city tradition, people and nature, civilization and Earth.

Modern Times

With the emergence of the city, the functioning structure of the village — the house, work area, private or semiprivate compound, cistern, granary, street, public gathering place, shrine — expanded into large buildings or whole quarters of the city for weaving, brick making, leather working, woodworking, food handling and storage, waterworks, trade and marketing, religious, artistic, social, and political events, defense and administrative control, and the education of the elite. Residences ranged from palaces with gardens for the rich to hovels for the poor and the slaves and stables for the animals.

War was a major shaper of cities in the Middle East and later in Europe, dictating walls, which defined and compacted cities, until firepower could breach the thickest defensive shell. When nation-states emerged out of city-states, they pushed the walls of defense out dozens or hundreds of miles to the national borders. The Great Wall of China was the ultimate expression of this. When city walls dissolved in the 1600s and

World Trade Center replacement, generic solution. *This basic scheme, with variations in form, represents all versions proposed, with 11 million square feet of commercial space — except for the submission by this book's author. Not mixed uses. Not community.*

1700s in Europe, the city frayed out a bit at the edges, encroaching in a minor way into natural and agricultural lands.

In the soon-to-be United States, the European colonist city-builders had so much more firepower than the Native Americans that walls were barely featured at all. By 1750, cities contained a mere 3 percent of the population despite their disproportionate economic and political power. By the early 1800s, a small number of urban people who could afford horses and carriages chose to live farther from the madding crowd, thus becoming early close-in commuting suburbanites. (We're not counting small farms and productive plantations and estates housing and sustaining their owners, managers, and workers as suburban.) In the mid 1800s, starting in England, trains made the suburbs accessible to more middle-income people, and small towns were linked with larger towns and cities. Significant commuting was beginning. In such villages and small towns close to cities most people lived a short distance from the train station.

In Scotland, starting in the 1700s, that is, concurrent with the early rise of industrialism, the *Enclosure Acts* tested the rights of the poor people raising animals on the commons against those of the rural landlords and found, as the law often does, that public rights were less important than private ones. The wealthy were then legally empowered to enclose much of the commons as private holdings and to evict the poor. This idea

spread to England, and thousands of people were excluded from self-sufficient agricultural life and rendered poor and desperate just at the time that throughout Britain expanding industry sought larger numbers of workers at the lowest possible wages. Former agricultural workers had to move to the city or starve on country roads. This "coincidence" of rural land use policy and urban history has been repeated in a variety of ways to this day as people in the low-consumption world (my term for the "Third World") are being forced off their land by governments representing the wealthy better than the local people on the land. Thus population growth has often been forced upon cities by policy decision. As industrialism gained momentum and Progress became the secular religion, slums appeared near the factories, and industrial pollution began spreading over thousands of urban acres.

In many cities, daily escape to the suburbs began to look like a very good idea for those who could afford it, and the single house with its own vestigial, foodless farm or slice of nature called the "yard," symbolic in America of the European Old World, became Home. In that mythical ever-so-private shrine lived the middle class family, at first "extended" and later "nuclear," which proved fissionable into yet smaller units: single-parent families and individuals at home alone.

By the late nineteenth century, improved construction methods and the logic of effi-

Thus population growth has often been forced upon cities by policy decision. As industrialism gained momentum and Progress became the secular religion, slums appeared near the factories, and industrial pollution began spreading over thousands of urban acres.

ciencies of proximity began expressing themselves in taller and taller buildings, creating economies of scale otherwise inconceivable. In America, home of pioneering big buildings, immigration was encouraged by policy: over 5 million immigrants arrived in the decade after 1880. New York became the first city in modern history to pass 1 million in population, and possibly the second in all of history, the first having been Rome 1,400 years earlier.

Technological development shaped cities profoundly. Elevators pulled cities into the vertical dimension while trains and streetcars spread them out horizontally. Hoisting devices had been in existence since the pyramid builders in Egypt, but in 1854 Elijah Otis invented the safety elevator: in case a cable broke, the elevator would immediately lock itself against a pair of ratchet tracks along the walls of the elevator shaft. This invention made passenger travel possible beyond the four stories considered the tolerable maximum for stairs at that time. Three years later Otis got his first order for two passenger elevators for a five-story china-and-glass company in New York City, and shortly thereafter buildings up to ten stories were being constructed in New York, Boston, and Chicago. By 1883, there were over 2,000 elevators in the apartments and offices of Boston. Thus, many years ahead of other countries, the United States took the lead in the vertical growth of cities just as, about fifty

years later, it would lead the way to sprawl. The cable car system introduced in San Francisco in 1873 was a brilliant mechanical device that used the weight of cars going downhill to help pull the other cars on the same single cable going uphill many blocks away, thus requiring an absolute minimum of energy. Cable cars spread around the country, but soon they were outdone by electric trolleys on rubber wheels and streetcars on steel wheels and tracks.

Thomas Edison invented the electric light bulb in 1879, and two years later his power station in New York City, the first in the world, came on line for the public. The electric light bulb soon brought inexpensive and relatively safe light to the deep insides of buildings far from windows, replacing the sooty gas lights of the Victorian era and increasing the potential depth of buildings just as elevators had increased their height. After the 1880s and 1890s the whole city, with trains and streetcars, elevators and electric lights, had the means to expand many times over, to a scale never dreamed of just a generation earlier.

The apartment building, common in Paris since the mid-1700s, was frowned upon by American society at the time and for the next one hundred years. "You know that no American who is at all comfortable in life will share his dwelling with another," said a typical 1880s developer.[11] But wooden tenements, built from 1840 through to World War I, shoehorned dozens of families into three- and four-story buildings while by 1879 some developers were ready to shift their experimenting to the wealthier classes, who also wanted to be near the businesses and cultural life concentrated in cities. In New York, one of the earliest of these was the eight-story Dakota Apartment House of that year (the one in which John Lennon was living at the time he was shot to death at its front door, just one block from the site of the first fatal automobile accident in American history on September 13, 1899). This spacious, stately building with its large rooms and high ceilings was anything but cramped. Almost

Trade Center replacement as a world community facing the Statue of Liberty.

A highly mixed-use community celebrating the best of Manhattan and the United States brings the best of trade to the site. Arrow to the right (south) points to the view from the "keyhole" or "view plaza" on the 20th floor (dark plane in the drawing).

upon introduction, these more expensive apartment houses became extremely popular, and in New York City hundreds were built in the 1880s.

New mass-produced high-compression brick took office buildings up to considerable heights, but the thickness of the wall at the base dictated a structural limit. One of the last of the tall brick office buildings constructed in Chicago, in 1891, supported sixteen stories on walls five feet thick at the base. As brick and stone were reaching their zenith, along came iron and steel, introduced in tall buildings in Chicago in 1884. Five years later New York built its first steel building, and the race between the two cities was on. In a novel set in 1890 called *The Cliff Dwellers*, these early large buildings are described as cities unto themselves: "A tenant could eat, drink, have a haircut, obtain legal advice or consult a real estate broker without leaving the building."[12] The tallest early buildings were the houses of commerce, especially insurance buildings, rising up to heights casting church spires into deep shadow, rendering insurance in the hereafter inferior to insurance for the here-and-now. By the 1930s, New York had gone from the cloudscrapers of the turn of the century to skyscrapers like the Chrysler building and the 107-story Empire State Building. Compacting forces of economic efficiency and cultural access, aided by vertical and horizontal transportation and lighting technologies, reached a pinnacle in New York in the 1930s.

The military pressure for urban consolidation was long gone, but the city had become compacted anyway, without walls.

At the same time, many forces pushed or pulled in the other direction. One was a simple reaction to the pollution of the industrial city and the intense competition and great expense of cosmopolitan living near the centers. Escape was an outward push, whether the reasons were ecological (soot, grime, noise) or sociological (crime, classism, and racism). Tuberculosis was rampant in sooty cities in the late 1800s and early 1900s. An architectural element that attempted to address the problem, the sleeping porch, became common; the idea was to breathe the fresh night air. It didn't work very well since daily contamination didn't settle out that fast, so this was also the time of rural sanitariums for rest and recovery. The first American national park, Yellowstone, established in 1872, was largely justified as a sanitarium of fresh air, an escape from the urban world for reasons of regaining health. The outward pull had elements of nostalgia for country farm living, which in the early 1900s was still only a generation or two back for big-city old-timers and a few miles out from downtown in any direction. Also, frequent and profound contact with nature was the dream of thousands of Americans who dreamt of the frontier while reading Walt Whitman and Henry David Thoreau and adventure novels of the Wild West.

Probably at the deepest level there was also a certain unconscious urge for none other than the ecocity, which, after all, is nothing but a refined urge to live both city and country life simultaneously — something I call the "ecocity impulse." Perhaps this urge is also a profound nostalgia for community in harmony with nature, remembered almost mythologically, even genetically, from the early village. For Americans, especially urbanites greedy for life and with marvelous new technologies ushering in a world of all possibilities, the question became "Why not have both lives simultaneously?" There had to be a way. For this unstated dream the car would be the instrument of deliverance. But first came rails.

If the train-created suburbs, as contrasted with the farming, logging, fishing, and other productive country towns, were functionally disconnected from the land, they at least created residential areas on a close to walkable scale, with diversity serving homey purposes if not economics grounded in the countryside. Some towns with strong ties to the land, when linked to cities by efficient, frequent trains, were to varying degrees transformed into dependent bedroom commuter communities.

Smaller than trains, streetcars make more stops and have a shorter practical range than trains. Rather than seeding new towns in the country or connecting earlier ones at a distance, they tended to expand the fringe of existing modestly large towns. "Streetcar suburbs" were a short ride out from the center and were generally compact, often with tastefully designed small apartments and moderate-sized to large two- and three-story homes catering to the entrepreneurial, professional, and upper clerical classes. But these first inklings of sprawl were nothing compared to what loomed in the future. The automobile, allied with its usually unacknowledged co-conspirator, the bus, became the definitive shaper of urban form.

The automobile was invented and taken out for its first test drive by Karl Friedrich Benz in Mannheim, Germany, in 1885, and by 1909 Henry Ford had set up the first automobile assembly line. Thanks to Ford, in the United States the vehicle went from being a luxury plaything for the rich or obsessed to being commonly affordable in less than a decade. By democratizing the car, Ford become the chief designer of the urban, suburban, even rural American landscape, and internationally his influence continues to spread literally as cars induce sprawl everywhere.

Ford declared that the most beautiful things in nature were the highly efficient creatures from which all excess weight had been stripped. With this natural model in mind, he conceived the assembly line, which was a marvel of efficiency but resulted in a product that ended up being the most inefficient transportation invention in history. In retrospect we can

Claremont Hotel addition in three successive zooms.
The Claremont Hotel on the Berkeley/Oakland border is a stately old building much revered locally. Here, it becomes the centerpiece of more development dug into the sloping bench on which it sits. The addition would create diverse living, work and shopping space, almost an in-town small town. The sequence drawings end in a street scene at fine grain scale.

The changes wrought by the automobile in one generation were epochal. In all but the strategically, commercially indispensable centers like Manhattan and downtown San Francisco, town centers withered.

see the positive lesson of the assembly line as well as the negative lesson of the automobile: minimizing motion maximizes efficiency.

The idea that efficiency is a problem of modern life is largely a misconception. Much of what we think of as efficient is just getting things done quickly that are pointless or dangerous and therefore shouldn't be done at all. The lesson as far as city structure goes is that if we end up covering long distances, moving about constantly, we will be leading wasteful lives, wasting resources and time and laying waste to land, air, and water. But all this was not yet evident when the Model T's and A's were pouring forth

from the factories, a car for every garage and a chicken for every pot. Prosperity was for all, and nature was so endless in America that it could soak up any punishment people could dole out.

The changes wrought by the automobile in one generation were epochal. In all but the strategically, commercially indispensable centers like Manhattan and downtown San Francisco, town centers withered. Very quickly the "doughnut" phenomenon and the bedroom community became realties. Doughnut: at night and on weekends everybody took their dough home to the encircling suburbs and left a hole in the

middle. Civic irresponsibility thrived. Bedroom community: a place where people slept and occasionally entertained and, from the early 1950s on, watched lots of television, but seldom lived in the cultural, urban, or even village sense. Downtown: a place that, if not derelict and boarded up, was, with rare exception, a money machine by day and a ghost town by night. As the dominance of the automobile became established, the hole in the doughnut became ever more literal as parking lots, new and used car lots, and gas stations pushed out most land uses other than daytime business. Maps of cities came to look like Swiss cheese, but with square and rectangular instead of circular holes, and this pattern was crisscrossed by broad and unforgiving bands called arterials and freeways.

"But," say the defenders of the American Way, "people love their cars." Well, they loved their streetcars, trains, and ferries, too. I have

to point out that auto love was not enough to satisfy the auto industry, and neither was intensive advertising or glorification in the movies or national economic policy and patriotism ("What's good for General Motors is good for America."). What would it take to get enough? Conspiracy. During the period from 1927 through 1955, General Motors, Mack Manufacturing (trucks), Standard Oil of California (now Exxon), Phillips Petroleum, Firestone Tire and Rubber, and Greyhound Lines — in violation of anti-trust law — shared information, investment money, and management activities for the purpose of maximizing profits and eliminating ground transportation competition in the United States. The net effect was to destroy urban rail transportation in over a hundred cities. To do

this, the conspirators set up several front companies in which they invested their money. These companies bought up and then tore up the streetcar lines of a gigantic nation, leaving citizens stuck with no transportation alternatives of note other than cars and buses.

Those who were shocked to see their favorite streetcar lines suddenly replaced by diesel-fuming buses were told by the conspir-

ator-owned bus companies about the virtues of buses over trains and streetcars. They were modernizing, they said: buses worked better, more "flexibly," than streetcars because they could follow the car out over the flatlands.

Shocking though it may sound to fans of public transit in general, the reality was that the buses were losers from the start, requiring major subsidies from government for the simple reason that it is impossible to cover large, thinly populated areas with frequent, convenient service. This is a problem of geometry and human numbers on the surface of the Earth. Efficient, economical public transportation can't work in sprawl. Ever since those first days of conversion from rail vehicles to buses, the subsidies to the car have been astutely hidden while ever-cleaner smog devices have been celebrated with fanfare. In the meantime subsidies to the buses and disparaging comments on their incorrigibly bad breath have been trotted out so that people can throw pies in public transit's face.

But if National City Lines, Pacific City Lines, and American City Lines, the three major front companies, and their investors couldn't make big money on buses, why did they tear up and scrap rail lines that they had paid good money for? Because that money was a pittance compared with how much they began making on cars, trucks, tires, gasoline, asphalt, and highway building. And, of course, they also made the buses and the bus tires and the diesel fuel for the buses that they sold to themselves (their bus companies) — and later, claiming intolerable losses, they sold the bus companies to municipalities trying to help their lower income people get around. An internal memo at Mack Truck that surfaced at the conspiracy trial explained it tersely: the "probable loss" for the investors in the bus companies would be "more than justified by the business and gross profit flowing out of this move in years to come."[13]

Finally, in 1955, after this conspiracy had been going on for almost three decades, federal prosecutors figured out there were less than honorable motives behind this strange disappearance of almost an entire country's urban rail system, identified the conspirators, and took them to court. The companies and their representatives were charged with violation of the Sherman Anti-Trust Act and found guilty in Judge William J. Campbell's US District Circuit Court in Chicago. Each company was fined $5,000 plus the court costs of $4,220.78, and each individual was ordered to pay $1.00 for his role in the conspiracy. Thus did big-time urban planning proceed to a successful if guilty conclusion. Thus America's car dependency was intentionally designed into our lifestyles to maximize profits for the car/sprawl/freeway/oil system. Thus the flat urban form called sprawl was not just the drift of public desire — far from it — but planned.

The Esthetic Tradition

Ecocity esthetics go back to the Hanging Gardens of Babylon and earlier to the wall paintings of Çatalhöyük depicting soaring birds and rampaging wild bulls. They go back all the way to roots in the villages where folk crafts took up natural themes in decorations of individual or collective dwellings, granaries, ceremonial lodges, open enclosures, and shrines, with buildings and monuments purposefully, reverently oriented toward the warmth of the sun, the signs in the stars, and the changing of the seasons. In Europe, on the eve of the colonization of most of the rest of the world, however, any esthetic akin to that of ecocities was nearly nonexistent. Kirkpatrick Sale describes the European attitudes of the times, quoting passages that illustrate the prevailing lack of esthetic sensitivity to nature:

It is there in such celebrations of urban form as Piero della Francesca's Ideal Town, an entirely lifeless human construct without a single blade of grass or shadow of tree, dominated by that most controlling of all inventions of Renaissance art, perspective ... But of all the images of control, the most pervasive and most revealing is that of the formal Renaissance garden, whose style was perfected and popularized ... with such careful artworks as the gardens of Compton Wynyates in England (1520) and Tivoli in Italy (1549). Here is the hand of man and not the grace of nature that is ever-present: bushes and small trees trimmed in rigid geometric shapes to look like wedding cakes or perfume bottles, closely clipped hedges along geometric walks, blocks of flower beds in uniform colors, carefully edged lawns and artfully distributed statues, benches, fountains, pools and bridges.[14]

Here there was no possibility of the "unweeded garden that grows to seed" that Hamlet describes with typical disgust, where "things rank and gross in nature possess it merely."

The purveyors of Western civilization learned a little from those they conquered. The British picked up the shady wraparound porches that kept buildings cool in hot climates, the "verandahs" of India, and took them to other hotter parts of their empire. The Spanish adopted the mud brick or adobe building from the North Africans, where the material moderated temperature admirably, and brought the idea with them to the New World, where some of the natives who had previously built in stone adopted it from them. But generally very little having to do with living and building sensitively with nature was learned from the natives. Instead, whether they fit well or not, European building styles were imported. In North America north of Mexico formal gardens, hedgerows, and fences were put up demarking northern European ideas of land ownership and esthetics simultaneously. Lawns were introduced to make an Old World of the New.

In America, the industrious habits of the Colonists and the relative defenselessness of native flora, fauna, and people translated into a rapid industrialization that placed buildings in any arrangement that seemed expedient. Some architects, craftspeople, and artists adopted some elements of native culture or exhibited appreciation of the land they had acquired, but not many. The land was tamed into usefulness, not lived with and understood. To the high-powered juggernaut of Manifest Destiny, as the forests were cut down and the soil of farm after farm was used up on the march west, as the raw energy of industrial exploitation, steel tracks, and iron horses pressed on, esthetic refinements garnered from the original inhabitants or nature herself would have seemed close to ridiculous. Inside towns there was the inward pull of the mechanisms of exchange — access by proximity — in economics and culture. The opposite and outward pull of available transportation put space between people almost reflexively, as did the psychological desire of a people longing to breathe free, free from other people of the sort they were leaving behind along with persecution and hard times in Europe. These factors, more than art, and much more than any attributes of nature other than its beckoning vast openness and rich resources, were the shapers of the village, town, and city.

The military crosshatch gridiron layout of streets, usually north-south, east-west, made it possible to build with minimal time-consuming thought. As Kenneth Schneider says, "The gridiron assured easy access to every property and set forth the image of democratic equality. No plan could have met the unknown requirements of industrialization and rapid urbanization as flexibly. The straight streets forming square or rectangular blocks could be set out and surveyed without agonizing decisions."[15]

There was, however, a minority tendency in the other direction. Thomas Jefferson and others tried to adapt an architecture to America's needs, borrowing from the democratic Greeks and the imperial Romans. The architecture itself and the arrangements of order and landscaping it fit into had less to do with nature than with the golden mean and other geometric abstractions and classical traditions. At the same time, some buildings were placed carefully in settings that were close to natural. In the Jeffersonian distrust of cities and idealization of rural agricultural America there was more than a glimmer of nature-appreciating esthetics. The campus of the University of Virginia, designed by Jefferson himself, was centered on a quadrangle of lawn surrounded by classrooms and student and faculty housing along two sides, a library at one end and an opening to nature — an expansive view of beautiful rolling hills — on the fourth side opposite the library. It is hard to be more architecturally explicit about celebrating both culture and nature than by placing a library and a natural view opposite one another on a central green. I'll describe this later as an important ecocity pattern I call variably a "keyhole plaza" or "view plaza," which celebrates both nature and culture by framing nature with the buildings. In the usual way of academic administrations, this idea sailed over their heads and, long after

Jefferson was gone, the University of Virginia placed a new building right in the middle of the view, completely enclosing the central green and losing the powerful original meaning of the arrangement and whatever it might have had to do with acknowledging nature.

Working with nature was expressed in the small, locally viable towns, in the farmhouses attached to barns in New England, in the two-story saltbox houses with a wind-breaking roof sloping down from the second floor across the single story on the cold windward side. Evident in these cases are the clustering of functions for practical reasons and the orientation of the buildings for climate moderation. Often the design of these towns and modestly complex buildings attained a level of real esthetic sensitivity to the natural environment.

The City Beautiful movement, launched by the World's Fair in Chicago in 1893, took the words of the fair's chief planner-architect, Daniel Burnham, to heart: "Make no small plans, they have no power to stir men's souls." After that the monumental arrangement of streets, large public squares, imposing buildings, colossal sculptures, triumphal arches, and grand reflecting pools became the desire of cities the world over. The movement codified and refined the themes of the monumental city going all the way back to Mesopotamia and Egypt. Its palette included broad avenues for impressive displays of autocratic and military power, tempered by the Renaissance garden, filled out with buildings in the American Roman or Imperial Federal style, focusing on splendid open spaces demarked by dramatic colonnades sometimes surrounding formal pools. Typically these buildings and spaces were festooned with Isadora Duncan-like statues and enormous vases overflowing with cast concrete, or in the case of temporary expositions and fair grounds, papier-mâché flowers. The columns, arches, sculptures, and reflecting pools were esthetic products of a domineering culture, but the mirror pools caught the clouds and blue sky, framed views and trees, and accentuated something of nature in that almost excruciatingly sensuous and nubile way of a Maxfield Parrish painting — but too large. In its exercise of contrasts, the City Beautiful style was an odd, rather oblique celebration of nature. Or perhaps it was a way of using nature to set off man's architectural glory.

That these grand designs were not more universally adopted by larger cities had mainly to do with construction costs and the designs' disregard for the dynamics of gathering people and functions together in small areas. The patches of open space and broad bands of avenues and boulevards were just not cut to the smaller size of the human body. Great for military displays designed to discourage revolution and stir pride in vicarious ways, in national grand designs, they were neither inviting nor very functional for culture and economy. The contemporary technologies of tall buildings — strong structures, elevators, and electric lights — which turned the action toward smaller areas of land and up into the sky distracted from the panorama of spacious boulevards and large, building-defined spaces. In the United States, the tall buildings literally rose up and over the esthetic and psychological effect of the City Beautiful. One kind of grandeur contradicted the other as two kinds of skylines vied for the same esthetic space. Ironically, at the same time that Burnham was saying "Make no small plans," the tall-building planners were making yet bigger ones that cast shadows over his creations and influence.

The Garden City idea, originating at about the same time, was an attempt to create what might have been called an ecocity if the term "ecology" had been in general usage then. With its requisite greenbelt, the Garden City was developed from the turn of the century until shortly after World War II, and it is still studied in architecture schools and influential in new town and large development planning today. The idea in Ebenezer Howard's 1902 book *Garden Cities of To-mor-row*[16] was "to wed city and country." Garden City advocates rejected the suburban idea of bringing city and country together by placing private houses in "yards." They were thoroughly concerned with creating a lively community, economy, and culture, not with escaping from them. Howard and his many associates and followers hoped to have a nat-

ural and agricultural zone — the greenbelt — around a town and planned for densities about twice the average of today's American suburban communities. Housing was planned to be a short distance from a city center in which commercial jobs were located and not too far from civic amenities and internal parks. Industries were to be located on the fringes, adjacent to the greenbelt. The general average size was proposed to be 1,000 acres of development in a natural and agricultural reserve of 5,000 acres owned by the city and supporting 30,000 inhabitants.

A number of such cities, among them Letchworth and Welwyn in England, were built. Although reasonably lively, they have lacked the vitality of larger or older cities and their charm, blight, art, corruption, cultural flowering, and intense pollution. Both of these cities are somewhat smaller than planned, and they seem to do well according to many criteria, including financial ones. While housing in England is chronically in need of government subsidy, Letchworth and Welwyn have never needed assistance. The concern of the founders for equity resulted in a community ownership plan whereby the residents buy shares proportional to their living space and limited to five percent dividends per year, with any excess going to community uses.

Lewis Mumford points out that there has been the stiff criticism of the built Garden Cities and the many claims that they have

Smallest case eco-community design.
A composite house for two or three families in the country. This illustration assumes home-based work and considerable food production for it to amount to a community with some economic vitality, justifying the term "community." Featured in addition to solar greenhouses and terraces is a fun treehouse, bridge and slide into the small lake.

failed and comments: "This is surely a singular kind of 'failure.' What other new conception of city improvement has resulted in the layout and building of fifteen New Towns, in Britain alone, to say nothing of similar foundations, either achieved or in process, in Sweden, in the Netherlands, in Italy, and in Soviet Russia ... from Scotland to India?"[17]

In my opinion, the Garden Cities mark the moment when the idea of the ecocity appeared and was consciously, if partially, expressed and actualized. These towns sparkled with promise and cried out for further development and experimentation, but they were neglected in favor of stronger movements in economics and esthetics, and largely passed by.

The modern architects Walter Gropius, Le Corbusier, and Frank Lloyd Wright appropriated the technology of their times and aspired to create marvelous new realities. Drawing from Cubist painters like Braque and Picasso, from industry, from the expedients of engineering's straight lines and urban gridiron layout, from their fascination with new construction materials like glass and steel at a scale and strength never seen before, and from their own imaginations, they created a world of crisp right angles and generous fields of simple, uncluttered, efficient, honest form and surface.

Gropius, a leading spirit of the Bauhaus school in Germany in 1920s and 30s, wanted to harness industrial productivity to house workers honorably and inexpensively, liberating their time for leisure and creative achievement and reducing the social inequities in evidence everywhere. In many ways, though often dreary and repetitious, the resulting buildings did serve hundreds of millions of people, placing them near inexpensive transit, giving them access to jobs, parks and whatever cultural and economic benefits were provided by their particular cities. This International Style of architecture became the most pervasive reality of new higher-density urban development after World War II. Its monotonous ubiquity was condemned by people of higher income and more refined, parochial, traditional, and pastoral tastes. The undifferentiated scale of its larger buildings seemed dehumanizing, and, partly because of this, many people have rejected development from medium to high density ever since. In redevelopment areas such buildings were repetitious and monotonous in design, cheap in construction, and stuffed with more than their fair share of desperate unemployed people, frustrated youth, and occasional outright criminals. But then the buildings were blamed for the social problems that such low-cost and volatile concentrations of poverty and pain produced.

These buildings were the model for most subsidized housing, but even some that were not so loaded up with social problems from the start featured, for example, roofs with no eaves, covered porches, or awnings to shelter pedestrians from the weather. The modern style was supposed to look "clean" (without pointless decoration and superfluous detail) and functional (in terms of elegant simplicity and economy of materials). But those clean lines with no overhanging eaves were good for cleaning the sides of buildings with rain — and the clothes of people in the streets while they were still wearing them. As for being "functional," this often worked better in theory than in practice. Yet the intention was socially responsible, and remains the reason why millions of people still live in such "apartment block" housing: some economies of scale are genuine. Whether ugly or not, such dense housing — I hate to say it — provides shelter and access to transit, jobs, and social life at relatively low cost. True, it often lacks the fine-grained detail of rich ecological and social relationships and thus probably lacks resilience and full health, but it might regain these with some serious redesign and a little more judicious investment shifted over from subsidies to cars, oil, and sprawl.

Le Corbusier, with his plans for enormous pristine slab-and-block buildings fifteen and fifty stories high rising from the sylvan landscapes of the archetypical African savannas where humanity descended from the trees and learned to walk upright and free, imagined a superworld for the superman of the future. He drew stocky, slim-waisted, square-shouldered figures to decorate his walls, shapes of the New Man and New Woman who could crack granite with judo chops. His women, like Michaelangelo's, were actually men with breasts appended, as if with Velcro. His buildings were likened to male erections rising through sensuous though barren female lakes and grassy landscapes. He imposed a vision of buildings on nature to make the monumental plazas of the City Beautiful shrivel and blanch into parochial inconsequentiality. The linear city he designed for Rio de Janeiro was a single long, winding building dozens of stories high, with a freeway on the roof. It would snake through the primordial upwellings of stone that rise through the jungle and along the curving beaches. Fortunately, it wasn't built. There is a place to be big and splendid in constructed edifice, I am sure, but Rio's location inspires awe and requires care. No apartments looming up to compete with Half Dome in Yosemite Valley or sports arenas on the side of Mt. Fuji, please.

Le Corbusier's schemes were not, however, simplistic. His Marseilles Block, a sixteen-story apartment building with a shopping "street" halfway up and rooftop sculpture gardens and children's playground, and many of his other buildings consciously brought together many aspects of full community life. You could say his designs were sometimes tending toward community design. If the same elements were assembled in his buildings with more consideration of nature's workings in that particular place, they might

Village on creek and bay.
A slightly larger community than in the previous illustration features, in addition to solar and intimate pedestrian access, orchards, vineyards, agricultural fields, sustainable yield forestry and fishing with considerable wilderness adjacent in an environment like some bays in Northern California or Oregon.

have constituted a step toward the ecocity. But in general, his effect was to promote the more sophisticated, not necessarily helpful, version of Gropius' boxes.

Frank Lloyd Wright used the horizontal expanses of the American Midwest and Plains to organize the lines of some of his buildings, and borrowed Aztec themes, abstract geometries, Art Nouveau shapes, patterns from nature, and building materials produced in economical industrial repeti-

tions for creations that sometimes worked in poetic counterbalance with nature. In his Broadacre City plans, according to which everyone could afford to have a house on an acre of land and get there by car or helicopter, were the seeds of an intellectual and esthetic justification for sprawl. He meant helicopter, too, and we'd be narrow to deride the notion knowing what we know of jammed freeways, nervous breakdowns of air controllers, and Murphy's Law, which says that if something can go wrong, it will. In those days the idea was bold, brilliant, and cheery. Now it's time to learn from it that some things worked ecologically and some did not. Ultimately, his work remained the one-building-at-a-time approach, whether in the city or the country. His version of working with nature, while often among the most harmonious designs to be found in architecture, will always be closer to the suburban or exurban ecocity impulse than the real thing.

If we draw out the lesson of the efficiencies of the assembly line, the striving for industry in service to all the people, the building of whole communities cognizant of access by proximity, if we select the best of village social creativity and urban personal creativity, thus reuniting conscience and consciousness in the enterprise of building cities in balance with nature, we can develop a whole new esthetic and begin weaving together the essentials of the ecocity.

CHAPTER 5

The City Today

INDIVIDUALISTIC CIVILIZATION IN THE United States was launched upon perhaps its last great binge, which is still going on, at the end of World War II, when people felt they had a right to a separate house — usually rather nondescript, even humble, with a moat of grass and picket-fence battlements — and a car to get where everything else in life went on. Americans also assumed endless oil — their country was the Saudi Arabia of the 1940s and 1950s. Atomic energy was going to start delivering electric energy too cheaply to meter any day now. True, this dream of comfort and ease didn't have the grand sweep of history's great religious and political causes, economic depressions, and struggles for empire or freedom, but the damage to nature and to the promise of salvation through what we've unleashed in the way of cars, sprawl, freeways, and related energy systems may ulti-

mately prove to be even more immense than that of any conventional world war. With about twenty million people killed in car crashes in the twentieth century, and a similar number by car pollution, the casualty list of the Car War ranks it right there among the other two big ones — and its casualty list continues growing into the 21st century. Its effect on nature in terms of species loss and climate change already ranks it as apocalyptic. From my perspective, the period from World War II to the present has brought us to the point of finally having to begin a conscious, worldwide effort to rebuild our civilization. (I wish more people agreed!) The first step in this effort has to be the rebuilding of our physical, constructed environment on the basis of ecological principles.

Our cities may now well be at a point in time in some ways analogous to that of the

Neolithic village about to be transformed into the city. I doubt that anyone at that time had any clear idea what the city would become. They were almost certainly unconscious of the larger pattern that would be developing into the future. Today, with stores of important tools and information in random disorder, we are fumbling our way toward the next reorganization, which, in this case, may yield the ecocity civilization. Just as the Neolithic villagers were unaware of the process of city building they were about to spawn, we seem to be unaware of the ecocity we might well create in the next few generations. This time around, however, if we aren't very conscious of what we are doing, the consequences will be dire. Perhaps even more important, if we don't explicitly recognize the role of conscience as well as consciousness in this process, we won't be able to harness the process for anything close to what we generally think of as good. For our next step in the pattern of healthy evolution, the two modes of thinking — consciousness and conscience — need to be seen as a whole.

For our next step in the pattern of healthy evolution, the two modes of thinking — consciousness and conscience — need to be seen as a whole.

Rebuilding and Rethinking — In That Order

World War II was over, all its horrors retreating into the past. Americans suddenly realized that they had about six percent of the world's people, 50 percent of its wealth, and 50 percent of its cars. What was to be done with by far the largest unscathed industrial machine on the planet, and with America's eleven million men in arms about to be liberated? They set out to build the ecological city. Theirs was the ecocity impulse to have both city and nature, both of them healthy and prosperous, simultaneously. But because this impulse was as yet uninformed by urban ecological principles, it ended up producing sprawl on a stunning scale. Within eighteen months of the end of the war, the American military had shrunk from 11 million to 500,000 and the biggest war-to-peacetime economic conversion program ever had begun. It was the construction of suburbia, facilitated by the National Defense Highway System, the limited-access "freeways" pioneered by Adolf Hitler's autobahns, and Henry Ford's mass manufacture of the automobile.

Where did all the soldiers go? Largely, they built cars and suburban houses and roads and went to college on federal money to learn how to do more of the same. They were consciously building a new and modern country and world, and unconsciously attempting to build the ecocity. Very few of them knew anything about the Garden City movement, but in their own way they were striving for a similar ideal. Everyone was to be humanized by grass, trees, roses, robins, butterflies, clean skies, and quiet starry nights. The vast majority thought it was a good idea. Linking semi-rural suburbs to the city with roads, cars, and television seemed harmless enough. The air was still crystal blue from sea to shining sea, with a few industrial exceptions and an occasional unfortunate incident in the Los Angeles air basin. If everybody could have a car and a house at a reasonable price, they would just solve the next problem as it appeared.

That Americans had not yet processed their own destruction of the Native Americans, who were far better connected with nature, had not yet thought about the dynamics of the city and the lessons of ecology, had not heard of the Garden City movement or of other relevant strands from recent history, prevented the ecocity venture from moving from impulse to insight and something healthier than suburbia. In addition, most people were personally compromised, bought off. By the mid-1950s, tens of millions worked for the auto/sprawl/freeway/oil industrial complex or were in debt to it or hooked on its products. This made them blind to the pitfalls and contradictions of the development patterns enveloping them. No one seemed capable of comprehending the implications of the rapidly multiplying number of cars and acres of sprawl. Even today, few people understand them.

The Marshall Plan was busy rebuilding European cities while America was building suburbs on its undeveloped, inexpensive land on the urban fringes. The first attack of Auto Sprawl Syndrome, that circular disease of cars creating sprawling suburbs and suburbs demanding more cars, began to produce serious degenerative symptoms. Slums spread and city centers began slowing down economically. Defining that first episode of Auto Sprawl Syndrome in the 1960s was the pattern of commuting from suburb to city center. Defining the second episode, coming in the 1980s, was the pattern of commuting from suburban home to suburban work place, both randomly scattered about a vast landscape.

Ecological consciousness was rising as problems became too large to ignore. Rachel Carson's *Silent Spring*, published in 1962, launched an activist reaction to particular industrial and technological assaults on the environment. DDT was killing birds, beneficial insects, and other organisms from pole to shining pole. With a politically difficult but relatively easy technological shift in production, DDT was outlawed. Industry cranked out new substitutes amid much noisy complaint. No lifestyle change or consideration of the structure of communities was required for this early environmentalist success.

While *Silent Spring* was making history, awakening ecological conscience, and saving a considerable number of species, back in the city the simple, big-box megabuilding emerged as the villain of enormous slum-clearing urban renewal projects. The expensive, tall apartment buildings of New York, San Francisco, and Chicago had their happy residents, but their more homely slum-replacing cousins were widely maligned. In a

Town in a semi-desert environment
Night view of a village that is not too different from the compact towns of North Africa and Yemen, but with slightly wetter climate and rooftop oasis, and linking pedestrian bridges.

few cases, as with the enormous Pruitt-Igoe urban renewal project in St. Louis, buildings were dynamited at the request of their own residents. This was not an auspicious time for Paolo Soleri to be introducing the concept of the large building as city. The fact that his proposed structures contrasted radically with the megabuildings in terms of diversity and their relationship to nature and community was lost on a society thinking in terms of 200-horsepower cars and enormous areas of uniform zoning. Kenneth Schneider offers some insight into the nature of the times:

> Both the faith and the sense of historic suddenness were vaguely with me in 1952 when I chose to join the relatively new profession, city planning,

which I felt could be a renaissance profession destined to shape urban history and help maintain the faith. A few years taught me that the profession was far from performing acts of renaissance. Alas, it became clear to me, city planning was destined to push paper for the established way of exploitation .... Planners operate without a conception of an ideal city .... We look vainly for a social ideal. Despite a 50-year debate about the neighborhood unit, the concept remains incomplete and exists only partially in a few locations. Urbanity, the quality of environments that stimulate a rich cosmopolitan life, is almost absent in the literature, except as a reflection of particular architectural settings .... While medicine operates with a vision of health and law with a distant image of justice, city planning exists without such a goal.[1]

Schneider adds that the process engaged in by the city planner helps hardly at all: "Any process which does not serve a deep cultural ideal inevitably serves the raw power of those who wield it. Most planners try to make planning serve a public purpose by focusing upon the process of planning"[2] and inviting public participation. In practice, this gives representation of a sort to interests beyond the financial, but this representation is almost inevitably lacking in knowledge of city func-

tioning, environmental implications, and an overall vision, that is, a whole systems perspective or ecology of the city. The planners' process broadens the range of input somewhat — by adding more, usually self-serving, influences. The democratic "methodologies," says Schneider, "allow urban ideals, but do not envisage or promote them. Urban processes, no matter how democratically based, are dangerous until a commonly accepted ideal of the city gives them specific content and direction."[3]

Though Soleri's ideas were based on setting and pursuing an ideal — the city built to play its proper role in evolution — and though his theory of miniaturization/complexification/quickening laid powerful theoretical foundations, his single-structure cities were otherwise unfortunately timed. His models and drawings of arcologies looked foreign to almost everyone. Environmental issues were chemical and conservationist at that time rather than issues of urban land use and abuse, architectural design, community restructuring, and associated lifeways. Also the main thrust of his writings represented the new-town approach: building a whole city at once. He had little interest in transforming existing cities where they stood. In any case, very few embraced Soleri's work when it was new, and, lacking an understanding of its foundation concepts and evolutionary context, critics have expressed amazement, thirty years later, that he still advances his arguments when "other megastructure architects" have long since given up. But there is an enormous qualitative difference between the megastructure building and Soleri's whole-city arcology, and people have been incapable of understanding that his exploration of a basic life principle is beyond fad and fashion.

Misunderstanding on the architecture and city planning front notwithstanding, by the 1960s and early 1970s good things were happening. True, they were largely a reaction to the Vietnam War, grinding poverty, racial injustice, and political assassinations, but the peace, civil rights, and women's rights movements were all making rapid progress. Environmental consciousness was rising, and Gandhian principles of non-violence and civil disobedience were being methodically applied in the peace and civil rights movements. Millions found cooperation with the machine-like establishment so "odious," in Mario Savio's words, that if they didn't "put their bodies upon the gears and upon the wheels ... upon all the apparatus ... and make it stop",[4] they at least tuned in, turned on, and dropped out.

With Earth Day 1970 and subsequent environmentalist actions and legislation, ecologically informed policies, programs, and personal actions were gathering momentum. Within twenty-four months after Earth Day, the *Clean Air Act*, the *Clean Water Act*, the *Environmental Education Act*, and the Occupational Safety and Health Act were passed. DDT was banned, and the supersonic transport was stopped. The youth culture, with its wanderings into metaphysical realms influenced by Eastern mysticism and psychedelic drugs, health food, and yoga and its wanderings back to the land in hippie communes or organic farms and out to the world in Peace Corps programs, created a condition of readiness to confront major lifestyle changes. Some came to Soleri's experimental city, Arcosanti, to live, work, and contemplate building a better world. These people were prepared to experience change. Later expressions of this experience took the form of "simple living" lifestyles, urban homesteading, and a more honest assessment of the impacts of their lives on society and nature. The words of cartoon character Pogo, immortalized in 1970 Earth Day events, were taken to heart and acted upon with good humor: "We have met the enemy and he is us." To this the young people who were willing to experiment with their own lives could add: "We have met the ally too, and he is also us."

It was perfect timing, then, one might think, that almost precisely as the OPEC oil embargo slammed into American experience like some of the front-page fist fights that erupted in the gas station lines of 1973, a powerful document from on high hit the press. A study called *The Cost of Sprawl*[5] was undertaken by three US agencies — the Council on Environmental Quality, the

Department of Housing and Urban Development, and the Environmental Protection Agency — and released in spring 1974.

Gathering data from all over the United States, the study compared low-, medium- and higher-density communities and measured their impact on schools, fire and police services, government facilities, roads, and utilities. It demonstrated that higher-density communities required 50 percent less land and 45 percent less investment cost in infrastructure (buildings, roads, landscaping, and utilities), caused 45 percent less air pollution and a similarly reduced amount of water pollution runoff, and used 14 to 44 percent less energy and 35 percent less water. Costs of fire, police, and other government services were similarly reduced in higher-density communities.

The high-density model, as the report defined it, was a mix far from the extremes of Manhattan, Hong Kong, or Paris. In fact, it included two- to six-story buildings and nothing taller. It left out the cost of the automobile entirely, nor did it mention the savings from using transit in higher-density areas. It also omitted any architectural and technological features — from ones as simple as enhanced insulation to ones as potentially flamboyant as solar greenhouses and green

Nature in the city - symbolized by large trees high in buildings. *Is it worth the extra strength of the structure — about enough to support one more floor? Is it worth it to finally take strong steps to declare a new relationship that is not exploitive but instead reverential?*

terracing on the sunny side of tall buildings — that might make real contributions to energy savings. If these features had been considered, the costs of cars factored in, and the extrapolation to yet higher densities calculated — if, in other words, the next logical step had been taken — it would have been apparent that the agencies were really on to something. Namely the whole-cloth new ecocity concept.

The Cost of Sprawl had profound things to say. If not the last word on the subject, it was one of the earliest and best reports. However close it came to a breakthrough, it nonetheless slid quickly into oblivion. The report on urban density and form, the opening into ecocity thinking, gathered dust in the Jimmy Carter years while attention and support went to initiatives for developing specific technologies that would have been far more useful had they been attached to ecocity thinking and development. Then, under Ronald Reagan, most of that initial effort was scrapped and the crack in the door to ecocities was shut at a potentially historic juncture. Thus, in the late 1970s and the 1980s, the big message about the city and the environment was ignored, and urban form and whole-community awareness drifted deeper into the back of the mind. Simpler changes came on line: car mileage went up, along with energy efficiency in buildings and appliances. And now, decades later, the shape of American cities and those of most of the rest of the world is worse than ever.

Appropriate Technology

A new willingness to experiment in technology, agriculture, and architecture emerged during the time from the oil embargo through the early 1980s. This willingness found a particularly strong expression at the federal government level. Though President Jimmy Carter failed to follow up on *The Cost of Sprawl*, he helped with renewable energy, putting solar collectors on the roof of the White House and establishing the Solar Energy Research Institute. In California, Governor Jerry Brown set up an Office of Appropriate Technology and abandoned the Governor's mansion and the attendant commute for a modest apartment within walking distance of Sacramento's downtown offices. He appointed Huey Johnson, founder of the Trust for Public Land, to head the State Department of Resources and the innovative architect Sim Van der Ryn to head the Office of the State Architect. These moves helped create a powerful wave of experimentation, putting alternatives like solar and wind energy on the map.

The British economist E.F. Schumacher, who had been studying small-scale industries and Buddhist economics for years, published his book *Small is Beautiful*[6] in 1973, and his timing was as good for his message as Soleri's was bad for his. The establishment of the first ecology center in the United States in Berkeley in 1969, Earth Day in 1970, the United Nations Conference on the

Environment in Stockholm in 1972, and the oil embargo of 1973 all prepared the soil for his message of small-scale, decentralized, self-reliant, and relatively simple technologies, businesses, and lifestyles. He did not live long enough to exercise his rising influence and feisty leadership, but his ideas affected policies and practices in the United States, Europe, and, by way of efforts for healthy "technology transfer," other parts of the world.

"Appropriate technology" was, in Schumacher's words, technology with a "human face," and it came to include solar, wind, geothermal, and biological fuels such as waste wood chips and alcohol; energy conservation measures such as weather stripping and insulation of water heaters, pipes, and buildings; recycling of all sorts; soil building through composting and human waste recovery; agricultural practices such as integrated pest management and companion planting; more efficient transportation systems such as bicycles and public transit vehicles, transcontinental railways, and electric cars. If the technology seemed appropriate to a healthier world, it qualified. Could there be an appropriate technology embracing all or most of the others, giving people shelter, access, order, and healthy relationships? It's the ecocity, of course.

As for bridging from appropriate technology to ecocities, we had an example that clarifies the continuity: the Montgomery Ward Building, built in stages from 1923 through 1927 in Oakland, California. Though it was exemplary for working well with public transportation and for energy and land conservation, it was demolished due to ego-infested local politics. It should have been an ecological landmark building illustrating how working intelligently in the third dimension solves numerous problems in one compact and elegant design.

At eight stories, the Montgomery Ward Building covered most of a two-block area and was the largest and tallest industrial building in Oakland until January 2001. Oakland resident and Berkeley city staff planner Sandy Grube, leading an effort to save the building, won listing on the National Register of Historic Places. This designation was granted largely on the basis of the building's Art Deco styling and Arts and Crafts detailing, but the department store's story is much bigger than that.

Goods arrived on the first floor by way of a short spur off the regional rail line that came right into the building, and were transferred to the upper floors by freight elevator. Customers arrived at the second floor showroom, looked over the displays, and placed their orders at the sales counter. Then their orders were written down, rolled up, and slipped into a canister that was placed in a pneumatic tube and sucked up to one of the six stories of warehousing above. Next, a clerk picked up the order and, on roller skates, zipped off to the correct shelves to locate the item and bring it back to a spiral delivery chute, which in turn delivered the package to the show room by gravity, generally in less than two minutes. The customer paid for the item and carried it out the door a few feet away and, usually, went home by transit. If the package was large, it could be delivered by a truck making the rounds.

Today, instead of goods being warehoused on a relatively small piece of land over both display room and arrival docks, they are stored in giant one-story sheds covering hundreds of acres of land ten, twenty, or even forty miles from Oakland over two ranges of hills and in the California Central Valley. They are serviced by a fleet of trucks clogging the freeways and burning millions of gallons of fuel a year. The customer arrives by car, parks in a gigantic land-gobbling lot, and walks to the one-story big-box facility. There he or she haplessly wanders the storage shelves alone, trying to catch the attention of clerks, pushing a cart, spends considerable time locating the item then carries it back to the checkout line. Finally, it's back out to the car and the long drive home.

Permit me to say that this is an insane waste of fossil fuels, asphalt, concrete, and the manufacture of hundreds of trucks, accompanied by massive quantities of air pollution and the consumption of millions of hours of people's time and millions of dollars

Below: Learning from nature – vertical layering of micro-environments. Valleys are cooler, wetter and less breezy than high places, and slopes facing the sun are warmer and drier than those facing away.

SOUTH-FACING SLOPE

NORTH-FACING SLOPE

HOT, DRY, BREEZY

WARM, MODERATE MOISTURE AND AIR MOVEMENT

COOL, STILL, MOIST

of people's money to move products in today's manner as compared to taking transit to the old Montgomery Ward Building and having goods delivered to your hands by a clerk on roller skates and gravity. Compounding the insanity, as Grube pointed out in defending the building, was the destruction of thousands of tons of building materials and of all the energy that went into creating the building. The structure could have been remodeled successfully, as, in fact, shortly before the demolition of the Oakland facility, was the case with the old Montgomery Ward buildings in Portland, Chicago, and Baltimore. It could have remained one of the best examples of urban ecological design and efficiency, of appropriate technologies united by appropriate architecture and urban layout.

Paradigms and Permaculture Principles

While appropriate technology was dawning, systems theory was coming on strong with Thomas Kuhn's *The Structure of Scientific Revolutions*[7] in 1962 and Fritjof Capra's *The Tao of Physics*[8] in 1975. Their "new paradigm" involved experiencing and understanding life as the interrelation of whole systems — organisms, businesses, conceptual frameworks — as part of larger whole systems: societies, ways of life, economies, bioregions, and ultimately the whole Earth and universe. A paradigm is a set of perceptions, concepts, moral convictions, and habits, activities, and lifeways functioning in the environment in its largest sense. Largely, what a paradigm sees is what it allows itself to see. A paradigm is not just the way we think and experience, but also a screen erected by the mind using previously ordered experience that, like the receptionist outside the doctor's office, admits or rejects the newly arriving experience or information.

The new paradigm amounts to lifeways and patterns of perception and thinking that recognize and relate to whole systems, not just isolated items and events, over very long time periods, not just for the duration of something's usefulness for limited immediate purposes. As human consciousness is aware that it is conscious, so new paradigm thinking is aware that paradigms exist and can be understood and worked with. We can see, for example, that as new information builds up suggesting that an older system of thinking or "old paradigm" is no longer answering questions, a breakthrough to a new understanding, a new set of theories, may be imminent. For example, as technology and experience advanced in the sciences, Newtonian physics seemed incapable of explaining ever more phenomena. Then at a certain point Einsteinian relativity matured and coalesced into a consistent body of information, perceptions, and systems of order that created a whole new paradigm.

In this book I am suggesting that this is exactly what we see happening in the realm of human habitat design. The old ways, ideas, perceptions, and information screens are failing us, and we need to go beyond trying to repair the car and its infrastructure and instead build the ecocity. It's noteworthy that just before the dawning of a new paradigm, resistance typically reaches its peak and pitched battles are fought to keep people from even considering the new information. But then the weight of evidence becomes overwhelming and rather suddenly, everything makes sense in a new way and the new paradigm not only blossoms forth in its own right, but acceptance becomes almost universal. As they say of a new idea, even more so a new paradigm: it is initially derided as ridiculous, then opposed violently as dangerous, and finally embraced as self-evident.

The old ways, ideas, perceptions, and information screens are failing us, and we need to go beyond trying to repair the car and its infrastructure and instead build the ecocity.

A whole-systems approach to agriculture and lives integrated into food production that goes a long way toward fleshing out a new paradigm is called "permaculture." It also is a design approach very similar to the ecocity design approach. The term was coined by the inventor of the concept and founder of the permaculture movement, Bill Mollison, and his definition goes like this:

> Permaculture (*perma*nent agri*culture*) is the conscious design and maintenance of agriculturally productive ecosystems that have the diversity, stability, and resilience of natural ecosystems. It is the harmonious integration of landscape and people providing their food, energy, shelter, and other material and non-material needs in a sustainable way. Without permanent agriculture there is no possibility of a stable social order.

Permaculture design is a system of assembling conceptual, material, and strategic components in a pattern which functions to benefit life in all its forms. The philosophy behind permaculture is one of working with, rather than against, nature; of protracted and thoughtful observation rather than protracted and thoughtless action; of looking at systems in all their functions, rather than asking only one yield of them; and of allowing systems to demonstrate their own evolutions.[9]

Though Mollison started by sculpting landscapes to create ponds, planting trees, bushes, and vines, and building rural homes, some of his associates have been impatient to expand the concept to embrace more aspects of human environmental design. According to the architect Declan Kennedy, the "term *permaculture* originally described a *perma*nent agri*culture*. Nowadays, the concept encompasses much more: a planning and design method with the aim of creating stable self-supporting systems, a sustainable culture based on ecological principles, which not only supply wholesome food for people but energy, warmth, beauty, and meaningful pursuits."[10]

So too in urban architecture.
Here we see a very small plaza built on the edge of a town with an almost completely natural environment a short walk away. Bridges are purely schematic.

Permaculture emphasizes fruit and nut trees and berry and vegetable plants that bear year after year or reseed themselves rather than annual crops that are sown and harvested every season. Arrangement and proximity are essential: trees are planted to shelter buildings and gardens from wind or to avoid shading areas that should get warmth and light. The plants that are most frequently used, such as salad greens and herbs, are located closest to the kitchen, with occasionally used main-course vegetables farther away, seldom harvested grains, berries, and fruits farther yet, and the distant woodlot, harvested only rarely, farthest of all. Beyond the semi-natural wood-lot, nature is mainly in charge: a healthy reservoir of life that, in the form of bees, butterflies, humming birds, and many other animals and plants, delivers the service of pollination, occasionally provides game for the farmer's dinner, and supplies completely untended wild fruits, nuts, and berries for the picking. Beyond that there is natural wilderness where people are non-interfering observers. Right on the windowsill, for color, fragrance, and inspiration, is the window box with flowers. We see here again the principle of access by proximity. Dogs and cats are not particularly useful in the eyes of permaculturists; at the permaculture village of Crystal Waters in Queensland, Australia, they are banned, and as a result the village is alive with hundreds of natural birds and frequently visited by native bats, reptiles, and marsupials.

The permaculture homestead follows the same principles in architecture, with the home sited for shelter from the winds provided by trees, hills, or berms, and oriented to collect or avoid direct sun, depending on the climate. To the already complex house with attached solar greenhouse, the permaculture designer might also attach a chicken house and composting bin that share heat with each other and the main house. Worms may be busily composting kitchen waste in the basement side by side with a dry toilet that is preparing an annual dose of fertilizer for appropriate parts of the garden. When it rains, the roof collects water with its gutters and downspouts, and cisterns store the water beside the house, ready for use inside or for watering the garden. A pond may be carved out of the yard on the sunny side of the house to reflect added light and warmth into the greenhouse and provide for ducks, which delight in eating the snails that are doing their best to eat the crops before the people do. The permaculturist can provide hives for bees, for honey as well as for pollination services. All this, again, represents access to diversity at close proximity.

A key principle, according to Mollison, is "Trust the chickens." It's impossible to predict how all the elements of your house and garden design are going to interrelate, he says, but you can try it out and carefully watch how it works. The chickens will deliver eggs and meat. They will rid your garden of certain grubs and other pests, but they will also pick at some of the plants you like, and they may need to be fed something in addition to kitchen scraps, local bugs, occasional weeds, and so on. If a clear pattern of benefit emerges with any element of the whole — chickens, worms, greenhouse, wood stove — many other benefits are likely to emerge unexpectedly. When building my own solar greenhouse in Berkeley, I was expecting food production and heat for the house, but to my surprise I also got peace and quiet: the greenhouse glass virtually eliminated the considerable street noise that had formerly plagued our place. We are familiar with the vicious cycle, where one bad thing begets another. This, in contrast, is the virtuous cycle, where good things beget other good ones. Putting beneficial things into mutually supportive relationships often has this delightful effect. So trust the chickens.

Permaculture amounts to the mixing of appropriate technologies and agriculture according to design principles that accomplish high orders of integrity of the parts of complex whole systems. Urban permaculture amounts to the application of these same design principles to urban habitats, to produce a yet higher order of integrity.

One of my favorite examples of urban permaculture was designed by Margrit Kennedy, Declan's wife, as part of the International Building Exhibition in Berlin in the mid 1980s. In the very dense Kreuzberg neighborhood — typically 900 to 1,200 people per urban block, in buildings averaging six stories — there was a large city parking structure used by people who worked or had appointments in the neighborhood. Relative to those living in the neighborhood, these motorists were few, and, living elsewhere. They were not local voters. With cars threatening the lives of the residents' children every day, there came a time when the locals decided to close down the parking structure and do something more socially useful with the building. Margrit Kennedy drew up a design with the parents in the area, and three years later the house for cars became a nursery school for children.

In good recycling tradition, the basic structure was preserved. The sloping ramps were left intact, with a staircase built into the shallow angle: two steps forward, one step up in a kind of limping gate, awkward but very inexpensive and practical. A masonry saw was taken to the three levels and a central atrium created by simply slicing out and removing the center of three floors of concrete slab. Classrooms were built by inserting divider walls perpendicular to the central opening, and the atrium itself was glassed over with a skylight and filled with dozens of plants. On the roof of the building a playground and garden were built behind a secure fence, and a small pond was added to support aquatic plants, insects, fish, and frogs. Around the outside of the building a trellis was constructed, and when I visited in 1989 vines had just started to wind their way up the fence that surrounded the rooftop playground.

Mixing architecture and town planning three-dimensionally in this fashion with this complex agriculture in relation to the bioregional realities of climate, weather, and sun angles becomes the context for a real thriving of appropriate technology and the human community. All of this was technologically and conceptually available by the early 1980s, but until this day only the smallest steps have been taken in this direction anywhere.

"Improved" Suburbs, Revitalized Downtowns

One of the things that does get built these days is the "improved" suburb, and among its builders are the "New Urbanists." Andres

Collin at Codornices Creek. *On the Berkeley/Albany border, Ecocity Builders and Urban Creeks Council "daylighted" (opened) a creek that had been buried for 50 years. A child named Collin Davis here enjoys the newly opened creek with sunflowers planted as first habitat — becoming native habitat in subsequent years.*

Duany and Elizabeth Plater-Zyberk designed the 80-acre Seaside project in Florida and the 352-acre Kentlands project in Gaithersburg, Maryland. Peter Calthorpe designed the 1,100-acre Laguna West project south of Sacramento, California. Other New Urbanist projects are built or under construction, and the building codes for the whole country are being adjusted somewhat to allow the sort of features they promote, such as older-style, narrower streets with smaller intersections. Says one of their prime movers, architect Peter Calthorpe:

Urban Development is ideologically a strong and compelling alternative to the suburban world, but it doesn't seem to fit the character or aspirations of major parts of our population, and many businesses. Mixed-use new towns are no alternative, as the political consensus needed to back massive infrastructure investment is lacking. Growth therefore is directed mainly by the location of new freeway systems, the economic strength of the regions, and standard single-use zoning practices. Environmental and local opposition to growth only seems to spread the problem, either transferring the congestion to the next county or creating lower and more auto-dependent densities.[11]

In response to this problem Calthorpe has proposed the "transit village" or "pedestrian pocket,"

...a balanced, mixed-use area within a 1/4 mile walking radius of a light rail station. The uses within this zone of approximately 50 to 120 acres would include housing, offices, retail, daycare, recreation and open space. Up to 2,000 units and one million square feet of office can be located within three blocks of the light rail station using typical condominium densities and four-story office configurations. These pockets could be implanted into the existing suburban fabric by the creation of a new light rail line and a corresponding up-zoning at each of its stations .... The goal would be to create an environment in which the convenience of the car and the opportunity to walk would be blended. The Pedestrian Pocket would accommodate both the car and the economic engine of new growth, the back office [that is, the office in the suburb remote from the city center]. Parking would be provided for all the housing and commercial space.[12]

In the early 1980s, New Urbanist thinking began with the recognition of the healthy aspects of the "traditional" small town with modest density, which we tend to remember as more typical around the 1940s, a time when our culture was not yet overrun with cars. Such towns had total densities two or three times as high as suburban developments. They had residential areas of modest houses with porches, private back yards, public front yards, ample side yards, often an alley in back with a garbage can behind the house. In the center there was a town square or plaza surrounded by buildings one to four stories tall, sidewalks everywhere, and cars parked along the curb. Rail access — regional heavy rail and local streetcars — was common. Originally called "neo-traditional," this approach embracing the car accommodating small town arrangement became known as "New Urbanism" as it evolved, although it was neither new nor really urban. In line with my own thinking about best urban patterns, these designer/advocates proposed a center-oriented development, often referring to such centers as "transit-oriented development" or TOD. An association called the Congress for the New Urbanism was founded to refine and promote such work.

But how well can slightly more dense suburban centers work if the assumption is that cars will be needed in addition to whatever transit is encouraged? Calthorpe has called the New Urbanist strategy a bridge strategy. De-emphasizing the car to some degree may be the beginning of a bridge to a better urban structure, but barely a start, and it will be helpful only if it is understood that way — which hardly seems to be the case. A few generations of people could easily live and die in such developments, swearing it's where they had arrived, and not a step to somewhere else, all the while consuming 80 or 90 percent as much gasoline as the people in the conventional development next door. In the New Urbanists' own manifesto, which they called the "Charter of the New Urbanism,"[13] the point that they don't want to cross over that bridge is, in fact, made quite clear. "In the contemporary metropolis," the Charter says, "development must accommodate automobiles. It should do so in ways that respect the pedestrian and the form of public space." Elsewhere the manifesto says, "Communities should be designed for the pedestrian and transit as well as the car,"[14] as if the process of design starts with the car, but we should remind ourselves not to forget people. Now why on Earth would they talk like that if some of the most well loved urban spaces in the world and even in the United States are car-free areas, such as some plazas, malls, parks, waterfront boardwalks, amusement parks, historic districts, Disneyland, and so on? The New Urbanists have adopted the idea that most people, and certainly Americans, need the excitement of cars passing quickly by and feel secure when there is a row of parked cars acting as shields against the cars moving in the street, and thus they advocate for curb parking.

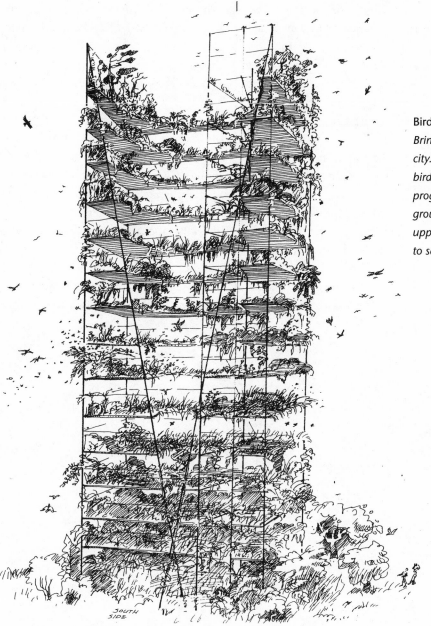

Birdhouse – no people allowed. *Bringing nature into high density city. To promote a variety of birds, the higher "floors" are progressively less accessible to ground-based animals, with upper ones even inaccessible to squirrels.*

SOUTH SIDE

Calthorpe was off on a better line of reasoning in the 1970s when he wrote an article called "Beyond Solar Suburbia."[15] In it he noted that an average suburbanite using every solar energy design feature and technological device available would spend ten times as much energy driving the car than could be saved with all the solar design and technology available for his or her house. In other words, appropriate technology in an inappropriate context was an illusion. Or, as Calthorpe himself once said: "Development must be seen in a social, political and environmental context in which solar and other alternative technologies are tools for new settlement patterns rather than compensations for the faults of the old."[16] Something more fundamental than add-on hardware was needed; the built environment had to be reshaped. The New Urbanists did go on to do that in some places and helped popularize the idea that we need to build around centers with rail service, but mainly in the suburbs or even farm country — and with so much acquiescence to the automobile one wonders if the result was mainly to convince people they had done enough when in fact very little was changed.

Improving suburbs and creating pockets of charm may be a step in the right direction, but what about improving the whole thing? Approaches to the whole city as a rule must take into consideration the whole city's land use patterns, the architecture built upon those patterns, and the way in which these are con-

Bug escape.
This feature at the top of windows, especially greenhouse windows, lets bugs escape — along with an inconsequential amount of heat. There could be a hatch timed to open and close every so often to save more warm air on cold days.

nected with the transportation system. Part of the land use element will be the restoration and preservation of agriculture and nature.

And in some places this genuine whole-systems approach is actually being put into practice. In fact, in essence, this is exactly what the planning departments of cities are attempting, and most of their tools, such as General Plans, zoning ordinances, and building codes, have significant potential for ecological benefit if written for that purpose. Many city planners wish they could do more in this direction but are hampered by a lack of ecological insight in the community as reflected in rigid rules and neighborhood associations that stop them dead in their tracks. In a few cities, however, many healthy ideas are deeply imbedded in the community and supported by skilled leadership, and therefore their plans are beginning to solve the larger ecological and urban problems.

Communities on the Cutting Edge

Whole cities, among them Curitiba, Brazil, and Waitakere, New Zealand, are taking on a wide range of ecological issues and are self-consciously called "ecological cities" by their own leaders. Many have established or are establishing greenbelts, constructing new urban rail lines, restoring natural areas, and building "greenways" or nature trails. Portland, Oregon, has removed a major motorway along the Willamette River to create a public promenade and park. Many are promoting bicycles and are doing ever better at recycling. Waitakere has a program for planting several trees for each new baby — a "baby forest" that the parents plant themselves. On a smaller scale, many intentional communities that call themselves "ecovillages" are being established, and many of them are united in a loose-knit organization called the Global Ecovillage Network that vigorously promotes their values and strategies.

One of the most inspiring experiments in urban design anywhere has been taking place over the past thirty-five years in Curitiba, Brazil, the country's eighth largest city at 1.6 million in the municipality and 2.5 million in the total metropolitan area. Under the leadership of Jaime Lerner and other recent mayors, this city became a model for solutions to intractable urban problems afflicting cities on every continent. Lerner was an architect and city planner, and when he became mayor in 1971, he brought along a number of innovative colleagues — and a team of inspired professionals launched into action.

Curitiba's success had a good deal to do with the considerable imagination and contagious enthusiasm of the mayor and his remarkably creative crew from the newly opened School of Architecture at the Federal University of Paraná. He launched a new institution, the Institute for Urban Planning and Research of Curitiba (IPPUC), which has endured through several other mayors'

Improving suburbs and creating pockets of charm may be a step in the right direction, but what about improving the whole thing?

terms as well as Lerner's, lending stability and maintaining pioneering ecological innovation in the city for thirty years. Happily, many of IPPUC's founders had visited Europe and recognized that pedestrian areas and public transportation can work beautifully. Unhappily, they saw their closest urban neighbor, São Paolo, clogged in cars and fumigated by smog. They were ready to try almost anything to avoid such a plight, even sensible ideas like creating pedestrian streets. With the influx of agricultural workers who were being replaced by farm machines, they expected the city to grow rapidly from its population of 600,000 — the stakes were high to get it right.

Their rare insight at the time was to make the land use/transportation connection work from the very beginning. They planned for future expansion of the city along five high-density corridors stretching out like arms of a giant starfish and served by frequent and inexpensive buses traveling quickly down streets dedicated to them alone. Other cities in Brazil were using federal money and their own to build subway systems. Curitiba decided the benefits of a subway just were not worth the extreme expense — 300 times more costly, said the planners, than simply paving five dedicated streets and encouraging density along those corridors with zoning incentives. Private companies would get to build the tall buildings along the corridors using their own capital and making good

profits while the bus companies would have the riders living and working within a short walk of the main "speedy" lines. Full and frequent buses meant profits for the bus companies too, without subsidy from government. Bus riders got access to jobs, housing, and open space all in the same package. The low cost of building the bus system also meant that a great deal of money was saved to be invested in many other things: purchasing land to restore rivers and expand parks, launching of social programs of a wide variety, and building neighborhood libraries called "Lighthouses of Learning." These have four- or five-story towers that look like lighthouses and inspire the affection of the citizens. "In a way they don't make any sense in Curitiba — the coast is sixty miles away," said IPPUC spokesperson Maria Rosario, "but they are the source of 'light and learning' and a good idea. People love them."[17]

The imagination and enthusiastic civic spirit of Curitiba's ecological and social policies have been its hallmark. It's not just that the city has achieved 70 percent recycling, the record in South America, but that old and broken-down buses have been towed out to the slums of new immigrants to become schools for entry level skills useful in the city, thus recycling buses as well as trash. Recycling hit a high point when several old quarries were filled with water, creating beautiful lakes, and major public institutions were built there, including the Free University for the

Environment and the Wire Opera House (called that because the graceful metal tubing and glass structure almost looks like an open wire structure).

By whatever means — sometimes taking it to the television audience in personal appeals, sometimes drafting school children to lecture their parents on environmental issues — Lerner's administrations have unclogged traffic, greatly reduced residents' travel time, established a 300-mile-long bus system and a 125-mile bicycle and foot path system. Curitiba's fuel consumption has been reduced by 20 percent compared to other similar-sized Brazilian cities. It has the country's lowest traffic-accident death rate per capita, and air pollution is remarkably low. According to Lerner, there is "little in the architecture of a city that is more beautifully designed than a tree,"[18] and with that philosophy his administration initiated a program with the help of which more than four million trees had been planted by the time I visited in April 2000. Street children are finding jobs through city programs, garbage collection and recycling are being accomplished in shantytowns, food shortages are being alleviated through city programs to trade agricultural gleanings for recyclables, and a twenty-four-hour health clinic for the poor has been opened.

Early in his first term as mayor, Lerner built Brazil's first pedestrian street despite protesting business owners — who later flour-ished financially as the project spread and turned into a 27-block system of pedestrian streets. That famous first Monday after Lerner's Department of Public Works put up the barriers at the ends of a downtown block, protesting car drivers arrived on the street intending to remove the barriers and keep the street for cars. But city employees were ready with long rolls of drawing paper. Hundreds of school children laid out the paper on the street and began drawing. The drivers backed off, the shop owners decided to wait a few days, and sure enough, it worked out just fine. For the young and restless one of Lerner's administrations built the country's first "24-hour street," a glass covered pedestrian street sheltered by one of Curitiba's characteristic metal tubing and glass structures of big semi-circular arches, similar to the Wire Opera House, which provides space for round-the-clock snacking, socializing, and courting. In the shantytowns where narrow footpaths wind between ramshackle hovels improvised from recycled materials, garbage collection and municipal recycling were beyond the reach of garbage and recycling trucks. Lerner proposed a "Green Exchange": The people themselves separate recyclable materials and garbage and take them to pickup points where they can be exchanged for food provided by the city. The Green Exchange then takes the garbage to the dump and markets the recyclables to help pay for the system. To get the populace into the swing of things, Lerner promoted the slogan "Trash that is not trash," and went on television to convince residents to separate rubbish into organic and inorganic waste. With a mix of public relations savvy and leprechaun spirit, he had garbage trucks painted green. And he kicked off the campaign by donning overalls and doing a day of collecting recyclable trash. "Shantytown kids ran after the garbage truck, tauntingly calling me 'trash collector,'" says Mr. Lerner. "But that was my idea. If it's not beneath the mayor to collect trash, it shouldn't be beneath them to separate it."[19]

Next, the city recruited four actors, dressed them up in foliage, called them the Leaf Family, put them on TV, and then sent them out on a tour of the city's schools to teach ecology and trade pictures of locally famous comic book heroes for non-degradable trash such as batteries and tooth paste tubes. The kids were excited to see them and listened attentively, largely because they'd

Integral neighborhood.

Urban "fractals," or portions of the city that embody the essential functions of the whole city on a smaller scale, and relate them successfully to the natural environment as well, are keys to understanding the ecological city. We are talking about centers-oriented development here. On the small scale, integral neighborhoods, such as this one proposed for west Berkeley in 1977 (not built), bring together a full mix of land uses from housing and jobs to shops and small manufacturing operations. As in other ecological architecture on a community scale, organic agriculture and sun and wind play a large part, along with compact development, rooftop uses and pedestrian streets. The "interior" pedestrian street in this case, on the extreme right, is one with restaurants, café, convenience store, very small movie theater and several other in-city village-scale services and products

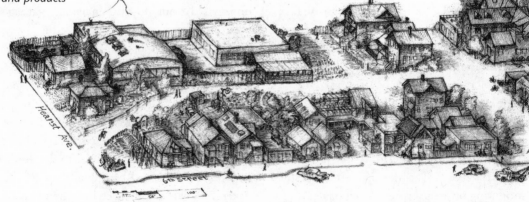

already seen the Leaf Family on television. That TV personalities would actually visit their school to talk about ecology was exciting.

Confronted with a car population growing faster than the human one, Lerner increased bus routes, frequency, and convenience. In a truly remarkable stroke of genius, he designed the spiffy "tube stations" that are now internationally famous, solving technical problems while inspiring the public with a stylish kind of urban feature somewhere between architecture and furniture. These glorified stations are in the shape of giant glass bottles on their sides with both ends cut off so that the bus rider can walk up a few steps, enter a turnstile, pay an attendant the equivalent of 45 cents US, and go anywhere in the city. No longer does the driver have to fumble with change or worry about who puts how much into the money box. The tube station celebrates the bus riders, its stylish design

announcing that the city cares about them. When the bus arrives, wide doors open and a wide plank drops from the side of the bus to the platform, allowing people to exit and enter the bus quickly, four abreast. Inside the tube stations, people don't get rained on any more. Upon installation of the tube stations, time for travel on the bus went down 24 percent and the ridership went up 20 percent, virtually overnight. Says Lerner, "If city dwellers used their cars less and separated their trash, half the city's problems would be solved."[20]

How was mayor Jaime Lerner able to accomplish all this, and in a country of modest means? With enormous imagination, flexibility, and determination. And, most importantly, as just suggested, he got the city reshaping project off to an excellent start: the compact development provided for efficient, inexpensive transportation and left room for recreational open space, natural restoration, and money for services to the people.

Not everyone is happy in Curitiba. Poverty is a problem that is being solved only slowly, despite innovative approaches. Cars are beginning to glut city streets as ever more people become prosperous — largely because they have been able to save money due to the excellent bus service. But as Marc Margolis has put it, whereas "in Brazil most mayors start out their terms as fabulously popular celebrities and finish them barricaded behind palace doors, dispatching press releases and disclaimers, and ducking the press, Lerner can walk the streets with impunity."[21] Said one of Lerner's critics, "the only thing lacking is for people to say that the sunsets here are nicer since Jaime Lerner took over."[22] But as the smog clears and the distant pink clouds show through, this may well be true. Vote for better sunsets. Vote Lerner.

Vancouver, British Columbia, has made major steps on the way to becoming an ecocity. It has two million people, whose ancestry and tradition are mostly British but are rapidly becoming more Asian. The third-fastest-growing city in North America, it has an intense, very three-dimensional but friendly downtown with sixty thousand residents. Larry Beasley, co-director of planning, says that they don't want a freeway, have never had one, and, if the present planning strategy and public mood hold, never will. Convinced that balanced development at high density works, his office is presently planning for forty thousand more people in the downtown.

The city has adopted a "special strategy to secure low-income housing, to foster housing for families with small children, and to ensure specialty housing for singles, mingles, seniors, extended families, alternative households, and any other household formation that presents itself," including a growing houseboat community. "Vancouver is now looking at both commercial and industrial live-work forms," says Beasley. "Both the shop house and the industrial loft are appearing as well as more multi-use high-rise, high-ceiling loft spaces that are snapped up by buyers and put to many uses in connection with residences."[23] Vancouver's chief asset, according to Beasley, is not any large global megacorporation hiring large numbers of people but "our good looks," meaning the mountains and the sea. To preserve those views while accommodating two million people, taller buildings are kept to a thin profile and spaced relatively far apart. The spacing also helps maintain a sense of privacy.

The planning department's major objectives for the city are "living closer to work, cutting commuting time, de-emphasizing the car ... slashing capital spending on roads, bridges and the complex accommodations needed for mass auto commuting." Vancouver is building an advanced transit system and networks of bikeways, pedestrian paths, ferry connections, and greenways. A major goal is to "replicate the rich street life that is enjoyed in older traditional cities," creating an intimate egalitarian world on the ground level but in a high-rise mode supported by the economic engine of the larger buildings rising without much notice beyond.

According to Beasley, "'Streetscaping' is the key design tool to tame the massive scale of the contemporary highrise, high-density development that characterizes Vancouver and most modern cities." To heighten the intimacy at street level, a variety of plantings are encouraged along intentionally narrow and crowded sidewalks, often with double rows of trees on each side of the street. In many cases the sidewalk has been "expanded and enhanced by setback requirements that, in essence, reclaim a slice of private space for the public streetscape. Remaining private, however, these slices of streetscape have great agility as places for outdoor retailing, outdoor restaurant seating, townhouse stoops and landscaping to soften hard urban edges .... Such setbacks have proven to not create economic hardship yet dramatically improve the public realm."

Creating a pleasurable environment at the foot of the towers, Beasley tells us, "is first and foremost the art of creating the classic streetwall. First, the space of the street is fully enclosed; second, the enclosure is strongly corniced, with taller tower elements set generously back; third, the height of the resulting streetwall is carefully constrained at 3 – 6 stories to secure a comfortable human scale." In many cases the terrace rooftops of these lower buildings along the street are planted with trees, adding a second screen of foliage between the people in the street and the taller buildings in the background; the first screen is provided by the trees planted at street level. The resulting effect, says Beasley, "'disembodies' the massive tower by screening its foundation and making it appear to float and then disappear beyond the immediate perception of bystanders."

As I listened to Beasley characterizing the streets as cozy and sociable, with the massive towers rising virtually unnoticed behind, it occured to me that a great many of the same people ignoring the towers and delighting in the small-scale at street level, live and work in the towers that this design concept treats as forbidding, massive, and out-of-human scale. Were there two attitudes in each person, one elevated into the views of mountains and sea, remote from the ground, soaring like birds and the other relishing another more terrestrial world as well? Despite the general reputation of towers around the world, are they really all that dehumanizing? Certainly nice folks capable of appreciating the finer ground-hugging details of life live and work in

the towers, and that includes tens to hundreds of millions of people at this point in history.

The State of the Theory

The Kentucky architect Richard S. Levine has proposed that "sustainable development" delivers piecemeal and gradual benefits but may actually prevent "sustainable cities" from coming into being. Building sustainable cities, however, would produce sustainable development. It's a one-way street, he claims:

The dominant model of desired change advanced within the movement is that, if only we can build up the arguments, tools, and resources in the sustainable direction, then at some

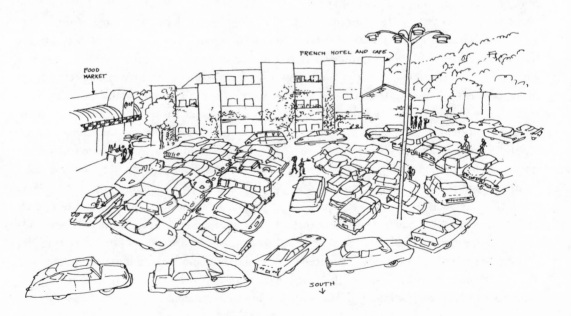

point ... the forces will shift and we will then be effectively on a sustainable path ....

These sustainable tendencies can at best provide certain necessary conditions for generating long term sustainability. Strategically, it cannot be the path itself. Sustainable development alone does not lead to sustainability. It may, in fact, by relieving some of the pressure, thus support the longevity of the unsustainable path. These many tendencies require synthesis and transformation into a new, dynamic process that overcomes the present system's tendency to produce and reproduce growth-oriented relations and processes

of imbalance, instability and decay. Such a change from the old to the new can only come from a catalyst — specifically, the design and institution of sustainable cities ....

To the extent that sustainable development agents move from crisis to crisis, using technological fixes to patch up larger structural problems, they tend to strengthen the systematic relations supporting unsustainability — especially when such "band-aid" solutions lead to instances where these deeper problems fall below the threshold of public attention and the political momentum for more fundamental change dissipates.[24]

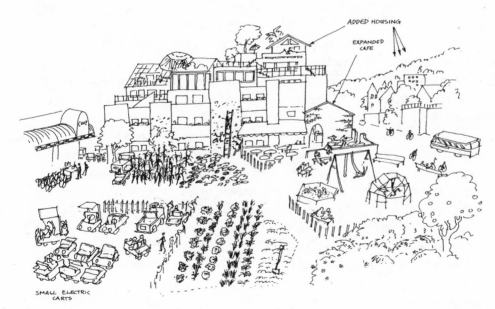

ADDED HOUSING

EXPANDED CAFE

SMALL ELECTRIC CARTS

Grocery store parking lot goes mixed use. *Parking becomes garden, orchard, outdoor café area and playground, plus small electric cart parking lot; community sees housing added vertically.*

Solutions based on understanding the city as a whole living system exist now, and, if put in meaningful perspective, so do approaches to reshaping downtowns and even suburbs.

The sustainable development path can actually be worse than that since most of it involves not patching up larger structural problems but ameliorating symptoms caused by structural problems not addressed at all. For example, converting Los Angeles's automobile fleet from gasoline to electricity would leave almost everything destructive about cars in place while transferring energy collection and generation and pollution abatement to power plants. In an era of rapidly declining oil reserves this will mean pressure for more environmentally destructive coal, tar sand, shale oil, or nuclear energy. Instead we could build communities that didn't require cars in the first place. Painting bike lane stripes on streets and posting "bike lane" signs, as citizens have done in Berkeley, mildly encourages bicyclists but does little to increase safety and nothing to change the city structure or to lengthen the distances that riders are willing to pedal between the origins and destinations. If those distances were lengthened, car trips would be reduced and be replaced with bike trips. In fact, if bicyclists feel that they are doing about all that's possible by taking a little more time to bike rather than drive, and the city remains otherwise the same, the bicycle lane markings may actually stall or kill the decision to make more substantive changes. Thus the good sustainable development step can be, as Levine argues, destructive to a much more meaningful "sustainability."

Scientists at the Lawrence Berkeley National Laboratory (LBNL) who study energy conservation on generous federal grants have repeatedly proposed reducing the urban heat-island effect with strategies such as planting trees to shade parking lots and even painting parking lots white to reflect light rather than absorb light and rediate heat. Again, why go to all that trouble greenwashing a parking lot and blinding the poor drivers with glare from white paint when the problem could be far better addressed with urban design changes that make parking lots completely unnecessary. One LBNL scientist — it gets this strange — has even proposed saving energy by taking down all the stop signs since they cause cars to stop and go, expending excess energy when they could just slow down and check for other drivers. That would save energy — and undoubtedly increase traffic and pedestrian fatalities and injuries many times over. Energy specialists, like fish in a small bowl, are failing to see the larger world around them. It's up to every-day citizens in large numbers to comprehend what needs to be done to reshape our cities and then to educate their fellow citizens and to promote and advocate the necessary changes.

Solutions based on understanding the city as a whole living system exist now, and, if put in meaningful perspective, so do approaches to reshaping downtowns and even suburbs. If we know what we are looking for, we will find that the city today gives us magnificent opportunities for genuine progress — for healthy, evolutionary progress.

More on theory later in the book, but here's a story illustrating something of the kind of basic thinking we need to do. When Cortez arrived in Mexico, the Aztecs had toy animals with wheels, but they had no carts, wagons, or other vehicles or objects so equipped. The first wheels they saw were on Spanish wagons hauling gunpowder, firearms, armor, and other supplies into their country. They had never extended the principle of the wheel beyond the toy. Similarly, countless details of the ecocity function admirably and pleasurably in our world right under our noses, and in a few places even its most fundamental and healthy land use patterns are in evidence. But we don't apply these features and basic principles to the entire society. Just as the Aztecs' toys were a source of pleasure, regarded as something special and enriching, ecocity features such as pedestrian areas, rooftop gardens, and bridges between buildings are enjoyed where they exist in small numbers, but are thought of as playful, special things of not much utility. Instead, they should be understood for the much greater value they would have if they were integrated into ecologically healthy land uses and infrastructure and applied to the whole society. They should be seen in their full usefulness and as part of normal economics. We tend to be amused that the wheel was neglected in pre-Columbian North America but we are doing the same thing all over again.

The old railroad dome car on
a thunderstorm desert night.

Access and Transportation

Access and Adventure

The shortest distance between two points is achieved by moving those points closer together. From then on, every trip is shorter. That's an efficiency multiplied thousands of times. Between cities and across long distances, approximating straight lines is a good idea. But within cities, access is most efficiently provided by proximity. You can either go there, or you can build for a diversity of activities so that "there" is practically next-door.

Animals move. When they cease living, they cease moving. In its poetic and darkly threatening tone, the Bible distinguishes between "the quick and the dead." If all else fails, we can at least have the illusion of being fully alive by moving about, and the faster and farther, the more powerful the feeling. This is no small part of today's problems. The

illusion is so pervasive that Mahatma Gandhi once felt moved to say that "there is more to life than increasing its speed." I think the only way to deal with this problem is to come up with an adventure that is as or more exciting: ecocity building.

Humanity's ancestors spent about 20 million years in the trees looking something like lemurs, leaping through the air in a three-dimensional living jungle gym. In this environment we evolved our grasping prehensile fingers with opposable thumb and our front-facing eyes capable of precise three-dimensional parallax vision. Deep genetic memories of this phase of our evolution bode well for a more three-dimensional life in the ecocities of the future. We are all gymnasts and trapeze artists to our double-helix cores. The three-dimensional forest environment helped launch us into a complex, exciting relationship with transportation: as we swung between and down from the trees and walked out onto the grassy surface of Gaia's broad tummy, we were as omnivorous for movement as we were for food.

It wasn't long before we started using our clever minds, tree-top-evolved hands, and 3-D vision to extend our range, comfort, and safety by fashioning moccasins, sandals, boots, and shoes. We leapt onto horses, built boats, lifted sails into the ocean sky, zipped down hills on skis, built our first carts and wagons, and eventually took to flying through the air in mechanized birds. It was an exciting story of wind-in-the-teeth sweet cowboy loneliness as the unsung hero got up and moved on, of the wonderful return to loved ones or the apprehensive reunion after a long voyage of exploration, of the adventure of unknown lands delivered to us on this planet's amazingly varied surface by no other means than simply moving across it, and of ever increasing speed. And now that some of us have traveled in the neighborhood of 25,000 miles per hour to the moon and back, perhaps it is time to reconsider the point of it all. Now that we have spent about half of the planet's full endowment of petroleum resources in this racing about, perhaps we should entertain the notion of overmobility.

The sense of adventure and that feeling of freedom it brings, are important issues. If adventure is many things to many people, one of the few things practically everyone will agree upon is that being stuck in a traffic jam is not one of them. Ironically, beyond certain not so obvious limits, transportation destroys the potential for adventure. If everyone is traveling regularly back and forth across the city, the everydayness of the trip reduces its pleasure. Too much movement through one neighborhood after another as they lose their ethnic feel, sprout chain stores and franchise restaurants, post identical billboards, and adopt uniform if scrambled esthetics, erodes the sense of excitement in the place and the alertness in the traveler. Excessive moving about crowds a place and, in the case of cars, contaminates the city with smog, noise, bad smells, inconvenience, irritation, and danger — all of an unintentional and pointless sort. No confrontations with ferocious animals, unpredictable tribesmen, or hazards of avalanche and prairie fire, no tests of subtle cultural understanding, knowledge of nature's challenges, or requirements for well-tuned navigation skills as in past times of real adventure. Just remembering to turn at the same old freeway exit is about the most that the sad-eyed commuter on the ten thousandth trip will be called upon to accomplish.

In the overmobile world of car culture the destination looks ever more like the source — roughly in proportion to the mobility. Similarly, in the world of jet travel, cultures and languages, the physical features of geography and architecture, even clothes and tastes in food are becoming homogenized. *I Love Lucy* reruns from orbiting satellites reach dusty villages in Mongolia and Peru. The growing extinction of species is caused partially by expanded markets made available by cheap transportation. Don't forget the depressing feeling of finding a bottle top on a remote mountain ridge, a cigarette package on what felt for a while like a lonely beach, or a vapor trail in what could have been a gorgeous sunset "hundreds of miles from anywhere." Transportation can go too far — it can change people's attitudes about travel and lower levels of respect for the places and people visited. To preserve a beautiful world, we need to travel not only with more respect but also, on the average, much less. When we arrive we need to stay a while, get to understand what's going on, become involved.

Inside cities, towns, and villages, very different rules apply, just as the rules within an organism — respiration, metabolism, balance, coordination, contemplation, dreaming, etc. — are different from the rules of intercourse with the world outside. Something happens at that skin, that edge, those city limits — something that changes everything. What happens is simply the protective containment of integrally functioning, mutually supportive subsystems and the selective filtering of materials and influences so that the organism can benefit from exchanges and maintain health. In both the organism and the city, that whole system is a system of subsystems.

Inside the city, the best transportation is the least: access by proximity should be the objective. The distances are potentially so miniaturized — "imploded," as some urban theorists would say — that within that semipermeable skin, that social/ecological boundary, fifteen miles per hour should be the general upper limit. Any higher speeds should be reserved for special transit uses and longer-distance bicycle commuting and sports cycling. Any speed above about fifteen miles per hour — a fast running speed or modest bicycling speed — becomes, on a geometric curve, more and more damaging as it increases. Below fifteen, even in a large vehicle, it is almost impossible to kill anything, even intentionally. Above that speed, safety erodes quickly. Speeding bicyclists occasionally cause serious accidents, though only rarely fatal ones. But when vehicles become motorized, pollution noise and exhaustion of fossil fuels escalate while fatal accidents become commonplace.

There is a simple law of physics involved here. The forces increase by the square as speed goes up: a crash at twenty miles per hour involves four times, not two times, the force involved in an accident at ten miles per hour, and a crash at forty miles per hour involves not four but sixteen times the force. The qualitative differences are even more extreme as certain thresholds are passed. At ten miles per hour two people bumping into one another bounce off with almost no damage. (My

Mills College, Lake Aliso.
Looking back from the restored lake to a new high-density plaza — again of highly mixed uses and full of community activity, though emphasizing education imbedded in nature.

daughter, at four years of age, once asked me if anyone had ever been killed in a "people accident.") At somewhere around twenty-five or thirty miles per hour bones act brittle and snap or collapse, and at about twice that speed, grizzly words reserved for insects hitting windshields apply to human beings, too.

When vehicles and their infrastructure (and here we can include horses, oxen, and burros, carts, wagons, stables, and so on, as well as cars, garages, and gas stations) are pres-

ent, the city is expanded by their physical size. When the city becomes motorized, it is significantly expanded by public transportation vehicles and infrastructure. But when the automobile becomes the dominant mode of transportation — move over people! The city blows out over vast areas, at extreme hazard to its citizens and nature alike. Ivan Illich says: "High speed is the critical factor which makes transportation socially destructive. The true choice among political systems and of desirable social relations is possible only where speed is restrained. Participatory democracy demands low energy technology, and free people must travel the road to productive social relations at the speed of a bicycle."[1]

The transportation system is much more than it seems. I have earlier described the role of the automobile in destroying the balance between the inward and outward forces of the city's growth and metabolism. I have portrayed the entire automobile transportation infrastructure — cars, sprawl, freeways, and oil — as truly monstrous in its effects, dominating life on Earth by becoming the single largest agent of resource depletion, habitat destruction, climate change, and species extinction. But this transportation infrastructure does not stand alone. It is supported by massive outlays of defense spending to keep cheap oil — for gasoline, greases, antifreeze, transmission fluid, oils, synthetic rubber, plastic car parts, asphalt — flowing our way.

Intimately connected with the car/sprawl/freeway/oil infrastructure is another of the more problematic technologies of our times: television. Television makes it much easier to combat what would otherwise be the loneliness of suburban isolation. It's a medium that tends to work best with constantly hyped-up, even furious change and provides a visual format, at one end of the bedroom, living room, or even kitchen, in which only a small number of objects work well at one time. A few fast moving cars are

ideal; cars and television were made for each other.

Consequences of Our Car Addiction

The automobile touches all of our lives, whether we own one or not, whether we realize it or not. It's personal: we breathe its excrement in the form of air pollution. It's dramatic: its alluring design details, its high speed and danger guarantee notice. It's sexy: millions use it as a place to escape prying eyes, as an impressive lure for a big catch, or, lacking a more comfortable and convenient accommodation, as a portable bed.

The car is probably second only to the house in status competitions the materialist world over, and often it even occupies the most room to boot. But the car is probably even more effective in such competitions because most of us can't even afford to own a house; in addition, the car can go out on the prowl gathering status anywhere while a house has to just sit there and wait for people to come to it. The car can announce who you are to hundreds, even thousands of people every day and broadcast your opinions with bumperstickers. And if this were not enough, our very identy is dependent upon it. Who are you, after all? "May I see your ID? Your driver's license, please."

The car is omnipresent. From tall buildings, hills, and airplanes people can't be seen, but the car is everywhere. The sprawling city layout indicates to the dispassionate eye that the city was built for cars, not people. If a Martian came to Earth and wanted to speak to someone in charge, he would first speak to a car.

As Ivan Illich says, the "automobile has created more distances than it has bridged"[2] — and, once created, rendered bridging those distances without the automobile virtually impossible. Thus we have become structurally addicted to cars. The structure of the city, even the whole national transportation system, has become thoroughly dependent upon them. Realizing there are 600 million cars in the world and about sixty million new ones arriving every year and then listing the resources destroyed or damaged makes the extent of the addiction evident:

- About half a million human beings a year worldwide and millions of mammals, reptiles, amphibians, and birds of various kinds are killed in automobile accidents.
- About a third of a million people annually die from air pollution, and cars are the largest single source of

Heart of the City project ideas - pedestrian/transit centers, aka city fractals, in six images.
Here we see one version of a larger "urban fractal" or "integral center" proposed for Berkeley, in which the main creek in the city's center, Strawberry Creek, is daylighted and restored, a pedestrian plaza is created, and the buildings are sloped to receive the sun's energy. The Berkeley Downtown Bay Area rapid transit station has several entrances immediately adjacent to this project, one visible in the picture. The drawing is proposed; the photo is what exists. The land is mostly owned by the University of California and is slated to become, at least mainly, a conference center, hotel and museum complex with possible new housing — all of which makes sense as part of this particular integral or fractal project.

pollution worldwide.

- Water is polluted by runoff from oily, sooty, sometimes salty roads, which, according to the Environmental Protection Agency, rivals sewage in damaging water resources.

- Precious time is spent commuting, sitting in traffic jams, waiting for car repairs, trying to sell or buy a car, changing tires, washing, polishing, fixing (or trying to), dealing with traffic courts, working to pay for the car, and visiting accident victims in the hospital or mortuary.

- A great deal of money is wasted on the purchase of the car, smog device, insurance, upkeep, grooming, accident repair, parking space, home garage and driveway, taxes for highways, registration fees, drivers' licenses, fuel (about 600 gallons a year for the average American motorist), parking and moving violations, property damage in accidents, hospital bills for accidents, and income loss during disability. Based on 2005 figures, in 2006 the total direct average costs of operating a mid-sized car in American cities, according to the management and consulting firm Runzheimer International, will run from $7,400 per car per year in Knoxville, Tennessee, to $11,844 in Detroit, an average of about $9,622.[3] Multiplied

by the number of cars in the US (very close to 150 million in 2006), the average cost climbs to one trillion, four hundred and fourty-four billion — or seven times as much as the damage done by hurricane Katrina if estimates of $200 billion at the time of this writing prove correct. The money — staggering amounts of money — that could go into building alternatives is thus diverted into the manufacture and support of cars, buying off the creativity we need to solve our problems. The millions of people currently working for the automobile in various ways could, with planning, industrial retooling, and retraining, be just as well gainfully employed working for the ecocity.

- Cars create noise with their engines, horns, and obnoxious car alarms. Train whistles and emergency sirens would need not be so loud if people were not riding in cars with windows rolled up and if cars were not already producing so much competing noise.

- Views are ruined by the car itself and its support systems of street lighting that blots out the stars in glare, billboards designed very large to be seen from speeding cars, gas stations, junkyards, sales lots, and freeways and their sound walls.

- Cars cause frustration and hostility in traffic jams and cultivate an interest in speed, power, violence, and aggressiveness ("road rage").

- Cars have separated people by causing the city to spread and walls of steel, glass, and speed to be erected; walls that breed isolation, alienation, and loss of human touch and eye contact in daily activity.

- The military-industrial complex builds its power partially on the manufacture and sale of cars, highways, and gasoline: Chrysler makes tanks, GM makes military trucks, oil companies make jet fuel. Originally, in the 1950s, the interstate freeway system was called the National Defense Highway System. A highly mobile society is one easily mobilized for war (logistically anyway), as the great freeway pioneer Adolf Hitler knew so well when he started building Germany's autobahns.

- The automobile is a kind of weapon in a class war. When it's a Cadillac in the ghetto, it's psychological warfare, blinding people and helping to keep them broke. Not being able to afford a considerable outlay of money (for a Cadillac or a rust bucket) makes you a member of a disadvantaged class created and maintained by the car.

- Cars are so heavy that, in any relative sense, they average out close to empty at any given time. In other words, what they are mainly moving is themselves! Cars are usually the heaviest things we own other than our houses, and since most of us cannot afford to own our own houses any more, cars usually *are* the heaviest things we own. To keep the heaviest thing most of us own moving almost every day over large distances costs us enormously. When they are not moving, they are occupying enormous space: one of the largest rooms in the house, and outdoor space that could be shops, parks, playgrounds, natural areas, farmland, and so on. If it seems a little absurd to think of cars as mainly self-justifying, it may seem a lot more absurd when we see how fast we are going. In 1972, Ivan Illich calculated that if the average American puts in 1,600 hours a year on behalf of or in his or her car (most of it work time to earn money to pay for the car) and gets 7,500 miles of transportation out of it, the car averages less than five miles per hour. Updating these figures to the early new millennium produces approximately the same speed.

- Cars lend themselves to intentional use as a weapon, a kind of freelance, low-budget tank in drive-by shootings and car bombings in Beirut, Bogota, Belfast, Baghdad, and elsewhere.

- And last on our list, enormous accidents plague the technology and its infrastructure. The Exxon Valdez spilled 11 million gallons of oil onto the coast of Alaska in 1989, and in 1993, the wreck of an even larger tanker, the Braer on the coast of the Shetland Islands, released 26 million gallons of oil. Car accidents can be truly catastrophic. On Interstate 5 in Fresno County, California, on November 29, 1992, a big dust storm descended on the highway, and cars and trucks sailed into the swirling sand at seventy miles per hour. The accident went on for ninety minutes on a quarter-mile section of road. Drivers steered out into the fields to be out of the way of traffic only to be hit by other cars swerving to avoid burning wreckage. Some cars were hit as many as five times. The toll: seventeen killed, one hundred fifty-seven injured, two hundred cars and trucks wrecked. Drivers blamed the California Highway Patrol for failing to close the highway, and the Highway Patrol blamed the drivers for driving too fast. Everybody blamed the weather.

Nobody blamed the sacred auto/sprawl/freeway/oil infrastructure.

- An even more bizarre accident happened on April 22, 1992 — Earth Day, ironically enough, the twenty-second anniversary of the burial of a car to symbolically put an end to the automobile era. That morning, in a thirty-block area of the downtown, the sewer system of Guadalajara, Mexico, erupted in a colossal explosion. Three hundred bodies were identified and months later rescue agencies reported more than nine hundred people still missing. The Mexican government declared a national disaster. The sewer, electricity, and telephone systems of the city of three million had gone down simultaneously. Dozens of buildings were flung from their foundations. As the smoke cleared, deep smoldering craters were slowly filling with sewage while cars and buses rested on nearby rooftops or hung from trees and drooping power lines. According to one reporter, "People were seen wandering aimlessly among the ruins, weeping, their clothing shredded."[4] Automobile gasoline had leaked into the sewer system from facilities owned by Pemex, the Mexican government oil company. No one blamed the auto/sprawl

/highway/oil system or suggested it might make sense to scale it back as quickly as practical, but in a very meaningful way, this was an automobile accident.

In my own neighborhood, the role of the automobile in the 1991 Oakland Berkeley Hills firestorm got almost zero coverage. The blaze consumed 1,700 acres of landscape, destroying 3,375 homes and killing twenty-five people. Most of the people who died were trying to escape in their automobiles and ended up in collisions at intersections or found themselves in traffic jams on narrow, winding, steep streets. The next morning the surrealistic landscape was littered with hundreds of burned-out cars with glass headlight lenses melted like tears weeping down bumpers. Gasoline tanks became the obvious centers of intense conflagration, as garages frequently appeared as source points of extreme radiant heat. Asphalt around gas tanks was completely burned away, leaving grains of sand and crumbly pebbles.

Although five thousand people were made temporarily homeless by the fire; not a single store was located in the fire zone. It was a single-use residential area and thoroughly automobile-dependent. Speaking on Berkeley's listener-sponsored radio station two days later I was, as far as I know, the lone voice pointing out that we were dealing with two fires here: the obvious catas-

trophe at hand and an immensely larger if far more subtle one, namely the fire under the hoods of the Earth's half billion cars. Without the latter, the catastrophe would not have taken place. This unacknowledged fire, created by poor community physical planning and automobile addiction, which pushed the development fringe out into the dry-brush chaparral in low-density housing, is the much larger disaster.

I tried to cultivate community interest in rebuilding at least a small part of the firestorm area in a more compact, mixed-use, community-centered type of development connected to the rest of the town with regular bus service or perhaps a streetcar line. One busy street, Broadway Terrace, heads straight toward the center of the burn area, connecting to downtown Oakland. Transit boosters had already advocated for a streetcar there. Fully 30 percent of the people who had lost their homes wanted to sell and leave. In addition, you have to realize that the views from up on the hill down to the bay and the distant Pacific through the Golden Gate are truly spectacular, and worth money to developers, and represent a "quality of life" to potential future residents. In other words, it was a major opportunity for reshuffling the land use pattern and building a spectacular model of an ecological neighborhood center.

I proposed that one to two dozen properties could be assembled to create such a

Berkeley past with eco-features. *Another Heart of the City Project version in a style favored by local architectural traditionalists.*

center. Saving four or five properties to *not* rebuild upon could secure open space for a neighborhood plaza. Three or four properties on the downhill side of the plaza could have been kept open to maintain the panorama for visitors to the plaza and residents alike. Properties adjacent to the future plaza could have become mixed-use, with shops on the ground floor, offices on the second, condos and apartments on the third and fourth, and cafés with a view of the bay on the rooftops and around the plaza. The streetcar could pull up to the edge of the plaza. The neighbors from three or four blocks around could conveniently walk over to enjoy the center themselves. Two or three hundred units of housing in several attractive buildings could

be created on what used to be those dozen or so properties of the old uniform single-house lots. An equal number of residential units on the farthest, most dangerous, and most automobile-dependent fringes could have been left unrebuilt — their owners could have first option to move into the new housing — and turned over to the regional park system. A large swimming pool or two could be placed on the side facing the fire winds — the hot dry winds always come from the same quarter there, the northeast — and the community could, by sharing its resources, have bought several pumps, hoses, and nozzles to direct a curtain of water from the swimming pool on any future fire that might come bearing down on the neighborhood center.

Nobody was interested. After about two months of effort I gave up and went on to other things. The lesson I failed to communicate at the time was that the car had created and maintained that fire area and most of the people who died were in or next to their traffic-snarled cars, stuck in jams, trying to escape — it was a car accident. Now the area has been rebuilt on exactly the same land use pattern, dependent as before upon gasoline for its very existence and for every day of its maintenance. It — and all of Car City — burns.

The only way around such accidents in the future is for people to have a much deeper understanding of what they are building and the roles transportation and access play in it. If plans are made in advance for rebuilding after disasters, informed by ecocity awareness, future disasters will be small in comparison. Funds need to be available for ecocity planning and rebuilding rather than as disaster grants and loans after the fact — which rebuild the same problem. Without such plans and funds, disasters caused by car-dependent design will not lead to ecologically sane rebuilding.

To wind up this litany of automobile problems and disasters, I'll give you a sobering statistic: according to the respected Environment and Forecasting Institute of Heidelberg, Germany, "each car [in Germany, but this is probably typical for wealthier countries] over its lifetime is responsible for 820 hours [34 days] of life lost through road traffic accident fatalities and 2,800 hours [117 days] of life damaged by road traffic accidents."[5] Just think about that for a moment. Each among that swarm of cars you see out there clogging our cities and jamming the highways every day is responsible on average for killing a person more than a month before his or her time. Violent, bloody, sometimes flaming deaths. Plus three months of other people recovering from such personal disasters. Poor countries have up to twenty-six times the fatality and injury accidents per mile traveled because of antiquated equipment, poor vehicle maintenance, bad roads, signs, and safety programs and sparsely spaced, poorly equipped hospitals and clinics. A closing oddity: of the first five California condors released into the wild by the captive-breeding program that is desperately trying to save the species from extinction, one died from drinking automobile anti-freeze and a second was hit by a car.

Better Cars?

The better car makes a worse city. This counterintuitive notion requires some understanding of systems thinking as mentioned earlier in my discussion of James Miller's book *Living Systems*. The more energy-efficient car means people can drive farther for less money, buy homes on cheap farmland, and extend sprawl farther yet. That new sprawl development then promotes more

Eco-modern. *Not likely to have been built since the 1950s, but functioning like the other designs, with minimum energy consumption, maximum biodiversity including rooftop gardens, public access to roofs and terraces for the great local views, and featuring the creek restoration and pedestrian plaza.*

driving, more cars, and more energy consumption — while making people feel good about it.

When it comes to better energy sources for the "better car," of all the hundreds of charges against the car, only one would be lessened — and only partially so — by a switch from gasoline to electricity: there would probably be less air pollution — unless the energy sources for electricity were shifted to coal, shale oil, tar sands, or nuclear (which causes air pollution during production as well as virtually permanent radiological pollution after production). There is a law of physics that says that accelerating and moving any-thing of a particular weight to a particular speed against a particular resistance of inertia and friction requires a particular amount of energy. In the case of the gasoline-engine car, the fuel is burned and the waste material distributed as it moves. In the electric car, energy is stored in batteries that have to be recharged by being plugged into a power plant, which has already burned the fuel and produced the waste. Batteries don't create energy, they just store it. Essentially, they are gas tanks for electrons. If we put cars and trucks on the grid, we will require staggering increases in electric generation capacity, and this energy will most likely come from rapidly declining oil and gas or — far worse — fossil fuels or nuclear energy.

But what about solar- or wind-electric cars? We will get cleaner energy shipped in from desert locations or windy places on gorgeous new electric power lines draping the landscape with yet more metallic spider webs. (By the way, the other three condors in the first batch released into the wild died, too — hitting electric power lines; fortunately, condors released later had better luck after receiving power line aversion training from their human handlers.) As much as I, for one, promote solar and wind energy, together they delivered 0.1 percent, or only one thousandth of US energy use in 2005. The renewable energy investment has not been made, and there is precious little there to build upon. Mull that one over for a while, environmentalist electric car fans. What about generating the electricity yourself, if you live in an appropriate location? The price of gasoline will be subtracted from the cost of owning a car, but the cost of the electricity-generating technology and its maintenance, battery replacement, and disposal will be added. Gasoline costs are, after all, only one-tenth the cost of car ownership in the United States. *Utne Reader* says, "Gas guzzlers are harder on the environment than electric cars, right? Wrong! A conventional car creates 26 tons of hazardous waste for every ton the vehicle weighs. A battery-powered automobile produces twice as much, 52 tons, including a witch's brew of lead and toxic acids."[6] What

about all those other problems: violent accidents, dismemberment of the city, alienation, full cost of ownership, paving paradise, destroying wildlife habitat, and so on? At least the electric car will be refreshingly quiet, but because of this pleasant development to the ears, it will kill and injure more pedestrians and more animals on the road — unless it is made to produce loud noises artificially and re-rendered obnoxious in that regard, too. What next?

War! Matthew Wald in his *New York Times* article entitled "A Military-Industrial Alliance Turns Plowshares to Swords" writes: "The press to develop an electric car has become a holy quest for environmentalists, who seek an antidote for smog, traffic noise and the nation's reliance on oil. But lately, this peacenik technology has found an unlikely patron: the Pentagon, which sees battlefield uses for electric-vehicle systems." Wald says these will be "scout vehicles that slip almost silently over enemy terrain and armored personnel carriers that have no exhaust pipes and so cannot be spotted by infrared detectors." Says Major Richard C. Cope of the Advance Research Projects Agency, the marine in charge of electric vehicle development, "At night, they can't see you, they can't hear you, they don't know you're there."[7]

I am not saying that if any cars exist at all they should not be extremely efficient and maybe electric. There will probably always be specialty uses for some. Instead, I am saying that in view of what cars do physically and the role they play in society, ecology, and urban structure and functioning, it is damaging to promote even the best of cars as if they were a major part of the solution. Car addicts welcome such excuses to continue their habit and avoid ecological city planning.

Defenders of cars mount one last defense: "True, there are problems, but if there were just fewer cars, the problems would not amount to much. In that case, there would be nothing really wrong with cars."

What they are actually saying is, "It's O.K. if a lot of people don't have cars and a privileged elite does, including us, of course." That's the undemocratic and unjust ancient city of hierarchy and strong separation of classes speaking. The disaster in Guadalajara is what happens when that small, car-driving elite fails to maintain the automobile technology. Those who do not contribute to the problem suffer. The smaller the number of cars permissible, the more privileged the few who retain their cars and thus the greater the injustice (assuming that owning a car in some final way does make life happy and full). Right now, among 6.5 billion humans only

Below:
Composite of design features popular in Berkeley.
This version, of the four shown here, comes closest to the planning objectives of the University of California.

one in ten has a car, maybe one out of six in the age- and physical-capability group able to drive. So for reasons of social justice, seeing that cars are a good thing, should we multiply their numbers on Earth by six? It would be better to realize that we're doing something colossally destructive and start building the city for people, not cars.

There is another angle here, too: "If the total population of humanity were much smaller than it is today, everybody could have a car." I've heard this one a few times. Remember, too, that only one out of ten of us presently owns a car and that we do most of the damage to the climate, to the biodiversity on Earth, and to its reserves of fossil chemicals. Does the idea of trimming the population down to match the Earth's carrying capacity for cars suggest that someone likes cars better than people? Even though I think the population should be smaller myself, I find this people-phobic, auto-philic attitude a bit hard to take. To do half the damage to the Earth we presently do, would the carrying capacity for a planet of car owners require eliminating 19 out of 20 of us?

The final excuse is the acts-of-God excuse — sweeping the problem under the rug because it's "just the way it is." The Hindus have gods for many inexorable forces of nature. One was specifically a god of smallpox, so unstoppable was that affliction — like hunger, poverty, storms, earthquakes, tsunamis, social injustice, indifferent fate

itself. So it is with the car. When a child, spouse, lover, friend, parent, co-worker, or admired stranger is struck down, we act as if little could be done about it, as if it were an act of God. Former President Bill Clinton lost his father in a car crash and former Vice President Al Gore witnessed an automobile hit and almost kill his six-year-old son. But in March of 1992 Clinton and Gore sent US trade representative Mickey Kantor to speak with leaders of GM, Ford, Chrysler, and the United Auto Workers to assure them that the President's administration was a booster of their product. Said Kantor, "I came here to Detroit because it's the number one priority not only on my list, but on the president's list."[8]

Senator Alan Cranston's son Robin was a friend of mine. He was killed in 1977 by a car before his father's eyes. A few years later I asked Alan if it had ever occurred to him that as senator he should take serious steps against the car dominance so that cars could be largely eliminated. He said he'd thought about it but it was just too big to tackle. Too big for the US Senate! Our leaders promote the car, as if helpless before the fates, incapable of doing anything else, but, in fact, many things can be done. Smallpox was eradicated, the dreaded god evaporated, the Earthly and cosmic landscape forever altered. Like the god of smallpox, the god of cars must be banished, too. We will be richer and happier without him.

Freeway Battles

I will address incentives in some detail in Chapter 8, but comparing freeway subsidies to rail investments at this point in the book seems timely. You noticed I called incentives for rail "investments" and incentives for freeways "subsidies." We are used to hearing the reverse because the dominant paradigm that values cars highly promotes rail transport as if it were an inefficient charity, a waste of tax-payers' money. But let's get it clear that one line of rail moves as much freight as eight lanes of freeway. So we shouldn't be talking about "subsidies" here but rather, without the negative, value-laden terminology, simply about "investments" in the kind of world in which we want to live.

So let's see what you get for your transportation dollars and look for a moment at the Amtrak budget. I know Amtrak has prob-lems. The last four times I rode it, the train was six, one, six, and twelve hours late at my destination. Pretty outrageous, isn't it? Luckily, I like reading, writing, wandering, watching the country side pass, eating, and drinking beer or wine on the train, of which I can do driving the car. So I just plan for time-consuming surprises. But it's not Amtrak's fault. It doesn't get public investment. The freeways do. In February of 2005, the Bush Administration proposed cutting Amtrak's national budget investment (called a "sub-sidy" by the Administration) from $1.2 billion to $.9 billion. For comparison, when the 1.5 mile Cypress Viaduct of Interstate 880, an elevated section of the freeway, col-lapsed in the 1989 Loma Prieta Earthquake just two miles from where I now live in Oakland, California, it was rebuilt for $1.2 billion. That's right, 1.5 miles of freeway got

Windmill, tree and bridge. *Close-up of the design from the previous drawing, looking up the street to bridge building with large native tree and birds.*

as much money as the entire passenger rail system of the richest country in the world.

Now for a glance at what you get along with freeways. Marshall Berman tells the following story:

> The South Bronx, where I spent my childhood and youth, is the site of one of the greatest recent ruins today outside Beirut. The physical and social destruction of the area began with the construction of the Cross Bronx Expressway in the late 1950s and early 1960s spreading gradually southward from the highway and northward from the emerging Bruckner Expressway in the late sixties.
>
> Then in the early 1970s the disintegration began to spread at a spectacular pace, devouring house after house and block after block, displacing thousands of people like some inexorable plague .... In the South Bronx alone, more than 30,000 people fled in the 1970s as their homes were being destroyed. Many of these people were forced to run more than once, trying to stay ahead of the blight that kept catching up with them. Thousands more in Manhattan and in Brooklyn went through the same ordeal. In fact something similar was happening in working-class neighborhoods in older cities all over the US.[9]

The city escapees not only fled the city but destroyed much of the city and open space between their homes and downtown with their transportation system for commuting, that is, cars and freeways. Since Berman's words were written, in the late 1980s, freeways have staged similar assaults in practically every corner of the world.

Rising to our defense we have such crusaders as Peter Newman and Jeff Kenworthy, introduced earlier in these pages, who have studied over one hundred cities around the world and came to the following conclusion:

> Traffic engineers still claim freeways are better for fuel emissions, but the results of our study do not. Economically, they also appear to have failed. Our data show that, instead of people in cities with freeways saving time, and hence being more productive, they just spend more time in their cars. Freeways space cities out and hence overall travel is increased. Those cities which do not go for freeways but instead built up transit and bicycle access have gained economically and environmentally .... Some short-term pain will be experienced as businesses and developers adjust their plans to a more transit-oriented city, but experience shows that the transition is worthwhile.[10]

Kenworthy and Newman report a direct relationship between density on the one hand and pedestrian amenities, bicycle and transit use, and energy consumption on the other. Almost always, the higher the density, the better for modes of transportation other than cars and systems other than freeways. They also report statistically on the relationship between the diversity of land uses and those transportation modes and energy use. The closer together different functions congregate, the better for pedestrians, bicycles, and transit. These are proportional relationships for city ecology of the same level of importance as the area-species diversity ratio is in island biogeography and evolution theory. In both cases, as far as I can tell, they are laws of order for living systems everywhere.

Kenworthy and Newman report that the highway planners simply do not look at the most important information. I found this hard to believe until I attended the hearings on the replacement of Interstate 880, the elevated freeway that had collapsed in Oakland during the Loma Prieta earthquake of 1989. I saw the immense pile of documents generated by the California Department of Transportation (Caltrans): several columns of paper three feet tall and placed on the stage next to the rostrum, a monument to thoroughness, intimidating to any potential opponent. With this much information, I presumed, the planners must have thought of everything. But not so. They turned out to have absolutely nothing to say about future impacts on land use in the suburbs.

Their argument was simply that expanded freeway capacity reduces congestion and increases automobile speeds, that faster-moving automobiles burn fuel more efficiently and cleanly, and therefore that widening freeways cleans up the air. End of argument.

The multiracial panel of men and women representing Caltrans sat there at their table as straight-faced, intelligent, and earnest-looking as parents and teachers at a PTA meeting. In the audience, the opponents of the plan were saying that, given a few years, the added freeway capacity would encourage further automobile-dependent sprawl development, and that as more people moved into these areas and started using the freeways, the freeways would soon be as clogged as they were in the beginning, only with more cars. (This is exactly what did happen in the subsequent several years.) This extraordinarily well-documented record was clearly articulated by the well-off and the poor, by neighborhood activists, environmentalists, minorities, transit users, and even car users looking for something better. The members of the panel did not question this land use/transportation connection at all. They simply ignored it, pointed again to the towering wall of reports, and repeated that their studies indicated that making room for more cars allowed them to move faster and burn their gasoline more efficiently and cleanly and therefore cleaned up the air. Is this record stuck or what?

Student at the Sixth Floor Street Café. *San Francisco and Bay in background.*

Another argument that the freeway planners brought forward was the cost to the city, the people, the whole region of not having a freeway "improvement" — the delays in delivery of products, the added time it takes to get people to work, and so on. In the case of the collapsed Interstate 880, Caltrans compiled figures showing that companies were losing money because it took their trucks longer to deliver the goods and that employees were losing money because their time was worth money and it was taking them longer to commute on surface streets or by transit than it would on a rebuilt freeway. These fig-

ures were tallied up to several million dollars a year and submitted as the added costs of not having the freeway replaced.

Meanwhile, however, thousands of commuters had switched to the local bus and light rail systems and both the people and the systems prospered. Some people were stuck on the opposite side of the bay from their families immediately after the earthquake and could not get back to them until the next day. They were suddenly shocked by their powerlessness to help their families — startled by their automobile vulnerability and their distance from loved ones. That night, profoundly shaken in more ways than one, some of them decided to find work closer to home, saving virtually all their commute money to spend in their own communities instead of sending it off to car manufacturers, oil companies, and smog-device factories. The businesses they patronized close to home gained money by the freeway closure. None of this information had been compiled by the Caltrans planners.

There are plenty of economic success stories about not expanding freeways. Toronto, Canada, a city studied by Kenworthy and Newman, decided in the early 1970s to forgo further freeway development, and no economic disasters materialized. Car dealers may have made less money over the years, though maybe people who otherwise would have become car dealers simply entered other lines of work and didn't come into competition with them as they otherwise would have. The city has prospered. It is among the cleanest cities of its size anywhere. Traffic jams did not become massive, and rather than now having hundreds of miles of new crowded freeways and worse air pollution, Toronto's smaller highway system, expanded transit system, and mixed-use development have significantly reduced the time residents spend in their car[3].

For a long time, I couldn't understand how things so obvious could be missed by our well-educated planners. Now I can: they don't miss the obvious — they simply bow to the pressure of the public caught up in its structural addiction to the automobile! Citizens put great pressure on politicians and planners alike to supply their habit. Planners know that there are plenty of statistics on the other side of the critical issues, and they are aware that today's decisions build tomorrow's dysfunctional or healthy environment, but they refuse to gather or think about information that would slow down or reverse freeway development. And, their jobs depend on maintaining their apparent ignorance.

Kenworthy and Newman take these people on their own terms, assembling tall columns of data on the other side of the stage. They tell us to appeal not only to facts but also to emotions and intuitions since the ultimate decision is the result of a vague but powerful thing called "public opinion," which touches the politicians, the planners, everybody. And if Not In My Back Yard (NIMBY) sentiment supports stopping freeways, as it often does, they recommend using NIMBYism along with the more broadminded arguments for conserving energy, fighting pollution, and saving the planet.

David Engwicht provides a quick answer to a question he often hears. Is he anti-car? No, his family owns one, and he uses alternatives whenever possible. He is against "the inappropriate use of automobiles."

Am I anti-car? Absolutely! I own one, too, or rather did until recently, a small truck, which I used for work and occasional errands around this largely automobile-dependent urban area known unofficially as the East Bay.

And there will be appropriate uses for what we think of as automobiles today in the eventual ecological city — for about one car per thousand people I would guess, and serving specialty uses. Something that deadly should be used most carefully and sparingly. But in the context of vast numbers of people using cars in everyday life, there is no "appropriate" use for cars unless simply as an acknowledged strategic compromise, with no excuses. Use it as little as you can. But if you need it, as you often will in the city built for cars, use it to make the contacts, run the errands, and do the work to teach and plan for ecocity conversion. Don't feel guilty; be strategic. And if you own one, take full advantage of your first opportunity to get rid of it. The ultimate defense against freeways is making cities that render them useless.

For Love of Rail and Ferry

By now you know that I think our "love affair with the car" is, in fact, a distorted dependency, an addiction based largely on the physical structure of the city. I will add now that over time millions of people have loved taking the train.

When I was I teenager I found few things more exciting than waiting for the El Capitan or the Super Chief at Lamy station near Santa Fe, New Mexico, ready to head off down the tracks into the early morning or night, a cold wind bearing down over the dry plains or rolling off the snowy Sangre de Cristo Mountains, or, at another season, watching tumbleweeds bounce along under a dusty sun. I'd be flattening pennies beneath freight train engines to give to friends, or saying good-bye before another trip to school or to visit distant friends in a different world.

The trains were beautiful in those days. They seemed to polish them up especially shiny, reflecting a fresh, clean sky. There were Fred Harvey Hotel-style Indian sand-painting motifs on the turquoise-accented walls and carpets of the trains and stations. There were romanticized images of handsome, implacable Indians in front of clear blue skies with pure white thunderheads and zigzag lightning — which was exactly what you'd see from the observation dome car: a 360-degree panorama rolling by as the train slithered sensuously ahead and trailed behind like a great industrial snake.

Today's trains, the whole thing two levels high instead of one, with two or three observation cars but lacking the front- and back-facing windshields of the old observation cars, completely miss the point. Inspirationally speaking, they are ineffectual wimps. Whoever designed them? But from the old dome cars, at night the stars would come out, far away from the cities — billions of stars overhead and all around, in winter sparkling over the wind burnished snow, in the summer between the towering clouds, which were flashing lightening bolts from the inside out, sometimes looming overhead, sometimes beyond an almost infinite horizon. I can't tell you how beautiful taking the train through the southwest was in those days. For a young person, wandering from seat to dinning car to observation car, feeling the adventure of the unknown, meeting new people — it was absolutely wonderful. And on those long trips, as if that kind of spectacle were not enough, you'd roll back into a blanket rocked and soothed by that clickity-click and slide off into real dreamland, too.

And then there are the wonders of the ferry. When the Loma Prieta earthquake closed the Bay Bridge for a month, suddenly Berkeley got back its beloved ferry to San Francisco. Just for fun I took it with my friend Nancy to San Francisco to have clam chowder, to look at this beautiful bay we share with 6 million other people and countless fish, cormorants, sea gulls, crustaceans, seals, cord grasses, and seaweeds. The ferries between the East Bay and San Francisco used to have full bars and famous dinner menus; the world's best corned beef, and Irish coffee. When was the last time, while eating dinner, you took a dusting of salt spray whipping through a door suddenly thrown open, laughing with friends, and getting just a little tipsy — driving a car?

One image I will never forget: walking down the stairs from the upper deck of the ferry, I looked over at the table next to the big parallelogram window — one of those design elements angling into the direction of the machine. There sat or, rather, unconsciously posed a man in a gray three-piece suit reading the paper. While people wandered from table to table eating hot dogs and sipping beer, wine, and soft drinks, while a volunteer band played music to help raise money for the Ferry Committee, nobody could have been more placid before a newspaper, unruffled by his surroundings. On this distinguished commuter's table was a cup of coffee, steam rising in lazy curls, and in his hands a copy of the *San Francisco Examiner* as he leaned back, lost in thought, and behind him the most furious crashing and bursting of waves imaginable, smooth slices of water cut by the prow of the vessel, leaping and exploding into foam and careening by on the far side of the glass as the little ship leaned into the night. Now where on earth can you see that if you don't have the right kind of transportation device?

Public transportation in the San Francisco Bay Area was once rail- and ferry-based. One of our local East Bay writers, Dashka Slater, writes about the pleasures of commuting on the ferries and trains:

It's hard to imagine how chummy the world must have seemed back then before the building of the Bay Bridge ... Transportation was a social occasion; Americans, in contrast to the more reticent Europeans, were notorious for chatting it up with strangers on boats and trains. Both the Red Trains and the Key System had a commuting ridership that was so regular that commuter clubs formed on the San Francisco-bound ferries. Some commuter clubs held on-board parties as often as once a month and just about every club gave a Christmas party on the morning of the last working day before Christmas. The car by contrast, is the world's most antisocial invention. People are at their worst in cars — I know I am .... If the metaphor of the train is "We're all in this together," the motto of the car is "Me first."[11]

You might as well *be* somewhere interesting while en route. Cars are made more interesting with tape decks, CD players, radios, cell phones, drink holders, and ego-inflating design features. But being in what solar train promoter Christopher Swan calls "rolling architecture" is really being someplace while going someplace, too. The cyclist or pedestrian in the town, city, or country is completely enveloped by the experience of that place. By the time a vehicle has devel-

Remodeling a typical downtown. *A central business district with ample free parking becomes an ecocity downtown.*

oped a speed greater than the cyclist's and become as big as a streetcar, train, ferry, or large airplane, the vehicle has become an environment in its own right. It becomes a designed place akin to architecture moving through a larger place, an environment within an environment. It sprouts all the special appendages that fit its circumstances, from loops hanging from the ceiling to grab hold of to advertisements over the windows, from dining tables and bars to magazine racks and cork life saver rings lashed to the walls. The car is a neither-here-nor-there world, sealing out the natural and the built environment while being too small to be an environment of consequence in its own right. Just as suburbia is neither here nor there in the sense of being neither city nor country, the car itself is neither a real environment nor a means to genuinely experience the larger environment while moving through it. If the larger vehicles are something like rolling architecture, cars are more like expanded, high-powered, wheel-supported, steel-clad clothing, and probably, being so personal and so strongly tied to identity, they function psychologically like clothing — and armor, too.

Why not vest real care and love in the public transportation vehicle? Why do we build freeway interchanges for hundreds of millions of dollars and shortchange the environments that are public transportation vehicles, especially considering that public transportation is far kinder on nature and society than the car system? If, in the ecological city, the city and neighborhood centers were intensively lived in, the nature areas, green corridors, and creeks were treated like places of real value, and the vehicles that connected the people were designed, utilized, and enjoyed in the same spirit, then both in *being* someplace and in *going* someplace, we'd *always be* someplace.

Personal Transportation: Bicycles, Carts, Shoes

Planners call cars "personal transportation" since one individual owns and controls the vehicle and generally is its only occupant or there are a relatively small number of passengers. Bicycles are personal transportation, too. The world has more than a billion bicycles — twice as many as cars. According to Ed Ayres writing for Worldwatch Institute, bicycle production worldwide surged in 1970 and has been, by units produced, pulling out beyond automobiles ever since. By 1990, bicycles were being produced at a rate almost three times that of cars.[12] In 2000, it was 101 million bicycles produced and 41 million cars.[13]

These figures are encouraging but a little deceptive — like saying that Howard has more because he has twice as many coins as Lucinda, ten Lincoln pennies to her five Satchegewea dollars. Bicycles weigh about one one-hundredth as much as cars, so in terms of the weight of materials used in construction, one car outdoes one bicycle by

Just as suburbia is neither here nor there in the sense of being neither city nor country, the car itself is neither a real environment nor a means to genuinely experience the larger environment while moving through it.

about one hundred to one. Twice as many bicycles, then, is still only *one-fiftieth* as much car weight. How about the energy required to accelerate and decelerate a vehicle weighing one hundred times as much as another — up to and back down from four times the speed? That would be fifty times four equals two hundred times the energy, but not really. We need other multipliers as well: one, because the energy required for acceleration and deceleration increases with speed; another to factor in wind resistance; and one more reflecting how often the vehicles are accelerating and decelerating and how long they stay at any given speed. I'd guess about four times again, which means that on a typical trip cars use eight hundred times more energy than bicycles. How about the fact that one is powered by a finite resource that damages the world profoundly and is used in enormous quantity and the other is powered by breakfast and lunch? How about the death rate in crashes, the full cost of the vehicle, its pollution? You get the idea. Cars have thousands of

In transition.

times the negative impact of bicycles, even if there are half as many.

Declaring the impending victory of the bicycle, then, seems a bit premature. Declaring we need to radically reduce car numbers is far more important, and to do that, redesigning the city for pedestrians and the people-powered machine called the bicycle is the big step.

But the bicycle-to-car ratio is encouraging in this way: the fact that twice as many people are getting service out of bikes than out of cars is hopeful. What if we had an intelligent world democracy and decided to vote for bicycles or cars? We could get rid of cars today and begin building ecocities as early as tomorrow. That only one out of 10 people on the planet owns a car shows the democratic potential here. The trend is especially hopeful because the city of walkable centers and the bicycle are best of friends. They support one another. For short-range commuting in from the edges of these centers and mid-range commuting from one walkable center to another not far away, bicycles are the perfect vehicle. They do take up about two-and-a-half times more space than pedestrians and they do inflate the horizontal scale of the city slightly but they require only about one-tenth to one-fifteenth the parking space of a car and an even smaller percentage of maneuvering space. If they are used as a means to get to, rather than through and around the inside of, the centers, their impact on the pedestrian city will be virtually nonexistent — except in the positive sense of bringing a greater diversity of people together from the new urban fringes, distances measured in a few minutes a day of bicycling rather than a few hours a day of driving. And they are incredibly efficient, about eight times more energy-efficient than walking because, in walking, a considerable amount of energy goes into muscle action simply to keep the body erect. The bicycle and the reshaped, ecologically healthy city were made for each other.

V. Setty Pendakur, of the University of British Columbia, points out that the bicycle is the major or only vehicle for hundreds of millions of people around the developing world: "While 40 percent of the driving age people in developed countries own cars, only one percent of those in developing countries do. North America, Europe and Japan have 16 percent of the world's people, while owning 81 percent and producing 90 percent of the world's cars."[14] India has a seven-to-one ratio of bicycles to cars. Pendakur points out that in India and other developing countries the bicycle and its three-wheeled pedal-powered equivalent serve both as personal transportation and as a way of making a living: they are used to make deliveries, to carry passengers, or as an open-air market stall. In India, the average bicycle occupancy is 1.4 people, which is higher than the automobile occupancy on the Oakland-San Francisco Bay Bridge. And yet, despite the usefulness of bicycles to so many people in poor countries, policies are being adopted to motorize transportation and penalize bicycles and their owners. "In Jakarta," says Pendakur, "the government has announced a total ban on bicycle rickshaws. Yet 100,000 licensed and 50,000 illegal bicycle rickshaws ... help support the livelihood of 1,000,000 people."[15] In Singapore, bicycles and rickshaws are dumped into the ocean by a government oblivious to the interests of the very small entrepreneurs and those conserving energy.

This massive subsidy to the automobile and penalty to the bicycle a about for several reasons, Pendakur says. In an effort to attract wealthy investors and expanding business, developing countries believe in "modernizing" by motorizing, acquiring status and higher money flows at the same time:

Aid practices of western countries are being used to distort the transportation attitudes of people in third world countries. For example, $10 billion in aid given by the US to China will be used on car-related projects in Shanghai in the next 10 years. However, only 2 percent of all trips in Shanghai are made by car, while 43 percent of trips are made by bicycle. This aid will be spent on improving roads, and as a result, bikes will be taken off the main roads in Shanghai .... A total of 83 percent of the World Bank's urban transportation lending is for car-oriented facilities. [Another reason for this bias against bicycles and in favor of cars is that] planners are middle or upper income bureaucrats with access to use of private or government vehicles.[16]

It should be mentioned that Pendakur wrote this in 1990; since then, Shanghai and other Chinese cities have outright banned bicycles on hundreds of streets, as Pendakur predicted.

Why, then, would I agree with Ayres' conclusion that in the "human habitations of the 21st century, there will be no place for death-dealing vehicles powered by exploding gasoline. In thousands of cities ... the bicycle will be riding high long after the internal combustion engine is gone"?[17] First of all, the automobile's absurdity grows more conspicuous every day — its "death-dealing" being only the tip of the iceberg. Transportation violence would virtually disappear if we simply built ecocities. Ayres cites, in addition to economy and kindness to the environment, mobility. In an urban context, bicycles work far better than cars. In heavy traffic, police on bikes are frequently successful in apprehending lawbreakers when police cars are stuck in traffic jams. Also, "the bicycle shows more potential for improvement — and consequently, for expanded markets — than the gasoline-powered car."[18]

Recent advances in mountain bikes lead

Pedestrian city center.

the way toward sturdier, more comfortable, and more reliable bikes for varied terrain, *World Watch* says. I would add that in connection with redesigning cities and towns, bicycles have enormous flexibility, fitting into a radically rethought infrastructure including bicycle streets, and elevated bicycle trails, bicycle parking structures taking up one-fiftieth the volume of car parking structures, folding bicycles that can be carried onto elevators and transit vehicles, and banking three-wheelers for fast hauling and for the sporty disabled and seniors. Ayres presents an example from Nepal: "Green roads" made of a dense covering of plant growth instead of pavement have potential in a shift away from cars and toward bicycles. Such roads are very inexpensive compared with highways, soak up rather than flash off water in heavy rains, and exclude the use of automobiles. The biggest reason that bicycles will eventually supplant cars is, however, simply that the anatomy of the ecological civilization will match bicycles and not cars. Walkable centers and small, narrow, cheap, even living roads are supported by and in turn support bicycles. Those walkable centers will also work well with transit. "Green roads," rails, bicycles, and walkable centers will all gang up against cars. Those are the positive reasons bikes will prevail in the long run. The negative reason is that gasoline will first run scarce, and then out entirely.

Ayres says that perhaps some cars in the future could be competitive with the bicycle: "If the internal combustion engine is replaced with a radically less damaging power plant, and if other major changes are made in the size and intrusiveness of cars, that would be a whole new story."[19] It would *not* be if the car was conceived of as essentially what it is now: a vehicle traveling far above Ivan Illich's sociable and ecological optimum of fifteen miles per hour, carrying a private person and a few useful items or friends long distances in spacious comfort. But what about electric *carts*?

When we in Urban Ecology were planning the first Slow Street for Berkeley, California, there were many people on Berkeley streets in small electric golf carts. There was a dealership in the city selling to golf courses and retirement villages and the Center for Independent Living was advocating for and attracting a large number of disabled people who found the carts convenient. We thought of such very small vehicles as substitutes for bicycles for in-town travel and for dealing with the heavy loads of family shopping. We investigated these carts with

great caution, being more than a little skeptical of the "better" car, but were encouraged to find them radically less damaging. Their small electric motors were about one-tenth as energy-consuming per mile as the automobile and, since trips are much shorter in the neighborhood and the in-city context, used more like one-thirtieth or one-fiftieth as much energy per average trip. We noticed that the electric cart occupied about 20 percent of the room required for parking and maneuvering a car. At fifteen miles per hour, it was also radically less intrusive socially, ecologically, and in terms of getting injuring or killing in accidents. The small, slow, weak, clean, energy-conserving, and safe cart *is* Ayres' radically altered car — altered so much that it is no longer a car at all. It's a cart. Should we ever develop significant renewable solar and wind energy, its humble demands for energy should be easily met. And it fits well with both Slow Streets and the walkable centers of the future ecocity.

Being There Instead of Getting There

While attempting to stop the rebuilding of Interstate 880 after it collapsed in the Loma Prieta earthquake (there was a parallel freeway just one mile east and transit was thriving after the quake, so it wasn't such a wild idea as it sounds at first glance), I came up with the slogan "Build community, not freeway." The freeway was rebuilt, but building the community will emerge victorious in the long run. That one-and-a-half-mile strip of freeway cost over one billion dollars, an expenditure that would have solved most transportation problems, even if we ignored transportation altogether and instead just solved access problems there by building right in the first place. In the low-income neighborhood of West Oakland, where the old freeway on stilts swayed and then caved in, people could use jobs and workplaces, shops and far better services.

Out in the suburbs from which commuters poured in over Interstate 880, poisoning the low-income minorities and pounding their eardrums with noise until that final grisly, deafening crash, they could use a little community-supporting development, too. Some of the money could have leveraged new mixed-use development in future suburban walkable centers, restored natural features, or helped social service organizations. Some could even have gone to ecologically imaginative buildings, providing both jobs during their construction and enduring workplaces after their completion. It would all have been based on what I call "proximity consciousness" — the awareness that access can be delivered by building and arranging a diversity of experiences, services, products, environments, people, and natural features close together.

Michael David Lipkan of Albuquerque, New Mexico, put proper emphasis on the concept by invoking the slogan, "Proximity Power!":

> When you think of power to build a new future, what do you think of? Nuclear? Oil? Hydroelectric? What about Proximity Power? ... Proximity Power is the power of choice, the power of convenience. It reduces our need for energy resources while it saves us time and frustration. It is the power of complexity. It is the richness of being close to opportunities. Proximity power lets us spend less time getting to our lives and more time living them. It is a survival necessity.[20]

How do we apply proximity power? By establishing proximity policies:

- Hire locally. Whether as official or personal policy, other things being equal, if several people are suitable for the job, hire the one who lives closest to the job.
- When seeking a job, try to find one close to home.
- If you are a landlord, rent preferentially to people who work close to home or don't have a car, or both.
- Try to find a home to rent or buy — even if it costs a bit more — close to work and the rest of your life: seeing friends, shopping, taking classes, recreating, and so on.
- Shop locally. Don't drive to the

The sooner the automobile-oriented malls are replaced by locally oriented shops tuned to the fine grain of pedestrian town centers, the better for us all.

If the first rule for supporting pedestrian access is to plan and design for mixed-use neighborhoods and buildings, then the second rule is to make such places and buildings enjoyable.

big-box regional mall. In fact, don't shop there even if you live next door, since it promotes other people's Earth-damaging shopping habits. The sooner the automobile-oriented malls are replaced by locally oriented shops tuned to the fine grain of pedestrian town centers, the better for us all. Buy books at local independent bookstores and food from farmers' markets. Circulate your money at close proximity where it can work other pleasant changes in addition to the ones created by your initial purchase.

• Use delivery services. Communities with diversity at close proximity and modest to high levels of population support delivery services, and delivery services, in turn, make it easier for walkable centers to function. For the disabled, delivery services are crucial. In many cities such services are already not uncommon. If the centers are compact enough, delivery can be by pedal power. In Berkeley, there is a delivery service called Pedal Express founded by the teacher and local energy and transport historian Dave Cohen. It uses cargo bikes to deliver messages and mailings, lightweight and high-value breads and vegetables. In Arcata, California, the Alliance for a Paving Moratorium

picks up organic food at a farm outside town and pedals it to the weekly farmers' market. Ever imaginative, the alliance's founder, Jan Lundberg, has added sail power to the delivery of food and other items up and down a river in the state of Washington. These delivery systems work magnificently at relatively close proximity.

Shoes are not exactly vehicles, but skates and roller blades begin the transition that becomes vehicular with skateboards and bicycles. Shoes, bicycles, and a few electric carts, supported by rail transit and by ferries for fun and utility, are the transportation devices that will assist pedestrian centers. The present hierarchy of transportation modes needs to be completely reversed: Feet first. Now it is cars first, then transit, then bicycles, then, lastly, access by foot. When we build healthy cities, it will be the reverse. Then we will be designing to empower foot-power transport.

If the first rule for supporting pedestrian access is to plan and design for mixed-use neighborhoods and buildings, then the second rule is to make such places and buildings enjoyable. Cars and clothing are not sold on the basis of utility and modesty alone, and buildings have to be exciting and expressive, too. Public squares and greens, complete with shops appealing to a variety of people, need to be close to residences and workplaces. Seattle has an Arts in Public Places program whereby

No smog, no glare — stars at night.
A downtown street in small city center at night — closer to conventional layout, but with living roofs, gardens and solar greenhouses, and no cars.

one percent of the cost of any new development is required to go to public art. Among many other intriguing things, the program has paid for cast-iron manhole covers in beautiful designs of the local Indian tribes. One look at a manhole cover with a relief map of downtown and you know where you are — and enjoy it. The opportunity to sit in a puddle of sunshine on a step, short wall, or bench is a pedestrian-friendly feature helpful in luring the would-be driver into a different pace of life. Berkeley has a "Peace Wall" covered with bright-colored ceramic tiles from around the world, mostly made by children and representing their ideas about peace with flowers, suns, rainbows, people of every color, doves, peace symbols, crossed-out atomic bomb clouds, broken swords, and a chip of the Berlin Wall. People love looking at it, contemplating it, sitting on it.

Buildings themselves can be helpful in weaning people from their cars, not only by their location in areas of great diversity at walkable distances, but by their not providing or not allowing parking. Car-free apartments and condominiums by this definition are ones with lease agreements or deed restrictions that forbid residents to own cars. Millions of people live happily without cars even in the United States. The developer's or owner's job is to find and recruit these people. Car-free apartments and condos can be built more cheaply than urban residences with parking provided — for about $20,000 to $50,000 less per parking place in the structure — and therefore can be sold or rented at a lower price than apartments or condos in a building with parking provided. The owner of such a building will pay lower city taxes, since the building places less stress on the streets and city services than one housing car drivers. If, in addition, some or all of the units are smaller than in conventional housing, the prices can be reduced even more. This will allow people who do not need many possessions or tools for their work to live inexpensively. For people who participate actively in the community, whose dining room is often a local restaurant, whose living room is mostly the plaza, café, theater, classroom, park, museum, and gallery, whose garden is a community garden with a toolshed shared with others, car-free apartments and condos are appropriate and potentially very inexpensive.

Presently most cities require that parking be built for new housing in higher-density areas so streets will not be jammed with parked cars and businesses can be provided with lots of driving customers. Instead, governments should give tax breaks to building owners for guaranteeing that building residents will not own cars, and zoning should allow higher density so that more customers can live a short distances from the businesses. Laws can be written to ensure that a car-free apartment or condo will house only car-free people. Tenants who have a change of mind and just have to have a car can always buy or rent one of the millions of car-oriented houses, condos, or apartments. The car-free arrangement simply provides another choice — a good deal for people whose very-low-impact lifestyles are a good deal for the rest of us. In Missoula, Montana, in 2000, a small housing developer proposed to build extra units in its downtown apartments, called the Gold Dust Apartments, in spaces that would ordinarily be the required parking, but offered the city, in exchange for a building approval, to write a contract with the residents that made their rental units conditional on not owning a car. Car-free-by-contract housing. The developers said they'd provide a contract to that effect and the city government, without even a zoning ordinance on the subject, said, "Go ahead. Try it out." So far it's doing well as a negotiated arrangement consistent with the city's General Plan.

In some places, even more effective than building new car-free apartments and condos, since considerable housing infrastructure is already built, is a car-free conversion. This means converting buildings that now provide parking into car-free buildings. As people gradually move out in the usual turnover, new people are recruited to sign the car-free agreement. As more and more ground-floor and basement space is made available in this process, former interior parking can be remodeled into shops of all sorts, new car-free residential space, or storage space. Parking spaces once reserved for residents in open-air lots can become parks, gardens, farmers' markets, playgrounds, or new car-free buildings. City, state, and federal governments can subsidize (invest in) such conversions from their transportation budgets because they provide access by remodeling for diversity at close proximity and save the expense of providing more roads, police traffic services, and the like. Car-free conversions provide by design, planning, and major remodeling the access that transportation is supposed to deliver.

Passageways like those promoted by Jane Jacobs in *The Death and Life of Great American Cities* that go through or between buildings in the middle of blocks add enormously to the pedestrian environment, as do beautiful hallways in public buildings such as libraries, government buildings, museums, post offices, and train stations. Streets and sidewalks can feature beautiful tile work or inset stonework. The details here, too, can be just wonderful. In 1976 I discovered a "time line" about a hundred feet long painted on the sidewalk in front of the small public library of Prescott, Arizona, taking in all of evolution from "The Beginning," replete with clouds, light beams, and angelic cherubs, through the crescendo of recent history in the last few feet before the library's main entrance. Other elements of connection, such as exterior transparent elevators with views, escalators, and bridges between buildings can be part of a lively pedestrian transportation system. From Minneapolis and Cincinnati to Adelaide and Melbourne, there are bridges and walkways between buildings, some sealed and some open. Nine bridges, one above the other, soar between the two towers of the forty-two-story California Center Building in downtown San Francisco. People movers — the conveyor belts in air terminals that help get people and bags from garage to service counter to gate — could be used in larger, more three-dimensional downtowns.

Features that protect in harsh weather, such as shade trees or cloth stretched over marketplaces, broad porches and verandahs in hot climates, wide eaves on buildings, or colonnades and arcades over sidewalks in wet climates, all make the pedestrian environment work better. In British Columbia, roofs are built over walkways that traverse long distances between university buildings. In Hilo, Hawaii, I walked hundreds of feet one night between several buildings of the university campus with dense sheets of water roaring down inches from me on both sides, without getting wet. Whole pedestrian streets or special intersections can be seasonally covered. Awnings attached to the sides of buildings serve this function fairly well today. Bernard Rudolfsky takes us on

...a leisurely walk under Bologna's famous *portici* to the mountain sanctuary of the Madonna di San Luca. The distance is about four miles, the last part leading through rural country to a height of eight hundred feet above the plain, with fine views of the Apennines, the Adriatic Sea, and, with luck, the Alps. The point is that one walks, first through the town, and then up the mountain, under a continuous stone canopy, an astonishing piece of architectural extravaganza, even to the blasé.[21]

In winter snow or blasting summer sun, this structure with its roof and marching arches lavishes kindnesses on the pedestrian, providing a means of access almost unbelievably extravagant to today's mind, but only because we forget how much more expensive and extravagant freeways for drivers are than amenities for people on foot. Bridges between buildings, sometimes called skyways, are condemned by some urban theorists as dreary features taking life from the city streets, and in some cases that is true. But if we were to lavish a little care on these structures, for example, instead of the sealed tubes linking some buildings at the second, third, or fourth floor level made of glass and seasonally openable, the

questions of esthetics and pleasure would be addressed. Minneapolis has seven miles of glass skyways that link hotels with shopping and other public areas in the city's cold, extreme winter weather. If the density is sufficient that both the ground level and a level above can be simultaneously lively, the complaint about bridges deadening the street level is gone even in the best of weather. Shopping centers, malls, tourist areas, and high-density areas I know in downtown San Francisco all thrive with lively street level activity and the levels above united by bridges. Spending a little extra money on these can make them great. Bringing them into common usage, rather than keeping them rare, usage could take them from Aztec wheeled-toy status to high utility in a world moving away from car dominance and toward pedestrian convenience and pleasure.

All of these pedestrian-pampering features are integral parts of the advanced new urban transportation system of the future, but there is more. Good bus stops, with posted schedules and maps, help support the walkable centers. Covered and wind-screened bus benches make a big difference. On a cold night in Stockholm in 1972, I sat down on a bus bench — and reflexively popped right back up. It was warm! Comfortable and attractive public benches in the streets and parks make a town pleasurable and lure people away from cars.

Urban features that do not at first glance seem related to transportation and access can

Nature's diversity + cultural diversity =
more integral design.

*A range of principles apply directly to
designing ecological cities. Chief among
them is the fact that taking up a fraction
of a rich natural bioregion, illustrated here
on upper left, can, on the lower right,
result in an overlay of human ecology that
has its own rich diversity that doesn't
degrade the adjacent natural environment.
There might be a slightly smaller number
of individuals of particular species but the
same number of species. The guidance
here is that natural diversity and the built
infrastructure need to be designed and
planned in relation to one another.*

inspire, even enchant, the citizen moving about the walkable centers. Such pedestrian features deflate the allure of the automobile by outdoing it. The water art of German designer Herbert Dreiseitl takes streams — artificial and real — down staircases, through swirling "flow forms," and into ponds for biological purification or puddles for city kids to splash in. In Freibourg, Germany, air-freshening channels of water about eight inches wide and four inches deep called *Bächle* course through pedestrian streets. Step in one, goes the local legend, and you will fall in love and marry in Freibourg. The spectacular six-story waterfall inside the big atrium of San Francisco's Rincon Center also functions as part of the cooling system, and when viewed from the right angle, creates rainbows inside the building. I haven't heard of these yet but they could be built: giant hanging planter pots hovering over a public square moving in the breezes like Alexander Calder mobiles, holding butterfly bushes and humming bird-attracting flowering vines turning in the wind, with butterflies and birds following their movements. Why not prism aqueducts carrying water through plazas and lobbies, leading to waterfalls or fish ladders, casting shimmering spectral colors across open landscapes or atriums or into sun porches or rooftop arboretums?

The terracing of buildings, with rooftops used for planting, as found in some older resort hotels, has recently been incorporated, for example, in the Austrian artist Friedensreich Hundertwasser's colorful eleven-story low-income housing project in Vienna and the new Gaia Building in downtown Berkeley. Pedestrian bridges linking similar pleasant rooftops are scattered here and there around the world. Kathmandu, Pokhara, and Darjeeling, facing the great Himalayan range, have a large number of rooftop restaurants, picnic tables, and informal work and leisure places under the sun, sporting colorful umbrellas, prayer flags, and drying laundry. Rooftops there, sometimes connected by bridges, are used by locals and tourists alike. The rooftops with bridges become a transportation system in their own right, and at the same time a place like no other from which to contemplate and comprehend the city and its environs. Some buildings incorporate bridges not just as links to other buildings but as a building in its own right, with space utilized for a library, for offices, or for residences passing over active streets below. I've counted seven such bridge buildings on the University of California campus in Berkeley; part of the City Hall complex of Palermo, Italy, is another one; and Frank Lloyd Wright's Marin County Civic Center Building is, in fact, a triple bridge over wide driveways.

Although aqueduct-like structures for streetcars in hilly cities like the one flying through the air on stone arches high above downtown Rio de Janeiro are rare today, they can link the ecocity's parts in a wholeness experienced in natural environments but seldom sensed in today's cities. Utility aqueducts, taking their hint from Roman water aqueducts, have been built new in Malagueria, Portugal, where they carry water — but this time in pipes — electricity, phone lines, and shade in the form of shadows cast into the streets in the hot climate. Everyone in a built environment like this can sense that someone knows and very likely loves the place and encourages that feeling in others.

"Intelligent" Highways, Cars, and Buildings

Advocates of the wired village of the future tell us to move electrons, not people — to wire the ten lane freeways, make them "intelligent," and build "smart cars." "Intelligent" buildings, they say, will have energy conservation systems that take care of themselves.

"Intelligent" highways inform the driver of conditions ahead by radio alert or expensive electronic signs. Electronic toll plazas collect tolls by reading a signal transmitted by the car as it passes, automatically transferring money from the driver's bank account to the bridge authority's. Someday they might even move the car forward by pumping electricity into the car's electric motor with a magnetic induction field created by wires buried in the highway itself, while, once again, simultaneously draining the driver's bank account and shifting money to the highway owner's. The "intelligent" highway can let drivers know that there is a wreck in the left lane three miles ahead so that they can get off the freeway and clog up the local streets. Often what it can tell drivers is something he or she can see anyway. In other words, this kind of "intelligence" can add counterproductive or superfluous information at high cost to a polluting, energy-squandering transportation system while victimizing people living nearby, who may not even own cars.

Telecommuting, another electronic "solution," also promises contradictions. It could simply reinforce suburban isolation and preserve or even extend sprawl. The vision of everyone alone in the suburbs united by electronics is fundamentally different from the vision of people in face-to-face communities. Just as cheaper, more "efficient" automobile transportation gave people access to ever more distant and cheaper land, thus expanding sprawl development, so the computer makes it possible to travel to the job site less frequently.

Not all that tingles with electrons is dubious or damaging, however. In the service of intelligent ways of building, computers could figure in reshaping cities and suburbs around walkable centers. Large corporations could use computers to help break up their mega-offices and centralized functions and scatter them to satellite offices in the suburbs and small cities where they could form nuclei of ecotowns and even villages, drawing in more density around public plazas and pedestrian areas on which the corporations could help capitalize. They could invest in the long term by designing and building passive energy-conserving buildings and installing solar technology. These centers could attract a variety of services to revitalized centers or new ones. Housing could be built near jobs, shops, and services to rebalance development. Efficient transit services and transit stations could serve these new centers.

In another version of such a physical arrangement, large companies could be replaced by smaller ones located in future small ecotowns, ecovillages, and neighborhood centers. Customers could switch from large companies to small ones, shifting economic investments and habits closer to home. Smaller companies could be encouraged — through loans and other assistance to small businesses of the sort provided by local economic development offices and federal small business administrations, and especially through changes in zoning regulations — to set up operations in small centers with supplies, services, shops, homes, education for the kids, parks, and nature all within walking distance. Electronic technology would seem to make this alternative more practical than ever by helping to create more diverse business cores for vital, real community centers. Most of the profits in this model would circulate inside the community.

Depaving while Rebuilding

The inescapable conclusion about access and transportation is that we have built an immensely destructive transportation infrastructure and need to create access in a different way. What this implies is tearing up the asphalt and rolling back sprawl: organizing the community, buying the seedlings, breaking out the sledge hammers, warming up the muscles, and firing up the bulldozers — on our side this time.

Nature depaves. Why don't we? My favorite depaving-cum-restoration project is one that must have been completely accidental. On the freeway from Santa Fe to Pecos, New Mexico, traces of the old road are still visible. The new highway cuts a swath through the landscape, while the older road weaves in and out of the line of the new highway. The old road goes up and down and snakes about from side to side considerably more than the modern wonder, sometimes higher, sometimes lower, looking like the oxbows of a meandering river. It appears that the state didn't have enough money in the budget for the new highway to grind the old highway into dust and gravel. It just let weather and time take care of it, and weather and time are doing a good job.

The old highway's asphalt has partially dissolved and washed away under forty or fifty years of intense New Mexico sun, occasional cloudbursts, and the action of water freezing and expanding, melting and evaporating, many dozens of times each year, year after year. The old dashed white line and the yellow line are still there in thick paint holding pebbles and grains of sand together where the long-departed petroleum bonding material has washed away. Chamisa, a powder-blue bush with bright yellow flowers in the spring, has sprouted there, though it usually lives exclusively in the wet, sandy valley bottoms. On the moldering old road its seeds have found the cracks and enjoy the increased runoff concentrated by the road surface. Thus, the crumbling road actually adds to the hilltop biodiversity. Whole trees, looking extraordinarily healthy, now ten and fifteen feet tall, grow right out of the middle of the road, prying open the failing pavement like wedges. The forty or fifty years represented on this humble piece of highway with its sun-basking lizards and cicada-buzzing piñons is a very thin slice of evolution. Given half a chance, nature will recover.

Jan Lundberg, founder of the Alliance for a Paving Moratorium (a project of his Culture Change Institute), was once co-editor and publisher, with his father, of the *Lundberg Letter*, the influential oil industry survey and subscription journal. He often asked oil industry executives what they expected would happen when oil ran out, and they always said, "It won't

Understory, canopy and emergents.
Forests frequently layer their ecological zones as understory, canopy and emergent trees rising occasionally above the other. The built infrastructure of dense pedestrian towns could also be seen in this light.

run out. We'll find more." Jan considered this answer absurd and dangerous, so he quit the *Lundberg Letter* and organized the Alliance for a Paving Moratorium, reasoning that calling for a halt to paving would be a shocking notion that could alert people to the destructive effects of the car/sprawl/highway/oil civilization. I told him that he was getting close, but we need something even more shocking: the battle cry "roll back sprawl!" If we could be vigorously building ecocity centers, at the same time we could be building bicycle and transit systems with the people in place to use them. In no time we could be happily bulldozing large tracts of suburban sprawl. Employing the powers of nature and time in the way of the under-funded highway department of New Mexico — by benign neglect — would be the low-budget approach to depaving. We could let derelict shopping center parking lots turn into weeds first, then bushes, then trees all by themselves — or buy the real estate and really do the job right.

We need not only to stop new paving, but to begin depaving what should never have been paved in the first place.

A major motorway was recently removed from the side of the Willamette River in Portland, Oregon, and replaced with a riverside walk and park. A marsh full of native plants and wildlife has been reestablished where the airport runway of Crissy Field used to be in San Francisco's Presidio. In the suburbs of St. Paul, Minnesota, in an area called Phalen

We need not only to stop new paving, but to begin depaving what should never have been paved in the first place.

Village, a failed shopping center and its enormous parking lot have been bulldozed to restore a lake filled forty years ago. According to St. Paul city staff planner Al Torstenson, it was a Master's student at the university named Sherri Buss who proposed opening the lake again, adding new development that would define a vibrant neighborhood center. To everyone's surprise, the idea was large enough to intrigue the neighbors, and Harrison Fraker, then at the University of Minnesota and now dean of the School of Environmental Design at the University of California at Berkeley, helped secure the funding for a planning process that produced a convincing plan. Says Torstenson:

It always wanted to be a lake, even after it was paved. Water collected in the parking lot during rains and stayed for days, and by the end a few cattails were pushing through the asphalt even before the bulldozers came back. One day last fall I saw ducks floating in the rain puddle in the middle of the parking lot near those reeds. The lake was insisting on coming back.[22]

One old-timer in Phalen Village told him, "I was there fishing when the first dump truck arrived. Never should have filled that lake in the first place." Most of the shopping center is gone now, down to 40,000 square feet from 260,000 square feet, and new construction and remodeling are adding to the housing and creating a small commercial center along a new transit line and a new bicycle path beside the restored lake.

Around the United States a few other old shopping centers gone belly-up are being bought by developers and transformed into new housing developments. None I know of is as ecologically inspirational as Phalen Village. Many are utilizing the New Urbanists' medium-density housing formula. The Crossings in Palo Alto, California, a project by architect Peter Calthorpe's Berkeley-based firm, has accomplished this next to a rail road crossing and commuter stop. Small plazas and greens have been created as part of these projects but they are not enough in number to make a dent on trends in the opposite direction — pedal-to-the-metal sprawl. But they do constitute an important step in the right direction.

The depaving strategy needs to make freeways smaller, reducing them from six to four lanes, for example, then to two, as demand shrinks because of ecocity building and the rebuilding of railroads and humble, inexpensive rural two-lane roads. We would be downsizing, not destroying transportation, perhaps adding small roads while removing big ones. Just as the biosphere needs corridors for migration, ranging, and habitat continuity, so society needs social-ecological links with itself and nature. These are the bicycle and footpaths that meander out from cities and towns into natural and agricultural land

and off to other cities. They would pass under or over intersections with wildlife corridors. They could be greenways like the ones in Nepal, which help prevent rather than increase erosion — roads of firm crabgrass-like plants that can give support to people on foot, bicycles, and even light vehicles like electric carts, but do not interrupt the natural order. Similar small roads, though hard-surfaced, are being built along city waterfronts, for example around San Francisco Bay — the Bayshore Trail — and in an ambitious effort in Scotland, England and Wales in a system called the Sustrans (for *Sus*tainable *Trans*portation) Traffic-Free Network. This is a system of country paths for bicycle and foot traffic, not for cars, and it is approaching, in 2005, a total of 10,000 miles and growing. Tremendous! The organizers, based in Bristol, England, are working to consolidate a number of these routes from one end of the UK to the other, including Northern Ireland.

This is how most of the transportation of the ecocity civilization should be. Like most other "radical" features of our future ecocity civilization, it actually exists, like a humble but intelligent creature poised in an obscure niche, mostly unnoticed among the dinosaurs, ready, with just a little encouragement, to replicate and spread over the world.

Shortly we will describe means to systematize depaving and withdrawing from sprawl in an orderly, strategic manner. But for now, time to travel on.

Table land.

CHAPTER

What to Build

The Builder's Sequence

In 1990, as keynote speaker at our First International Ecocity Conference, Denis Hayes asked how it was environmentalists could be making such progress in so many areas and still be falling behind regarding the "really big issues": climate change, habitat loss, and species extinction. He sounded much like Philip Shabekoff (see Introduction), but he was speaking ten years earlier. Hayes suggested that we need more vision. Ten years later, the same year Shabekoff made his comment, Hayes said the same thing at another event. I was in the audience and asked him, "What is your vision? Do you have a strategy to provide that vision, to solve those problems?" If he didn't, as past director of the Solar Research Institute, organizer of Earth Day, and now head of a large foundation, who did? What's the

answer? To which he said, "Richard, I don't have a strategy. I try to support whatever seems helpful."

Shortly after, an analogy came to mind: you can't build a house starting with the roof shingles or the baseboards. It can't be done randomly. Building a house — any complex project — needs a strategy to get to the vision. There is a natural sequence to follow. You have to start with the foundation. If a fleet of trucks dumps a pile of building materials on your construction site and you start nailing the roofing shingles to the plywood, painting the sheetrock, and sanding the floorboards, you're going to have terrible chaos, and an immense waste of time, energy, and materials. In short order you are going to have a pile of mangled, moldering building materials rather than a house. To build a house, you have to start by laying out and building the foundation.

181

With cities, the natural order of things starts with the land use pattern. The analogy that compares constructing houses and constructing cities is so simple, logical, and even conspicuous in everyday life that it can, in its own right, lead to solving "the big problems." How could we imagine restructuring society for a healthy Earth without a reasonable approach to rebuilding our physical homes, the largest things we build, our cities and towns? How could it make sense to structure cities on vehicles on rubber tires that have no loyalty to any place at all instead of on people who plant and nurture things in the soil? And if rebuilding is in order, doesn't it make sense to follow a logical sequence?

We need a strategy if we want to make real progress on the big issues, and essential to that strategy is simply to get the sequence right. Foundation first, land uses first. If the vision is the ecocity, the strategy is the builders' sequence. That has a real chance of solving the lager problems.

Ecocity Principles

What, then, might be the principles for building ecocities? To begin, we might try the Ecological Golden Rule: Do unto others — including plants, animals, and the Earth herself — as you would have others do unto you. Dividing the golden rule into two, we might embrace the social-ecological commandments taught to every pre-kindergarten child: Be nice to others and clean up after yourself.

Refining this a bit more, we could say that there are three major environmental prescriptions into which most others fit: Conserve, recycle, and preserve biodiversity. This is still a bit broad, but specific examples immediately come to mind. We are off to a good start, and people are beginning to get comfortable with what these three things mean. Then there are the "four pillars" of the Green Party of Germany: ecology, social responsibility, grassroots democracy, and nonviolence.

So far, there is little here that prescribes how to build and live in a healthy physical structure, a built community, much less a whole functioning civilization. Without expressing the ecocity insight and actual design principles for ecocity building, we can't make much improvement in our built habitat. Much more is needed, and a few good people are in the process of thinking it through. There are Bill Mollison's five principles of permaculture; the six "guiding principles" of "social-economic-natural complex ecosystems" of systems thinker Rusong Wang of Beijing, China; the eight planning and design principles of developer Joseph Smyth and Citizen Planners of Ventura County, California; the nine design principles William McDonough Associates developed for the 2000 World's Fair in Hanover, Germany; the ten "guidelines for rebuilding the ecocity" of the Australian planner/activist David Engwicht; and the twelve "Ecopolis Design Principles" of the architect Paul Downton, the activist Chérie Hoyle, and the other leading lights of Urban Ecology Australia.

Without dissecting all of the above and in order to communicate a sense of the thinking on the subject, let me present the "Ecopolis Development Principles," as written in 1993:

- restore degraded land
- fit the bioregion
- balance development
- halt urban sprawl
- optimize energy performance
- contribute to the economy
- provide health and security
- encourage community
- promote social equity
- respect history
- enrich the cultural landscape
- heal the biosphere

One can extract several major principles from Paolo Soleri's and Ian McHarg's work while architect Christopher Alexander of Berkeley, California, offers no fewer than 253 "patterns" in his book *A Pattern Language*, ranging from pleasing ways of arranging furniture and lighting to grand principles of arranging urban patterns on the largest scale, from tentatively recommended prescriptions to virtual holy commandments: Thou must build thus.

Struggling with all of this, I've come up with lists of about twenty to forty "principles," but here I'll just mention the few I think are most important:

- *Build the city like the living system it is.* In other words, build it on a basically three-dimensional, integral, complex model, not flat, random, and in large areas uniform and simple. Like any living system, the city should be compact, and it should be designed primarily for a population of living things, mostly people, rather than for machines like cars or even buses. Its physical body and logistics must be based on the needs of flesh and blood, not steel and gasoline. It must be a pedestrian city. Paolo Soleri got it right about that shape and function — three-dimensional with well integrated complexity — and for doing virtually nothing to explore his discovery, we have been suffering more and more lost opportunities every day, for four decades. The "organic city" is a term some people are beginning to use, and so it must be.

- *Make the city's function fit with the patterns of evolution.* Not only must it not destroy its sources of sustenance — i.e., be "sustainable" in the current language — but it must also regenerate natural systems and support and express creativity and compassion. As the universe evolves, its entities like stars and planets spawn beings like animals and plants, and some of these invent wings like those of insects, birds, and bats that never were before. And so people invent, among other things, Justice, Truth, Beauty, and the techniques of law, philosophy, and art to embody them. These and future inventions not yet imagined are what cities should support.

- *Follow the builder's sequence — start with the foundation.* This means starting with a land use pattern that supports the healthy anatomy of the whole city. The land use/infrastructure — or "landustructure" as I sometimes call it — has to be ecologically tuned from the start, either the start of a new town or the start of reconceptualizing and remodeling an existing city. The ecocity map, which we will explore in some detail later, is the tool to get the foundation right.
- *Reverse the transportation hierarchy.* Transportation is so important in working with or against landuses and thus supporting or subverting practically everything else that could be healthy about cities that it is imperative to plan for pedestrians first, bicycles next, rail transit next, "flexible" transit (on-the-road busses) next, and lastly cars and trucks.
- *Build soils and enhance biodiversity.* This one is self-evident, but intimately connected with the above four.

Beyond this, I support the ecopolis development principles (though I would change the fourth one to read, "Halt sprawl and roll it back") but would like to add a last proviso, a principle of sorts: Ecocities are not everything virtuous. I just don't think anything, even the best things, can be all things good and nothing bad. But ecocities can set up a new game board, much healthier for nature and society, which will define a new panorama of problems while solving many old ones. With ecocities, at least the field of action is likely to be far more relevant and peaceful in terms of human cultural products and their effects upon the Earth. As we build ecocities, we should at least be able to greatly reduce our "collateral damage" to nature, and that's more than is promised by any other strategy I've ever heard of.

Bike Tour

With principles in hand, what to build? The ecologically healthy community, of course, from land uses on up, from small-scale villages to radically reshaped metropolitan areas and the whole ecocity civilization of the Ecozoic era.

Let's say we land on Earth about 100 year from now, and we're looking around. By now, ecocities are common. They are sparkling outposts in the deep green forests, oases in the desert, frontier towns in oceans of grain, islands

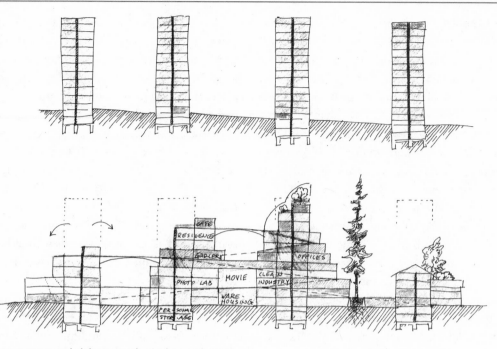

From towers and slabs to ecological complexity.

Currently warehousing is at truckers' distance from cities, where land is cheap. In the ecological pedestrian towns served by train, warehousing would be in the darker interior zones on lower levels with other uses where light is artificially controlled, such as movie theaters and certain production facilities. The mixed-use buildings then rise above the storage — similiarly as many houses are conventionally built over storage basements. Instead of millions of trucks burning hundreds of millions of gallons of gasoline to access storage every day across endless miles of highways, small electric fork lifts and elevators do the task in distances measured in dozens of feet.

on salty waters lost in clouds of sea birds as the ferries push through the waves and the fishermen come home. There are hundreds of them on every continent and dozens on offshore islands where people don't worry about the fact that there are no bridges connecting them to the mainland.

So here we are, down on the ground, approaching an ecocity right now. Forget renting a car, even though it's an electric one; they won't let you past the dinky parking lot behind the pharmacy anyway. We're taking bikes.

It will be a leisurely cruise, much calmer than driving a car, and you can be especially relaxed because in this town you're not going to be hit by a car. Despite our slow speed, we will be experiencing enormous change in short periods of time. In typical car-dependent cities, miles and many minutes can go by on the freeway or main streets with little change in that monotonous repetition so prized by the billboard advertisers and graffiti scribblers who capitalize on the nothing-much-worth-looking-at situation. Not so in the ecocity. We round a bend from out in the countryside and leave grainy fields, shimmering summer deserts, or secretive forests behind us. Suddenly we pop into a neighborhood center, one of a few scattered around a major ecocity downtown. Look over your shoulder — it's still there, the fluffy summer clouds, birds, breezes, loamy earth smells. But blink your eyes and you are also inside something that looks like a larger country village.

In just three or four blocks, this neighborhood center changes its scale and character dramatically. Two- and three-story houses and other buildings on the edge quickly give way to five- and six-story structures and bustling street life. The taller buildings step back in rows of balconies and terraces. It's a three-dimensional adult jungle gym. People are leaning over the railings of both balconies and pedestrian bridges to talk to other people two or three stories below in the street or on other balconies and bridges. They can hear each other, too. No cars rumbling through to drown out human voices.

Design motifs and local styles and materials as well as actual bridges link the buildings with consistency of style, echoes of traditions. The people here may love wood and detail, in other places simple whitewashed stucco on adobe or perhaps stately old brick or stone predominates. Consistency in design is suggested by locally available materials and — a felt presence in all ecocities — the sense that the whole thing is built on the human measure. In some places traditional design motifs may be interrupted by a modern insert of startling contrast, but the counterpoint is exciting, and these buildings too are linked functionally, often with bridges to the others.

In this neighborhood center, the one through which we are bicycling, there are varied rooflines, small towers, and plantings everywhere on porches and balconies and in window boxes. Fruits, berries, and flowers attract bees, butterflies, and an assortment of birds. It's a bit like a multi-species Mardi Gras, with windmills cartwheeling above the rooftop trees and shimmering light reflected off solar collectors and greenhouses passing through the moving branches of trees, bushes, and vines. Nodding sunflowers look down at us from several stories up. The neighborhood seemed sleepy at first glance, at the edge of town, but it buzzes with activity for its central two or three blocks.

In the neighborhood center, we glide to a stop in a small plaza. Looking around, we see handrails, trellises, stairs, ramps for the handicapped, cafés on street and roof level, an atrium movie lobby, a sculpture garden with panoramic views. At first it's a jumble, but soon enough things begin to make sense, continuities are established, linkages unite, cycles close, consistent design elements begin to be evident. It is almost as if small hills had been built to bring people up to higher vantage points with beautiful views, while at the same time creating a fascinating thing to look at: the neighborhood center itself.

Leaving the neighborhood center, in only two or three blocks we are cycling through what's left of the old suburban belt of homes that in the twentieth century went on for dozens of blocks in all directions. Now, garages have been converted to second units and houses raised a story or two, making way for third and fourth units. Then, quite suddenly, we are in open space, on a rural road — and it only took us six or seven blocks to move through the entire neighborhood area from open agricultural space on one side to restored natural land on the other.

Cycling from the neighborhood to the city center amounts to four or five blocks of country trail built for foot and bicycle. Motor vehicles — trucks mostly — in small numbers are on another two-lane route, mostly underground. On both sides, as we pedal along, we see the natural landscape as it might have been 500 years ago, but with the addition of several ecovillages, each with its associated cluster of farms plots. They stretch off down a curving valley about a half-mile to three miles away. Each ecovillage is like a small neighborhood center in its own right — which they all once were — but now reorganized in the traditional compact village form. Unlike typical neighborhood centers, however, many of their residents work the land and secure the convoluted greenbelt. The closest village now operates a community-supported agriculture project (CSA). Its subscription customers participate in the farming from time to time. Not surprisingly, this increases the city folks' understanding of and enthusiasm for local agriculture and, they all say, makes the food taste better. From a porch or terrace in town, you can occasionally see your friends across the open fields working the farm.

Midway between the neighborhood center and the city center, the edge of the downtown is only two or three blocks away on the far side of a bridge. That span carries us over a local creek in its shallow valley, with native vegetation on one bank and an orchard of fruit and nut trees and a bicycle trail on the other. Then, where a small hill rises, we dive into a tunnel for one more former city block, now open space. Inside the tunnel, several open skylights illuminate most of our path, with two or three fiber optic lights on the ceiling between the skylights connected to solar-powered electric battery lights. These, along with bicycle lights, are sufficient at night. Inside the passageway we are under a nature corridor that deer and recently even antelope have begun to use in passing between the neighborhood and city centers.

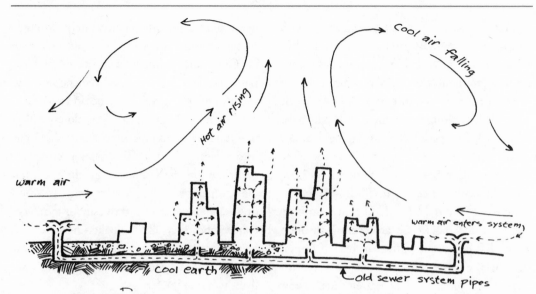

Passive cooling of downtown. *Warm air rises — it can be used to draw air into and under city centers, using the cool earth below to cool the buildings with zero or very little energy.*

In just another few blocks our small road, now on the surface, enters a kind of "gate to the city," with special buildings facing one another where our road becomes a main city-center street. From these gate buildings or, when they are especially formidable, "ramparts," the burghers hang out their banners announcing and celebrating civic events. Here we find ourselves entering one of the downtown center's "keyhole plazas," a.k.a. "view plazas." From above, the plan of one of these public squares looks something like an old-style keyhole, open in the middle with a slot on one side. Larger buildings surround the plaza, a public place filled with people, plants, art and water works. The plaza and associated buildings generously frame a cherished local view. The sky is crystal clear — no pollution. Sun and wind power the place, and there is

practically no motorized transportation.

In the distance the area's tallest mountain is silhouetted against the sky, framed by the town's most celebrated tall buildings. The buildings themselves rise up like the cathedrals of a religion of reverence for life on Earth, carrying trees and bushes up to high crags and crests and cascading vines and flowers down into the canyons. Butterflies and falcons live up top. Swifts and swallows slice the air at ninety miles per hour, missing residents, office workers, lunchers, and rooftop joggers by inches as they hiss by and break right or left, plunge, or shoot straight up almost out of sight.

From the street we can see people walking and biking over the bridges between public areas on the fifth floor. Pedestrians — cyclists, too — are way up there on bridges between the tenth floors of buildings, emerging high overhead, crossing over streets, boulevards, and alleys and disappearing again. In really large ecocity centers, there is often a set of bridges at twenty or twenty-five stories. Some have sets of bridges every five stories with express elevators set for stops on these floors — the idea is that nobody has to walk more than two stories up or down, getting very fast access and some modest exercise on the stairs. Slower elevators hit every floor. In our particular downtown, artificial waterfalls cascade down six, seven, or eight stories into small ponds, breaking up into mists that cool the surrounding buildings. Where the sun pours in, these practical flourishes introduce interior rainbows to the city canyons. All this is paid for with assessment-district funds that replace the air conditioning expenses formerly paid by each building owner. Now, windows are thrown open to the fresh air, except in mid winter when light is reflected by mirrors down into the streets to brighten and warm the public realm.

In cooler climates, the towers and domes of the city center are oriented toward the sun collecting energy and good intentions. Where it's hot and humid, patches of parachute cloth are

"Keyhole" or "view" plaza. *This feature represents a principle in its own right: and conspicuous the healthy union of nature and city by way of celebrating a special local view in the very layout of the public plaza. To place the most significant architecture in a supportive relationship with nature by framing a view from a public plaza declares our city's love of nature — a major step toward survival these days.*

stretched out horizontally and ripple on cables between large rooftop trees and metal or wooden poles. These urban "garments," as Paolo Soleri calls them, cast moving pools of shade into the light breezes of the streets below, saving millions of dollars in cooling machinery and energy.

Our bicycles glide by notches in the variegated walls of buildings that offer glimpses of bays, rivers, distant hills, and broad horizons. We bicycle over bridges spanning local watercourses right in the middle of town. There are ancient rock outcroppings and magnificent trees in settings that make them the focus of special attention. We move along orchard-lined streets on bicycle paths marked by different colored brick or tile; there are no curbs here to help separate cars from pedestrians because there are no cars. We stop to pick a fruit or two and pass street cafés, hardware stores, dentists' offices, banks, bookstores, entrances to underground warehouses, main-street movie theaters, pedestrian-mall computer shops, alleyway bird-watchers' suppliers, side-street food stores, and galleria fabric outlets.

Police use bicycles and operate out of closet-sized "neighborhood stations" on the fifth and tenth floors as well as on the street level, apprehending their suspects by darting across bridges, up and down ramps and elevators, and through streets and alleys, often radioing ahead for assistance or locating the rare fleeing suspect by looking down or up from a bridge or rooftop railing. Jane Jacobs'

famous statement about the safety of streets with people in them — "the streets have eyes" — can be expanded here to also include rooftops and bridges. Ambulances are well-outfitted gurneys pushed quickly by strong paramedics on roller skates to the scene of necessity in less time than it used to take any vehicle — traveling about eight times the distance. There are no sirens screaming through the streets but rather a bicycle bell: "Ching-ching, ching-ching. Coming through now, 'scuse us. Ching, ching."

These people, who would be ambulance drivers in other times, wear the sporty "lock and roll" skate-shoes, popular among ambitious business people, athletic types, and teenagers who call them "skoos." Flip a lever and they roll free; flip it back and the wheels lock tight for walking. With four or five small wheels per axle and five or six axles set in flexible plastic, they are so comfortable many people wear them all day in "lock" position at work, in classes, around the house. Then, for doing errands, they flip the lever to "roll" and take to the streets and bridges at twice the walking speed.

As we pedal along, we have pedestrians, lock-and-rollers, bikes, occasional streetcars, street musicians, playing children, and songbirds for company. The intricate weaving together of ecology and society is a pervasive impression. This New Synthesis Architecture rises to an unavoidable level of consciousness. Its purposes can't be missed in an environment with this many parts assembled, each integral to the fullness of the whole. Hummingbirds pollinate the flowers and delight the people. People plant the flowers and the fruit and nut trees, which give birds food and shelter. Salmon jump, snap, and wriggle their way up the fish ladder in the city center, eating the insects that eat the leaves of trees and bushes. The salmon's relatives at sea provide dinner for people in the restaurants. Dragonflies drink from side pools, eat occasional pesky mosquitoes and gnats, and inspire five-year-old future hang-glider pilots who, with their parents, plant fuchsias and trumpet vines for the hummingbirds.

In the middle of the city center we decide to take one of the oversized elevators, folding bicycle and all, to the fifth level bicycle and pedestrian network. As we are about to enter the glass elevator attached to the outside of a building, a strangely shimmering shadow spreads over the streets. People by the hundreds stop and look up. An enormous cloud of seabirds is passing over the city, and we fall into a hush to listen to the breathy sound of millions of wings. In patches and waves, the flood of primordial life flows over the roofs of some buildings and pours between the taller ones like waters around boulders in a mountain stream. The birds are on their way to the great wetlands that have been reestablished just two miles outside of town. They take fully ten minutes to pass over and through town, as we take the elevator to the tenth floor and into the midst of the birds.

The people in our rooftop café have carried their refreshments to the railings to look down upon this stately flood of living creatures and up at the few birds taking higher paths. Some birds come within two or three feet, and a few stragglers alight on the railings and tables for snacks as the flock begins settling down on the marsh in the distance. "Never fails to take my breath away," says one patron, replacing one of the café's pair of loaner binoculars. As people drift back to their tables, the conversations turn to the spectacular air shows produced for free by various species of migratory bird. Each species comes in its proper season. Great flocks dive, turn, split, and explode apart and swirl together out over the bay and marsh, turning swaths of sky from coal black to shimmering silver and back again, all reflected in the waters below.

Back with our bicycles on the fifth floor, we take a brief ride around the downtown on rooftop streets and bridges among the gardens, benches, and miniparks. The usual crowd is flying kites and radio-controlled gliders. Continuing on the elevated bicycle-pedestrian street toward the far edge of the city on the opposite side from our point of entry, we emerge from the city center in another three or four blocks, the density dropping rapidly as we descend gradually toward *terra firma* — or, more correctly, as *terra firma* rises toward us, since there is a hill on that side of town and our rooftop street/bridge heads horizontally straight for it. Within three blocks, city-life excitement is replaced by a kind of almost village calm, and in one more block we pop out into the magnificent views of the countryside where the elevated bike and pedestrian route finishes its trajectory through the canopy, branches, and tree trunks, touches down, and winds off into the countryside. Soon the path plunges under a close-in wildlife corridor. Not far away to one side, so does the main rail line. The bicycle path emerges one mile farther away, and the railroad tracks reappear far in the distance as they begin their stitching in and out of the landscape toward the distant mountains. There the tracks wind part-way up, then burrow under the granite into the next bioregion while the bicycle road makes switchbacks up the slopes, passing through a few more tunnels, visiting a few ecovillages, and finally running up and over a pass.

We have now experienced most of the major features of the ecocity in a matter of, say, a lazy thirty-minute bicycle ride, plus a fifteen-minute stop to watch the birds. At sixty miles per hour on the freeway bypass, we'd shoot around the whole thing in just three minutes or glimpse it stop-and-go through the billboards, sound-wall gaps and obscuring smog. We'd miss it all. Instead, staying within Ivan Illich's fifteen mile-per-hour maximum convivial speed limit, we saw a walkable neighborhood center, a pedestrian downtown, and open land between and around them, and we enjoyed the views from the keyhole plaza and a spectacular panorama from the tenth floor. In fact, a route through the whole city was walkable in perhaps twice the time it took us to bike it. We crossed restored creeks, dove under nature corridors, and returned to the wide open spaces.

Ecocity Layout

Let's look more closely at some of the features of ecocities. In contrast to the general urban pattern today — skyscrapers in the middle with tens of thousands of acres of groundscrapers all around — the general pattern of ecocity development is much more fine-grained. The center is still tall and dense, but reorganized in an enormous complexity of uses, while the area around this will, in most cases, be new belts of nature and agriculture. Islands of civilization in the form of dense subcenters and compact neighborhood centers are situated fairly close to the major city center. Farther out, at a distance seldom traversed by daily commuters, there are real towns of some scale. Neighborhoods turning into villages and hamlets reach out, scattering ever more sparsely across agricultural and natural country into the hinterlands. Probably most of these outlying communities will be agricultural villages producing much of their own food and selling farm produce to the more centrally located towns and cities. Of course, real geology, topography, and ecology will modify this pattern from place to place. Some towns, for example, may have no nearby ecovillages if, for example, they are situated on the edge between a desert and an ocean. There, people might fish the sea but have no production from the surrounding land and little reason for ecovillages located there. To this general urban anatomy all other ecocity features and technologies are attached, which empowers them to function fully.

Streets are the main circulation systems of the ecocity as they are in the conventional city. However, since they are designed for humans rather than cars they can, in many cases, be relatively narrow. Still, for variety, wide boulevards here and there provide real urban/bioregional views worth celebrating and streetcars call for occasional streets that are wider than those designed for pedestrians.

The layout of streets in today's cities is a topic of lively discussion among planners and architects, with some championing the right-

Pyramid-like larger complex buildings. *In this illustration, a gigantic flock of migratory birds flies through a city of such buildings, each like an island with its own character and great views, each island adjoining the others by bridges and the lively pedestrian streets below. Terracing maximizes availability of the city's surface with nature and a diversity of angles to the sun and panorama.*

angle gridiron pattern, others the meandering curving lines conforming to the contours of the topography. Still others champion the cul-de-sacs and dead-end streets on the suburban theory that the only good street is one with almost no traffic. Grid advocates derisively refer to the squiggly lines on many suburban development maps as "dead worms," and squiggly-line fans refer to the grid pattern as "rectilinear," as if it were a long version of a four-letter word. Others insist on streets radiating out from center points, parks, or circles, and still others like the ordered power of broad boulevards arbitrarily pushing their way through the urban fabric, usually connecting major centers. They are charmed by the light and air provided by these grand swaths of open views — boulevards like those of Paris and Washington.

But if we imagine cities with precious little automobility and near total human mobility, street layout becomes less important. Since pedestrians make fewer demands on the infrastructure, there is less pressure to accommodate transportation with wide streets. With footpaths cutting through blocks and no fear of being run over crossing the street, the whole town becomes permeable and pleasantly accessible whatever the street arrangement. These streets are especially good for children and people with disabilities. The sightlines of the grid are fine for a sense of location and for an esthetic appreciation of distances. Curved streets create a close-in inti-macy that can have its value. But the straight lines of San Francisco's hilly cityscape create their own excitement by producing views of the bay, ocean, and hills that would be lost with curved streets.

Higher buildings with elevated public access also provide the kind of grand views that boulevards do and that, if well designed, are great views in themselves. The supposedly democratic opinion that building heights should be limited to a consistent two, three, or four stories so that views can be maintained makes no sense. If all the buildings are the same height, everyone's view is of more buildings of the same height next door or across the street. While living in a two-story apartment in Berkeley, for years I had to walk several blocks from my home to see beyond the two-story houses and three-story trees to the mountains across San Francisco Bay or to glimpse a sunset. Uniform height limits give everyone the right to no view at all: democracy of the lowest common denominator — if I can't have a view, then no one can. The solution is to have public access to high places: terrace and rooftop restaurants, miniparks, and promenades. All that is a design challenge to be worked with, and one with endless good solutions.

Walkable Centers

Downtown is already seen as the heart of almost any city and could be seen as the heart of the concept of a vibrant culture in tune with nature. Downtown and neighborhood centers in larger ecocities will constitute a whole new urban reality, complete with neighborhoods and clusters of dwellings nestled into terraces on the upper reaches of tall buildings, floating high in the sky, as replete with natural (if transposed), life as many places on ground level. Co-housing arrangements — a form of cooperative housing in which residents, usually ten to thirty households of individuals, couples, and families, share some facilities but have small-scale private facilities such as kitchenettes and dining nooks and modest living rooms for entertaining — could be worked into larger clusters of buildings, for example around a large rooftop garden or plaza.

In the highest-density parts of these ecocities, uses that need artificial light or function well with controlled artificial or reflected light could be placed in the lowest floors or basements of buildings. These uses include certain kinds of laboratories, semi-automated clean industries, theaters, warehouses, rental storage space, mechanical and electrical systems for heating and cooling, and the like. Storage will often be just a few feet below or immediately beside the citizens going about their business rather than accessible only by automobiles and trucks heading out to the low-value real estate fringe.

Centers have traditionally been the cultural hubs of art, entertainment, civic education, and political discourse. In the ecocity, what I think of as the *agora futura* is here, not in the television studio. It's that wide place in the street or sidewalk or small open-air gathering place available for debating and sharing news. The city centers, large and small, are for moving from one activity to another, for just hanging out, for placing chairs in the sidewalk cafés and delis, for benches and walls at the right height to sit on, for trees for a little street-side fruit picking. In larger-scale centers, intensity of use by large numbers of people justifies the channeling of money and resources of imagination into special civic features. Investment potential is very high where many people congregate. More people per square foot equal more money per square foot, and this is where money should be lavished on public art and art that addresses ecological realities.

The land area of neighborhood and city centers is a topic of debate. The New Urbanists say that people will only be willing to traverse centers of activity one-quarter-mile wide, and their various plans show uniform circles of that size spaced along transit lines and centering on rail stations. The planner Dirk Bolt studied cities in Thailand and discovered that one kilometer, or closer to a half-mile, worked for Thais on foot. Car-free city promoter Joel Crawford substantially agrees with Bolt and imagines clusters of car-free cities surrounded by nature or agriculture along convoluted loop transit systems, with each center having a diameter of a half-mile.[1]

My own guess is that a variety of sizes of centers from fifty yards in a small village or neighborhood to about three-quarters of a mile or even a mile in an ecologically healthy downtown, makes sense for lively mixed-use pedestrian centers. Ecosystems are generally healthiest in great diversity, and hence one might think that diversity of city center, district, and neighborhood sizes would tend to be healthy for similar reasons.

An odd thing to my way of thinking is not only the uniformity of the diameter of the New Urbanists' and Joel Crawford's city centers but the fact that they generally set a four-story height limit as well. The size of the center is the horizontal multiplied by the vertical, and so a four-story height limit, though more dense than one or two, still forces horizontal development out several times over the area of, say, a center of ten- to twenty-story buildings. Tall centers are thus potentially — and often actually — far richer culturally and economically than low-rise ones and promote the kind of civic focus and concentration of capital that can produce grand plazas, parks, settings for sculptures, and other special features. Crawford and the New Urbanists seem oblivious to the realities of the centers of thousands of everyday downtowns in practically every country in the world and to what makes them "work" economically and socially.

Given the world's population, the efficiencies of the three-dimensional organization of urban activities, and the disasters of sprawl, it seems eminently appropriate to create urban centers of a higher density than a four-story maximum allows. In fact, given that building materials and construction techniques are easily able to support much taller buildings and that hundreds of millions of people are living or working at the fifth story or higher already and appreciating their urban benefits, it strikes me as nothing less than perverse for architects and urban theorists to refuse to explore the possibilities and to even try to keep others from doing it. Some of the most beautiful buildings on Earth are taller than four stories. Elevated gardens, art and public spaces on rooftops, terracing that could take rooftops up four stories at a step, bridges and rooftop streets that could make a real adventure of the third dimension — almost none of these are seeing serious experimentation, and yet exploring this world is full of promise. The glimpses into this elevated realm we do have, from every continent, are very promising and inspiring. To refuse to even experiment and to actually attempt to ban such explorations through zoning or imposing intellectual limits is evidence of a stunted architectural imagination, to say the least. In addition, it is just plain damaging to the efforts to solve our urban ecological, economic, and social problems. Cultivating low height limits amounts to cultivating intentional ignorance.

There is a reason that millions of people

in the low consumption world live in concrete-block or concrete and steel buildings five to twenty stories high: the low per-person costs permit decent shelter near jobs and transit, whereas waiting for more luxurious one-to four-story arrangements could sideline millions of people for their entire lives from any but shantytown shelter and grueling commutes on foot or on crowded busses. Sitting in a friend's cozy one-story house in Oakland, California — a very nice house that I helped build, in fact — a couple I recently met from Belo Horizonte, Brazil's third largest city, were almost apologetic in telling me that they lived on the sixteenth floor, explaining, "It's just the way millions of us live in our country. Not many people can afford to live in houses." They considered themselves active environmentalists and much looked forward to their occasional trips out into the countryside. When people like this are put into the smaller buildings scattered over ten times the land area, transit no longer works for them — they need to buy a car, and they can't afford it.

The city is the people.
At a conference in Berlin I once attended, a lecturer showed her audience a picture of a beautiful eastern European square with people in the foreground and repeated a popular thing to say in city design circles: "the city is the people." I reflected on that a while and drew this picture.

People will put a little more effort into crossing larger centers if the centers are diverse in uses and higher in density. Different districts and centers have very different moods and human energy levels. The sturdiness of the people and their particular traditions figure in. Bicycles add to the diameter of centers as do people-moving conveyor belts and escalators. Such mechanical devices as elevators and escalators in the intense city centers function at efficiencies easily justifying their construction as public works. The oldest movies ever, made by the Lumière brothers of Lyons, France, from 1895 through the turn of the century, show, among the many mechanical devices of the day that fascinated the film makers, parallel moving sidewalks: conveyor belts sliding along side by side that increase in speed as the pedestrian steps from the slower to the faster track. Such devices could likely increase the population in city centers by two or three times while decreasing the energy requirement per person. Considerations like these seem abstract now, but as we delve into ecocity studies and experiments, they will come to seem quite practical and common.

Extremely cold and hot, dry climates tend to compact centers, and so do waters surrounding the community. Mild weather and open flat land have the opposite effect. Elevators and bridges between buildings add to the vertical dimension and total population. Banning the car for all but emergency use and planning for diverse uses at close proximity will create a new condition. Then we can begin to refine our knowledge about the "natural" size and shape of a wide range of types of centers.

Complex Buildings and New Synthesis Architecture

The mixed-use buildings traditional in many cities before automobile domination still remain in some places, with shops on ground floors and apartments above. Urban critic James Kunstler calls this, somewhat tongue in cheek, "low-cost housing"[2] as he describes the smells of the pizza shop below lowering rents for the tenants above. In an ecocity future, this kind of building will be typically located in and immediately adjacent to ecocity downtowns and in transit villages and neighborhood centers, but it will have new ecocity features built in, such as multi-story greenhouses and rooftop and terrace public spaces. In the area immediately adjacent to the denser centers, one zone out, predominantly residential "zero lot line" development will be typical, with each building touching the building next door. Row houses, Indian pueblos, ancient Middle Eastern towns, and many two- to four-story Victorian houses from San Francisco to Melbourne exemplify this pattern. Adding more uses to this arrangement of buildings will produce a mixed-use block. It will look fairly dense from the street, but the style can be quite beautiful, and back yards will be peaceful pockets of green and private retreat because the buildings themselves will create, as they do presently in cities like San Francisco, a thick wall separating back yards from streets.

The complexity of the ecocity goes deeper yet. Buildings seem intuitively simpler to us than machines. This has something to do with identity. We animals believe that we are in many ways more complex than plants, and unlike plants we move around. Our houses remind us more of plants. They don't move around. In fact, many of them are made of plants — usually wood, sometimes of bamboo, with thatched roofs, with mud blocks held together with straw or other plant fiber. In the south central mountains of China near the coast there are enormous round clan houses four and five stories high, accommodating 150 to 300 people and made of rammed earth: mud mixed with short, hard twigs and sticky rice, the strongest, tallest earth/plant material walls I know of. Other buildings are made of stone, the very archetype of immobility, or stone-derived materials that don't even grow as plants do. Though doors and windows swing and slide, the whole building doesn't up and go. Neither do most buildings, besides factories, have big parts like electricity-generating turbines that spin at impressive speed, nor machines that stamp out product or conveyor belts that look busy.

This simplicity is changing, however, as houses are increasingly outfitted with all sorts of complex tools, from dishwashers to computers, from timing switches to control sprinkler systems, alarm systems, lights, thermostats, and garage door openers. Some buildings already regulate blinds and even the opacity of windows to

"The city is the people" is not enough. *Our creations have a life of their own; they are living organisms — that we have to be careful to consider.*

conserve energy for heating and for balanced natural lighting. Someday, we are told, we will have intelligent houses that will be programmed to take care of their own cleaning, provisioning, and bill paying, maybe even repair.

When biological systems are let in through the window box, rooftop garden, and solar greenhouse with butterfly bush and hummingbird feeder; when recycling chutes and compost systems are designed to be integral to the building; when the collection, storage, and distribution of rainwater is built-in with gutters, down-spouts, filters, ponds, and cisterns; when buildings become co-op, co-housing, extended multi-generational ethnically-mixed family residences; when the building becomes mixed-use in itself with jobs, shops, education, entertainment, prepared food, groceries, drugs, banks, and then is linked to other buildings with bridges, elevators, and bicycle skyways; when urban orchards, creeks, and prevailing winds come marching, flowing, and blowing through semi-interior atriums and corridors, and overhead canvas garments and spreading rooftop trees unfurl their shading seasonally; and, finally, when the entire city becomes essentially a single structure, tempting poets and mechanics alike to an analogy with living organisms, then we will have complex buildings.

We are talking about a new synthesis of ecology and society in architecture here, not about the synthesis of the individual or the family with nature. The hunter-gatherer's lone lean-to, the isolated farmhouse, even the permaculture farmhouse homestead, is not new-synthesis architecture — it may be eco-personal or eco-family, but not eco-social. Rather, when diverse community activities are gathered together in one physical container and that built community is harmonious with nature, we have the synthesis of the social and the ecological and a special kind of architecture deserving the name. New-synthesis architecture is just beginning to appear in some places with the shift toward "balanced development" and the more mixed-use buildings that are part of that shift. The Hundertwasser Haus in Vienna, with its trees planted on terrace roofs and brought right into some of the rooms and growing out the windows, is an example, a clear symbol of, and an actual living example of, bringing nature into city structure.

Traditional villages and towns, many of which are still with us, focused on compact centers, and sometimes those centers constituted the whole town. This traditional village structure featured buildings close together, often joined together side-by-side, demarking streets and public plazas. Residences over shops next door to pubs and across the street from the post office, and government offices and church a block or two from school were typical, and so were relatively high buildings. In Turkey, Italy, and Nepal I've seen villages only three blocks long with buildings five and six stories high arranged in this fashion. In many parts of the world, for reasons of conserving heat and village land, few buildings are one-story except in the smallest, poorest of hamlets. The Southern California contractor

When diverse community activities are gathered together in one physical container and that built community is harmonious with nature, we have the synthesis of the social and the ecological and a special kind of architecture deserving the name.

Michael Hoag has recently been campaigning to create traditional pedestrian villages, copied directly from northern Italian rural towns, as an alternative to urban sprawl. He is worried that the Great Central Valley of California will end up as one enormous Los Angeles. His idea, supplemented with ecocity structure and technologies, could result in strings of rail-connected towns and small cities instead.

Oddly, though, the new ecovillages, for which there is a worldwide movement of support, are not adopting this traditional village pattern. The Findhorn Ecovillage in Scotland, the site of the first international conference on ecovillages in 1995, is a scattering of trailers and one-story buildings, with only a handful of two-story houses with solar greenhouses and famously productive gardens. There are two major community buildings, one of them a beautiful meeting hall about three stories high. The other thirty-some ecovillages represented at the conference were made up mainly of scattered individual buildings with little sense of arrangement around streets and plazas and no zero-lot-line buildings at all. Their most common ecological features amounted to solar design and equipment, windmills, their own style of organic food gardens, and beautiful crafts work in local and recycled materials. Four miles away from Findhorn is Forres, a traditional old small Scottish town with solid, dignified three-story buildings of many uses around narrow streets and cozy wide spots serving as small plazas, a good example of the traditional village

form. Our host community had real ecological conscience and consciousness, but its planners were taken aback by my suggestion that they might consider adapting the traditional village structure of the communities all around them. They had never thought of it quite that way. I proposed that those traditional towns and villages exemplified a profound structural response to social cohesion and cooperation and some aspects of ecology, in particular access by proximity and responsiveness to local climate conditions and building materials.

Meanwhile, the dedicated and hardworking people at Findhorn are making impressive progress in restoring the nearly destroyed Caledonian forest of Scotland, but they and the other ecovillage dwellers and promoters are not alone in overlooking the virtues of the traditional village. Berkeley, California, where I lived for 30 years, with its world-class university, has no plaza or town square, no pedestrian street, and no clear policies to stem the growing dominance by cars. A few neighborhood centers have a bit of the feel of a traditional village but none have a small plaza, none approach the vertical ratio of height to width of traditional villages, and all lack relief from cars pushing in from all sides. Santa Fe, with more solar houses than any city I know of, and proud of this accomplishment, has developed a sprawl pattern that would make citizens of Dallas-Fort Worth feel right at home — and this with the model of the pueblos and even its own deep history as a pedestrian city right under its shoes. Our communities all have their histories and patterns of learning — and forgetting. The traditional village structure is one of the most profound inventions in history, and it applies to cities as well as villages. For this reason, the village can save the city, and then the city the world. But not if it's built without an understanding of its full historic and ecological role.

In ecocity transformations there is a direction, scale and form seen here: toward the centers, and smaller and taller. According to

Peter van Dresser, New Mexican solar innovator, one-time rocket scientist, science fiction writer (he coined the term "astronaut"), and author of *A Landscape for Humans*, the time has come for "small-scale recentralization."[3] Van Dresser's "small" is relative and applies at all scales. Ecovillage pioneers could just be the ones to lead the way for cities too, if they adopted the traditional village form. But, city folks, no need to wait.

Four story height limit not!

One of the more destructive influences on imagination and honest confrontation of urbanism in our times of extreme overpopulation is the notion that buildings over four stories degrade humanity and should be banned. A major principle behind ecological cities is that higher living organisms are complex, three-dimensional entities and an apt model for city structure, another one of those principles in nature that should be taken seriously. Here we see the four-story lack of imagination applied to the much-enjoyed Claremont Hotel, shown remodeled earlier in this book. Remove the top and you remove not only the services for that many people, along with the people, but also the love of architectural exuberance and creativity. And here we've even included part of the building that totals five stories.

Water

Cities should use water conservatively and occupy a small area of land, thereby interfering minimally with the intricate patterns of the water cycle. This is yet another reason for the generally compact three-dimensional city-building prescription: do not block the soil's permeability, do not dump storm waters off streets and parking lots at the wrong time, that is, when rain is heavy and water courses are swelling. Berkeley is about 35 percent streets, parking lots, and sidewalks, 25 percent rooftops on mostly one-and two-story buildings, and 40 percent "open space," mostly marginally permeable lawns. The sealed 60 percent and semi-sealed 40 percent, equaling approximately 80 percent (60 plus 1/2 of 40) of sealed surface, sends rain rushing off to scour out the plants we ecocity builders have carefully dug into the restored creek banks. The runoff grinds away most of the insect and fish eggs and flushes them out into the salty San Francisco Bay. Then, in the summertime, most of the creeks dry up because too little of the runoff has soaked into the soil to join the waters of our small, local aquifer, which otherwise would ooze its groundwater into the creeks all year round.

In times of flood, as became obvious in the Mississippi and Missouri flood of 1993, building sprawling cities out over flood plains is a disaster. The larger the land area diked and sandbagged, the more the rivers are pinched and their flood levels raised. Instead, cities and towns can sit as smaller elevated islands in or next to the flood plain, rising above the calamity, letting the flood cover larger acreage with shallower waters, as happens in Curitiba, Brazil, causing little damage and often some benefit through fertilizing stream-side parks.

We should lay down our cities on a much smaller footprint on the soil, catching and releasing water at rates similar to the landscape around those cities. There are places on which to avoid building altogether, and there are places to which we might want to introduce water for reasons both traditional (in fountains, say) and new (in interior waterfall mobiles or sculptural prismatic fish ladders, for example). In all cases, what we do with water is important.

In water, as on land, it is possible to add to biodiversity in the way we build our cities. Oceans just off sand beaches are often far less biologically diverse than along rocky coasts and coral reefs, especially if those rocks and reefs provide tide pools and convoluted, partially protected areas at many levels relative to the tides. A city that is built to become an artificial island or archipelago can act like this to increase biodiversity and the productivity of many species if its wastes are carefully limited and recycled, and if its people take responsibility for building in imitation of tide pool coastlines and brackish estuaries, providing canals, marshes, calm waters, and mixes of fresh and salt water in certain places.

On low-lying coastal lands, aquatic environments can be created by digging canals into the land for saltwater circulation. Interior canals and

The traditional village structure is one of the most profound inventions in history, and it applies to cities as well as villages. For this reason, the village can save the city, and then the city the world.

lakes can be excavated and kept clean by two artificial creeks or rivers with locks that open and close so that the rising tide fills the inland waterways with ocean or bay water through one "creek" or "river" and the ebbing tide draws the inland water out through the other waterway. On a larger scale, these tide-pumped waters could carry boats into and out of the new water system as the tides rise and fall twice a day. In Venice, California, a system of inland canals was built in imitation of Venice, Italy, but it doesn't have the benefit of tidal locks and circulation of water. The result is a colorful but smelly set of residential islands, with dead fish floating and being deposited on scummy backwater shorelines. Nobody swims or fishes there. The building of timed tidal locks on two entrances to such a system could turn a set of sterile canals into a thriving ecosystem serving fish, bird, and animal as well as human populations.

There was a time, before television made entertainment passive and cars scattered people to the boondocks, when cities had fabulous pools for the enjoyment of citizens. With a growing interest in ecology and larger populations concentrated in walkable centers, such pools can be designed back into cities and linked with existing rivers, streams, lakes, and bays. There is no reason we could not, in less than a generation, be swimming with giant sea turtles, sliding down water chutes on the coast with sea otters, joining in the activities of porpoises and dozens of other especially sociable species once we have allowed their populations to rebound.

We can design cities in close association with, and awareness of, the needs of life in the water.

On a more prosaic level, we can catch rainwater for our own uses in cisterns and rooftop and ground-level temporary ponds and lakes to gather water in dry areas and prevent flooding in wetter seasons. Thus we can also store close to home some of the water that is otherwise typically impounded by the flooding of distant valleys behind dams. Such seasonal or periodic bodies of storm water in and around cities, which are described in some detail in Anne Whiston Spirn's book *The Granite Garden*,[4] can create beautiful scenes, reflecting city and sky, mountains and stars in these ephemeral waters. Sprawl provides parking lots that can double as holding ponds in this manner, she says, but with open space replacing parking space come fleeting scenes of paradise already glimpsed in Curitiba. In cities meeting most of their water needs through conservation and capture of local precipitation and by reducing their land areas greatly through ecocity restructuring, local water tables could be recharged and dried up springs brought back to life. Adding this to energy generation by sun would allow the removal of many dams and let rivers run free again.

Natural Features and Biodiversity

The bottom line in evolution is the survival or extinction of species as they go about transforming themselves and their environments over tens of thousands to tens of millions of years. Humanity's confrontation with other species has been a growing disaster since even before cities appeared, but it has accelerated with the advent of urbanism and is now careening out of control. Yet, cities could be places where biodiversity outstrips almost anything found in nature. Already cities are botanical gardens of ornamental and food plants, complete with human pets and pests, imported from around the world intentionally and accidentally, bred and fed. But urban biodiversity can be deceptive, trading higher local diversity for lower worldwide diversity. When an introduced plant or animal drives indigenous ones to extinction, we may not be conscious of the loss. Our environment may appear lush with variety, but all those introduced species live (or lived) somewhere else, too. Their addition in the new place is not an addition to the world, but the extinctions they cause are a subtraction from it.

With a deeper understanding of the ecocity vision, however, there is no reason that biodiversity cannot be preserved at almost today's levels. The naturalist Sterling Bunnell points out that, ironically, cities serve well as places for reintroducing certain endangered species back into the wild. Endangered animals that people have been attempting to protect sometimes find predation, severe weather, starvation, and agricultural insecticides in the country far more dangerous than problems in the cities. Some animals, such as the least tern, a rare colony of which lives between runways on Alameda Naval Air Station in San Francisco Bay, are protected by the oddest of human habits in cities. When the airfield was operating the birds were protected from feral cats by — of all things — jet bombers and fighters taking off and landing. Since the base closed, sympathetic humans take care to fence and patrol the colony for its preservation. Cities can be good halfway houses *en route* to some species reestablishing themselves in nature.

Arcologies, Armatures, and Implantations

Earlier we took a bicycle ride through a "conventional" city transformed into an ecocity. Yet there is another type of ecological city possible, a "new town" built on ground unoccupied by previous urban development. It starts as small as the aspiring ecovillage and could include modified New Urbanist developments and whatever descendants of the Garden City that might come next, as well as the single-building cities proposed by Soleri.

Imagining the ecocity in the arcology model takes the idea of three-dimensionality and organic compactness to a hypothetical zenith. The intensely three-dimensional arcology could be envisioned as a space city on the land, open to its environment instead of closed like the completely sealed Biosphere 2 experiment near Tucson, Arizona. It would be one-thousandth as expensive per person as an orbiting space city and 1,000 times more relevant. We could build several such Earth-bound sci-fi cities for a few tens of thousands of people for the price of a nuclear submarine or two, with far better results for future security. Some

Aztec toy animal with wheels. *We already know the principles for ecological design of our communities — we just don't use them very much, in fact, hardly at all. When Cortez came to Mexico, the Aztecs had wheels on toys but never generalized the principle to utility or broad application. It seems puzzling to us looking back to the sixteenth century but we, too, are ignoring something, namely the principle of "whole systems" or "ecological" design. Now we need to pull all the pieces together in linked healthy parts and apply the results broadly.*

single buildings have more than twenty-five thousand occupants right now, without being anything close to ecocities.

In an arcology, some plazas, large and small, might be more like enormous rooms — maybe the John Portman hotels around America, with their big public interiors, are hints of things to come, though I see no reason not to open up his designs in various places to sun, wind, and weather. Jon Jerde's shopping and entertainment complexes are something like single-use versions of the kinds of spaces we could expect of arcological interiors. In both the reshaped city and the essentially single-structure new town, beams of sunlight will project down and slowly move their spotlights from west to east through the quiet interiors all day, and stars will sparkle through the openings at night. Birds will fly through, and some will nest in these large, semi-open rooms. In rainstorms, wild cataracts

could be collected from rooftops and sent coursing through sluices, falls, and swirling pools on their way to great cisterns on the scale of small reservoirs. The Memorial Publico Building of Curitiba, where the Fourth International Ecocity Conference was held, takes a small artificial creek through its towering glass foyer; the Scandinavian Air System headquarters building outside of Stockholm has a small artificial cataract running down the staircase on one side; the foyer of the Rincon Center, mentioned earlier, has a waterfall/sculpture with rainbows in the swirling water droplets. These could be more grand and subtle, and more numerous and varied in arcologies.

Recyclables will be gathered by gravity via spiral chutes of the sort we saw in the Montgomery Ward Building. Warm air will rise in vertical shafts by its own natural convection from the large solar greenhouses on the sunny side of the city. "Waste" heat from industry down in the basement will be directed up on cold days to the different neighborhoods nestled into a three-dimensional latticework or used to preheat industrial processes.

An arcology will take far less land, less energy to operate, less connecting materials such as pipes and wires, and less surface coverings and paint per person than any other building type. The building materials will be substantial, as enormous strength will be required. The arcology will not, however, need to be as strong per story as a conventional tall building since it will be thoroughly tied together structurally, resisting lateral forces, and "base-isolated" against earthquakes, floating like a rock or a strong ship on softer soils.

The distances to be traversed to lead full economic, social, and civic lives in an arcological city will be extremely short, and preserved nature and agriculture will be just outside the city gates as well as invited into the city itself in small pockets and passages. Life will be maintained here with great "frugality" (one of Soleri's key terms), with great efficiency and enormous cultural richness but little waste — in fact, with almost all waste folding into new resources for some other activity going on in the city or very near by.

From the top of such a city/building the views will — obviously — be breathtaking. Looking over your shoulder as you leave it, you will see the city rising out of the landscape like an ancient castle, perhaps, or a strange immense sculpture or small mountain, but with a skin almost translucent with life. All around, in sharp contrast, there will be nature, agriculture, forest, water, or desert, as appropriate to the city's particular location. Only if we build structures like this will we be able to actually comprehend and assess such intense experiences of compactness on one hand and openness on the other, experiences of remoteness from the land's surface below while nature remains instantly accessible by us simply stepping out the door.

Other new town approaches, building on the Garden City movement and the recent ecovillage movement, could carve out communities where giant industrial farms will be breaking up simultaneously with oil running short. Michael Hoag's "imported" rail-linked Italian villages for California's Central Valley, for example. Grasslands and forests could be rescued from desertification, and in these and some of the spaces already left vacant in many cities around the world, new town-style projects could be created to serve human purposes with little or no damage to natural systems. In such country, a city could be home to a College of Wildlands and Agricultural Restoration engaged in training while experimenting in the field. In industrial rust belts, "brown fields" could be recovered for new ecologically tuned development where integral neighborhoods, in-town ecovillages, and major developments, like the Halifax project proposed for Adelaide by Urban Ecology Australia, could be built.

A proposal by architect Herb Greene gives us an intriguing way of visualizing some larger patterns in a city being transformed into an ecocity. He has proposed that we build "armatures." Greene and a number of other observers of buildings over the ages have noted that new buildings are occasionally built on top of or added to the walls of much older buildings. In fact, it has only been with the excess of energy and wealth prevailing in rich countries today that we have gotten into the habit of expunging the ruins of older buildings before building anew on the same site. Çatalhöyük was built one floor upon the ruins of another for 850 years. A compacted hill six-stories high stood as foundation for the last generation of buildings there. Similarly, solidly built Roman coliseums and aqueducts have ended up as foundations for medieval and recent apartment houses. Greene proposes intentionally building such immensely enduring spinal structures — big decorative walls, really — on which future buildings could be added and existing ones subtracted as the needs of times change.

In Italy, the ancient armatures were made of stone. Today, they could be made that way again, but they could be stronger yet if made of concrete and steel. Holes could be placed in them for expansion-bolt attachments for changing new structures. Keeping such an armature dry and sealed with paint or another sealant would make much of the skeleton of the city virtually eternal. Using near-permanent materials adds great efficiency over long periods of time, minimizing building material requirements deep into the future and saving enormous amounts of money for other of society's investments. Armatures amortize.

Stewart Brand has written a book called *How Buildings Learn*[5] about the transformation of buildings by remodeling over the decades. Greene applies the idea to cities that are transformed over centuries, accommodating the same kind of flexibility on a larger and longer scale of space and time. In addition to their economic efficiencies, his armatures would contribute considerable esthetic interest to ecocities.

"Implantations," architect Richard S. Levine's concept, amount to larger-scale integral neighborhoods, or in-city villages and active neighborhood centers becoming in-city towns or even arcology-like structures. Levine believes that the larger neighborhood-community scale or small district is the smallest scale at which the solutions to ecological, social, and economic problems can be addressed. The most promising sites for implantations are those that have been abandoned because of major economic dislocations, such as the cleared former industrial areas of rust-belt cities. Levine has proposed to the city of Vienna that an implantation over a major railroad be constructed, which would be shaped something like — again — an Italian hill town. The streets and small parks and plazas of such a structure would provide views of the mountains and woods, the Danube, and the ancient cathedrals of the city center. Vienna, in fact, is used to burying transportation infrastructure. Not only does it have a subway, which my disability activist friends say is among the most accessible, it has bridged and buried under a city park two-and-a-half miles of the freeway that runs down the far side of the Danube from the city center. Something I thought was just an ecocity fantasy when I was writing *Ecocity Berkeley* in 1986 is thus built and operating in Vienna.

Each new implantation would amount to a car-free walkable center, a fully functioning community, but one located adjacent to a previously existing center — for the economic benefit of both. The implantation itself would be a kind of instant ecological neighborhood or town center and could be a powerful model of the benefits of this kind of building, influencing the whole larger city and other cities everywhere in the manner of one of Paul Downton's "urban fractals" — a part of the city with all the essential parts assembled together and functioning well on a smaller scale than the whole city. This urban fractal idea describes very well my own "integral neighborhoods" and "ecological demonstration projects", such as the Heart of the City Project proposed for downtown Berkeley that we will look at later.

Arcologies could be seen as implantations in the larger size range. They could also be seen as the ultimate armatures, with not only solid walls but also horizontal floors, something like artificial land stacked in layers, open for light, air, and circulation, and staying in place for generations. Imagine an arcology's vertical framework and suspended platforms, with twenty or thirty feet between ceiling and open buildable floor surface as the basic structure. The circulation system will include elevators, walkways, and bridges at many levels, chutes for recycling, and large tubes, sometimes made of colorful cloth, for air circulation. Utility cores will carry electric and fiber-optic channels, water, "gray water," and sewage or, if dry toilets are used, compost elevators, almost like dumbwaiters. Into these basic "city infrastructure services" residence or business divider walls and two or three stories of inside mezzanines and whole floors could be built, rearranged and remodeled on a time scale of decades or generations, while the whole arcology superstructure will last 2,000 years or more, attaining or surpassing the longevity of Roman concrete still with us today.

Here we have a picture of both permanence and flexibility — a construction project that is expensive up front but, amortized over the lifetime of the whole built community, many times more economical than the conventional city. Today's car city is endlessly demolishing and rebuilding itself while supporting the dominance of automobiles, freeways, and oil technologies that not only waste energy, land, and life profligately, but produce corrosive pollution that attacks and greatly shortens the life of building materials. With arcological armatures, such afflictions would whither away.

Breezy, shady hot
climate ecocity.

CHAPTER 8

Plunge on in!

Four Steps to an Ecology of the Economy: Map, List, Incentives, People

Economics and politics are actually one seamless continuum — the engine of productivity together with the rules of its functioning, including who benefits and who pays. Those rules are called natural laws and human policies. This whole system is made up of plants and animals recycling and rearranging the raw elements of solar and mineral wealth, eating one another and evolving on behalf of each individual, each species, and all of life. As if this weren't complicated enough, human beings work on behalf of families, religions, ethnic groups, professions, cities, nations, alliances of nations, and the United Nations as well as self, species, and the one and all of life in the universe. We seem to be so diverse in our deeper selves that each of us works for a different mix of those constituencies, some-times forgetting one or more altogether. Yet we are always building that edifice that supports us all, our civilization, which is made up of all our cities and physical systems functioning according to rules we made up ourselves, based in turn on the rules of nature. Given the order, with the human edifice built upon the natural one, it is clear that if our rules differ markedly from nature's, we are likely to run into problems.

Therefore some basics prevail. As Thomas Berry says, nature's economics are primary, humanity's economics derivative. Chérie Hoyle of Urban Ecology Australia puts it tersely, "No ecology, no economy. No planet, no profit." According to Hazel Henderson, the pioneering economist and futurist, the economy can be graphically represented by what she calls her "cake chart," a take-off on the pie charts economists use tirelessly to

211

express percentages of this and that.[1] The top layer of the cake is the "private" sector: production, employment, consumption, investment, savings. The next layer is the "public" sector: infrastructure, schools, municipal government, and various services. The third layer down is the underground economy including tax dodges, black market exchanges, and the like. Beneath these three "monetized" layers, in which cash is used as a means of valuation and exchange, is the non-monetized layer, based on bartering, home-based production, subsistence farming, "sweat equity," and what she calls the "love economy" of volunteerism: working to support the family and friends with vegetables, cleaning, baby sitting, medical advice, and so on. This base layer of the human economy rests, in turn, on the bottom layer of the cake, nature's economy: the natural "resource base," which not only ultimately provides everything basic to the human need for sustenance but also serves at no cost to clean up our messes if we don't get too far out of hand.

Since economics deals in the proportional valuation of resources, goods, and services for the purpose of exchange, use, savings, and investment, it is inevitably based on numbers. The most basic of numbers essential to the economy are the following: Humans "appropriate 40 percent of the planet's organic matter produced by green plants," says Erward O. Wilson.[2] And if the oceans are included, humanity takes approximately 25 percent of *all* solar energy-powered biomass production. Only three percent of all mammals, birds, reptiles, and amphibians on Earth's surface (in terms of biomass) are wild; the other 97 percent are here solely to serve us. In biomass, we humans *are* about a hundred times the size of the runner-up species in our size range on Earth in all its history. These are the basics about the foundation layer of the economic cake.

Jared Diamond in *Collapse: How Societies Choose to Fail or Succeed* as well as in his earlier book *Guns, Germs, and Steel*[3] makes a convincing case for what he denies is environmental determinism, that is, the overriding power of the economy of nature — nature's economics being "primary," society's being "derivative," to use Thomas Berry's terms. In addition to environmental influences in the rise and fall of societies, Diamond throws in human cultural factors in explaining those culturally all-encompassing ups and downs. But let's face it, the flows of energy from sun to soil, food, and wood to rain, and the temperature of the air have repeatedly sunk culture after culture with blind spots to what should be the obvious massive flows of energy and material that constitute our main environmental conditions. In our economics in the early 21st century, the 800-pound gorilla is the car/sprawl/freeway/cheap-energy city. This is the real economy drawing down not only most of the biosphere's solar energy for exclusive human benefit but also the fossil fuel savings account of the whole planet in an evolutionary blink of the eye. And almost no one but you and I know it! Not even E.O. Wilson who I talked to about it personally, drawing only a blank expression. I feel like some incredulous citizen of Easter Island in the history so well related in *Collapse* screaming, "You idiots! How can you be cutting down the last trees on our island to build those bizarre giant statues? Don't you know we need trees for houses, boats, and fire?" Today those bizarre giant statues are sprawl cities. That's the real economy.

Let's turn now to how things are distributed within our species: about 20 percent of us have about 80 percent of the resources at our disposal and the other 80 percent have about 20 percent. Actually, there are wildly varying estimates, but you get the idea. Things are a bit out of balance among us humans as well as between us and the other species. If we believe in democracy and justice, then we need a frontal assault on cold, greedy, heartless values. And we have to stop taking refuge in staying culturally lazy and comfortable with the *status quo*, like the Easter Islanders sticking with their wise cultural traditions. Like theirs, but more so, our sacred *status quo*, our "unnegotiable life style," to quote George W. Bush, is hurtling along and anything but static. What is, and is accepted, is not necessarily what's going to save us, or, as Einstein said, you can't solve the problem with the same thinking that created

the problem. Though ecocity building can't solve the problems of greed and lazy self-congratulatory acceptance of our culture as the great one all cultures have thought themselves to be, it can help by giving us a means to run the whole society on a small fraction of today's demand for land, materials, energy, and other living creatures. Perhaps, then, being frugal and considerate in its very conception, structure, and functioning, it can help create a climate of frugality and respect for life processes.

The clearest expression of a strategy to build ecocities and the kind of society they would make possible and at the same time one of the best descriptions of how such an economics would work is contained in what I call "Four Steps to an Ecology of the Economy."

First, the map. We need to determine what goes where. We need to draw up an ecocity zoning map — a map of the city's anatomy, its land use and infrastructure. We need to plot the areas to be developed for density and diversity, the future walkable centers, based mostly on existing lively centers. We need to identify the areas in which to restore nature and agriculture, the zones farthest from those centers and therefore the most dependent upon cars. We can undertake this mapping exercise knowing full well that it will need revision and refinement as we consider the many variables of any town. As Jaime Lerner, three-term mayor of Curitiba, Brazil,

Though ecocity building can't solve the problems of greed and lazy self-congratulatory acceptance of our culture as the great one all cultures have thought themselves to be, it can help by giving us a means to run the whole society on a small fraction of today's demand for land, materials, energy, and other living creatures.

counsels us, we can't wait until we have all the answers, because if we do that, we will wait forever. Waiting for perfection or certainty is an excuse for inaction. We have to plunge in and expect to make adjustments along the way. Once we have a good map, we will have a much clearer idea of all the details that follow. This map is the indispensable first step in "the builder's sequence." True, we can say to ourselves, "Reinforce the centers with higher-density pedestrian-balanced development and withdraw from automobile dependent areas and important once-natural features." But it's much clearer to have it also on paper.

Second, the list. We need to develop a list of technologies — those required for constructing mixed-use buildings and solar technologies; for producing bicycles, streetcars, and rail stations; for building greenhouses and rooftop and organic gardens, and so on — businesses, and jobs that, based on the ecocity zoning map, will contribute to a vital economy. Not taking this crucial step helps explain why we have been winning so many battles and still losing the war.

With the map and the list of technologies, businesses, and jobs, a vision and ways to build the ecocity are coming into focus. The map and list give us some clarity about what we need to educate ourselves for if we are going to have a healthy future. They create a context and a means for evolution toward a more creative, compassionate relationship between society and nature. They lay out where everything fits in the physical communities, what products, services, technologies, businesses, and jobs are required, and where they best fit.

True, there are many things that don't relate directly to building. A second category on our list, "Part B," we might say, contains all those technologies, businesses, and jobs that create products and services that are not particularly related to city structure but that are relatively healthy in their own right — recyclable or biodegradable, energy conserving, non-toxic. These serve us in providing healthy food, clothing, medicine, information products, and services.

Third, the incentives. We need to rewrite the "incentives package." The present car/sprawl/freeway/oil system is viable only because a long list of incentives, including enormous subsidies, supports it. It is regulated into existence. We need a new set of incentives to make it profitable to build a society at peace with nature. Developing the laws and policies, ordinances, codes, regulations, taxes, fines, grants, contracts, loans, and leases to support the community defined by the ecocity maps and animated by the businesses that build and maintain the ecocity civilization will make it so. Without the proper incentives it can't happen. A whole culture of support needs to be created here and expressed in such incentives. An ecocity civilization ultimately needs the imagination and support of people everywhere creating the incentives that make it possible to switch from one list of technologies and jobs to the other.

Fourth, the people. We need to *gather the people*. They are everywhere! In Berkeley, for example, there are hundreds of students and retired people who would love to live in reasonably priced housing downtown. The location is ecologically appropriate according to our ecocity zoning map. All sorts of functions and services are there: jobs, food, arts, entertainment, the university campus, good transit to the whole San Francisco Bay Area. Downtown Berkeley is a perfect place for car-free housing. Developers should build it and go out and recruit these key people, who are out there and ready and willing to sign car-free leases. A local developer named Avi Nevo recently offered to build a new, car-free apartment house a half-block from the downtown rapid transit station and do just that: recruit tenants who would do very little environmental damage while bringing new customers — themselves — to the downtown. His tenants would sign a lease agreement not to own a car and Nevo would make his building's operation legally contingent upon that agreement. He sought a use permit from the city to that effect.

To my amazement, the "progressives" on the Berkeley City Council voted the project down, saying they wanted the parking. The real reason, since they all claimed to be ardent environmentalists, is more likely the reflexive opposition several of them have toward developers who, from time to time, they feel obligated to cast as greedy exploiters of their constituents, thus maintaining clear party-line fights. The height of the building was cut down, too, making room for fewer people near transit and eliminating a rooftop garden that we in Ecocity Builders had been promoting and Nevo wanted to build. Claiming to support the citizens while voting against their housing and claiming to support transit, energy conservation, and CO_2 reduction in

Below: Ecological town in Northern California-like environment.

the atmosphere while voting against placing housing near transit gets more obviously inconsistent by the day. Hell, why not say it? It gets downright hypocritical.

Chambers of Commerce in most cities vigorously promote their towns, looking for conferences, tourists, companies to relocate, and so on. Cities with the ambition to lead society into a prosperous green future will need to adopt some of these boosterist techniques to seek out and round up talented people who are ready to relocate and carve out conscientious careers. People are needed for the technologies, businesses, and jobs on the list, and if there are incentives and if the people are invited, they will come.

I grew up in Santa Fe, New Mexico, and across the Rio Grande valley was Los Alamos, the "Atomic City" constructed to save the world from the Nazis. There the atomic bomb was designed and the first several constructed, including the two used against Japan. Whatever we may think of the use of the atomic bomb or its influence upon human events, the city itself had an *esprit de corps* based on its mission to defend the country and the "Free World." Today we could similarly initiate larger-scale community building projects with a mission — for example converting military bases to ecological towns. We could people these transforming communities with citizens who want to lend their talents and dedication to building a better world, to achieving peace between people

Cities with the ambition to lead society into a prosperous green future will need to adopt some of these boosterist techniques to seek out and round up talented people who are ready to relocate and carve out conscientious careers.

and the rest of the biosphere. I know that such people exist, because every year a few dozen find Ecocity Builders, coming to our organization from all over the world. They ask me if I can steer them toward exactly that kind of experience. They want to build their education and careers around ecological city building, and many of them want to live it. I am constantly frustrated by having to tell them there is no place where such work is being done in anything close to its wholeness. If an economy were being built around the four steps, there would be many such places.

The Four Steps Exemplified: Ideas for Ithaca

Joan Bokaer is not only building EcoVillage in Ithaca, New York, one and a half bicycling miles from downtown, but proposing to transform the city of Ithaca itself. She has an economic strategy that, first, assumes the reinforcing of centers and the withdrawal from sprawl pictured on the ecocity zoning map (Step 1). Next, her strategy defines the particular kinds of work needed for such a city (Step 2) and resolutely confronts the need for incentives to make the first two steps profitable to investors and the whole community (Step 3). Then she lays out an investment strategy based on what she calls the Green Fund. Watching it unfold, we are reminded of Jane Jacobs' engine of prosperity (the city) finding its optimal relationship with its hinterlands and thus becoming capable of assuming its unique place in the larger world economy. Gathering the people (Step 4) is where Bokaer started — by inspiring and helping to organize the residents of 90 households and their predecessors at EcoVillage at Ithaca.

Bokaer points to the cycles of money circulating in an economy, being reinvested, and accruing to owners of businesses while workers are being replaced by new technology. A relatively small number of owners benefit disproportionally while a larger number of workers set out to find new work, retrain, attempt to start their own businesses, or go on unemployment or welfare. Why not encourage them to get ready to build the ecocity? Why not actually help them by investing in businesses that contribute to that effort? She suggests that the city could be an active investor and thus join those who generally benefit the most in any economy, the owners:

I propose that each city create a Green Fund in which the city itself invests in numerous local entrepreneurs and holds a percentage of stocks in those investments, and its citizens become partial beneficiaries of the enterprises that succeed. The investments will stimulate economic growth by supporting local entrepreneurs, and it will foster a diversified economy by continuously investing in more of its citizens. The returns on the investments should be used to promote an ecological rebuilding program.[4]

"Density shift" – from sprawl to vitally three-dimensional. *Old style suburb of a small town is disappearing while the town center is growing to become the whole town, in three drawings.*

A city government embarking on such a strategy could count on itself for the zoning changes and ordinances beneficial both to those it invests in and to itself — that is, to the citizens of the city.

The city government could start small, with a modest-sized government like in Ithaca, setting aside, say, $2 million the first year. It could invest in a hundred of its businesses, choosing to fund the ones that are on the list (Step 2). Some cities offer loans and technical assistance to, for example, recycling businesses. Says Bokaer, instead of loans these funds should be investment stakes and should be extended far beyond recycling to support a much wider range of businesses for ecocity rebuilding — with top priority assigned to businesses that will flourish with changes in the land use and infrastructure made in the direction of walkable centers and restored natural and agricultural areas:

> The city, working in cooperation with the surrounding towns and the county (which should eventually become one government, city and bioregion together) need to draw up an Urban Growth Boundary (UGB) to define its edge .... The outside of the UGB should be zoned for a variety of forms of agriculture including agricultural ecovillages, and for the protection or restoration of natural areas. Future development would take place within the UGB along transit lines, filling in much of the area presently used for cars. Since the surface space of about one-third to two-thirds of most US cities is devoted to the automobile, that leaves a lot of room to fill in. As the population grows along public transit lines, there will be more people paying taxes and shopping in the city — without requiring auto commutes and parking — thereby stimulating the local economy without creating traffic problems.[5]

Bokaer makes the point that house-moving and construction materials recycling businesses should be among those most favored with Green Fund investments. The Green Fund itself could be seeded by a mere half a percent to two percent allocation of the city's general fund every year, supplemented each year by profits from its investments. It would be a growing revolving fund the profits of which would pay for its administration and for helping more businesses until the city was thoroughly transformed. Increases in tax revenues due to the rebuilding could be spent on furthering the ecocity transformation process, "turning life in the centers and along the transit corridors into a true paradise,"[6] as the city invests in beautiful downtown pools, fruit trees in the streets, ecological restoration and arts programs, and so on.

Consistent with the Green Fund, she also proposes a voluntary membership ecological rebuilding program administered by the city that directly addresses withdrawal from sprawl. The citizen freely chooses to join that program by "living within the urban growth boundary or, for those outside the boundary, participating in agriculture and giving up the privately owned automobile."[7] Pasadena once examined what might be called its "automobile balance of trade" and found it was losing $6 million a year in costs to the city that were not recouped in parking fees, fines, state gasoline taxes transferred to the city, and other car-generated revenues. Since members of the ecological rebuilding program would be saving the city money by not owning cars, they would qualify for free transit passes, coupons for taxi rides, no-charge swims at the pools, subsidized medical coverage — whatever the city decided was appropriate for the citizen who saves them so much money — in addition to enjoying the benefits available to everybody, such as tree-lined bicycle paths away from automobile traffic along the newly opened watercourses.

This ecocity restructuring strategy could be implemented anywhere. Rust belt cities need to reinvigorate their inner neighborhoods, and many of them already have derelict open space that could be the beginning of great new walkable centers and restored agriculture and nature corridors as well. Depressed lumber towns could use more diversified economies based on considerably reduced logging and increased other uses of land, from restoration and fishing to herb and mushroom cultivation, from tourism and small college campuses to hightech research think-tank as well as the usual practical services to any community: food, clothing, hardware, repair, banking. Macho loggers recast as carpenters would feel more at home taking dangerous risks in building the new taller buildings of the ecological country town than retraining for desk jobs. I know the style and feel of those two kinds of gratifying physical and practical work, logging and carpentry, and I know that they are similar because I have worked at both.

Economically "successful" suburbs cranking out dollars but suffocating in asphalt could find their centers and begin the transition. As we have seen, it was an incentives package, complete with GI loans for new houses in the suburbs, tax dollars for freeways to get there, and student loans to learn how to build and run cities like that, that helped create sprawl. If we add the "ecocity insight" to the impulse to have both culture and nature, which helped fuel sprawl, and if we adopt an economic strategy like Joan Bokaer's, we might just get what we plan — just as we did when we planned sprawl and freeways.

Ecocities and Rethinking Economics

Are we in a "post-industrial world," or does the notion simply indicate a blind spot of the office to the world? The United States, northwestern Europe, Japan, and a few city-regions in the developing world are becoming office to the world. From inside these places it may appear that we are in a post-industrial world, in the information age. That world is an information world, as are all business administration worlds, but it is not the whole world. "Post industrial"? Never before has the planet been more industrialized, more ravenous in its consumption of energy, resources, and low-cost labor. Never have we taken more from the soil, waters, and atmosphere. In the office to the world we are isolated in our cities and suburbs from actual remnants of nature and vast tracts of poverty and resource exploitation around the Earth. We are cut off from firsthand experience by physical distance and that weird one-way "communications" filter called "television." Frugality, as Soleri advocates, has to be designed into economic systems with the honesty and imagination to do, as Buckminister Fuller advised, more with less — the specialty of ecocities.

"Shrink for Prosperity" might be a slogan to help establish a better-founded set of economic premises. This notion has several dimensions. On the face of it, it looks contradictory since we are so used to growth appearing to be the very basis of prosperity. Our economists tell us this over and over. But they fail to discuss the fact that the planet is a finite environment and neglect the reality of the exploited peoples in the world who frequently see no real gains from economic growth and often suffer terrible losses. Are the rich getting richer and the poor getting poorer, or not? History is the record written mainly by the victors. Winston Churchill said to his colleagues that "History will be kind to us, gentlemen, for I plan to write it,"[8] and after making a fair share of it, he did "write it" in his study of World War II. Economics is the record written mainly by those with the money, by those directing the exploitation. Both records, historic and economic, tend to bolster the interests of their authors and their patrons and the illusion that the world can be an office oblivious of the effects of the productivity it manages. The notion that we live in a post-industrial world is comforting because guilt for the over-exploitation of industrialism is assuaged. It's as if everyone subscribing to it is saying, "That's just the nature of it. That's just what the reality is." But it is a false construct based on denial and is having dire results. By any normal understanding of it, growth simply can't be indefinitely sustainable on a finite planet. Long-term prosperity will require very

"Post industrial"? Never before has the planet been more industrialized, more ravenous in its consumption of energy, resources, and low-cost labor. Never have we taken more from the soil, waters, and atmosphere.

judicious "shrink." We will need to "pow-erdown," as Richard Heinberg puts it,[9] one of the authors of the current flurry of books predicting major economic disloca-tions after oil production peaks and begins its inexorable, irreversible decline.

Then there are people who are doing well enough who would rather buy some-thing hand-made even if a little more expensive, and shop at the corner store rather than Wal-Mart because they like the owners and their contribution to the com-munity or because they want to see their money circulate in the community rather than go to Wal-Mart owners out of town. They want to free themselves of reading time-consuming email and junk mail and sorting coupons because they'd rather relax in the garden than work penny-pinching to save $3.56. Some people adopting these essentially bioregional practices regard themselves to be part of the "simple living" movement.

In reality, there is nothing "simple" about living in a close-knit community or close to nature. Pursuing organic agricul-ture or permaculture? Very complex worlds are those. But if complex, ecological sys-tems can be fairly easily comprehensible, they are not that hard to understand once a little systems thinking is applied. Take any old bird, for example — it's incredibly complex down to the DNA deep in its microscopic cells, far more complex than the structure of a city. And yet, we can eas-ily grasp the basic function of its body parts and behavior: beak, wings, feet, eyes; feeding, nest building, mating, egg incu-bating, and so on. Similarly, understanding the ecocity makes it possible to greatly reduce the physical impacts of complex cultural structures, technologies, and activ-ities. The ecocity facilitates many forms of complexity, providing interconnections and high levels of efficiency honoring their foundation in nature. The ecocity is in itself a miniaturization/complexification of the city in keeping with the "miniplexing" dynamics of evolution, but an urban trans-formation that makes things much more clearly understood and comprehensible as well as healthy.

The whole city, shrinking from the sprawled giants of today with their contra-dictory internal functions, becoming complex, integrally tuned three-dimen-sional structures, should produce complexities linked to one another so effi-ciently as to produce enormous prosperity relative to resources consumed. We may discover that the kind of prosperity that enriches life the most is a prosperity of opportunity for untold enjoyment of time, creativity, and nature.

Buy and boycott lists are crucial. Why so many people assume the giant corporations have all the power mystifies me. The attitude gives up our democratic powers without a

We can disem-power the giant corporations immediately, just by not buying their products.

If energy were expensive we would assuredly "discover" genuine efficiency on the larger scale. To help rebuild the city, then, we should consciously, as part of the strategy, increase energy prices by taxing fossil fuels while we still have the time, shifting resources toward renewable energy and the building of ecocities.

fight or even a conscious whimper. We can disempower the giant corporations immediately just by not buying their products. We can use boycott and buy lists that reflect our values relatively easily with great effect, and there are organizations that provide such lists and ratings. So far there are no real lists available for supporting ecocity building with our dollars. But starting your own list is not that hard:

- No shopping at any place that has a gigantic parking lot or is served by freeway off-ramps.
- No shopping on-line for what you could buy in a walkable neighborhood.
- No more new cars, and as soon as possible no old ones either.
- No buying into gated, suburban communities with triple garages, and so on.
- Yes to buying so that your money circulates locally as much as possible.

From many different angles, thinking through our four steps is a start for buying like a conscious and conscientious ecocity/bioregional citizen would.

Ethical investing is another approach. The investor might begin asking questions that lead to supporting companies that should be on the positive list. No investing in car/sprawl/freeway/oil companies; no investing in companies with headquarters in car-dependent suburban office parks. Take this seriously! Divest now! "Buy and avoid" lists could be developed for municipal bonds. A city with a General Plan that was zoning-friendly to ecocity development and restoration would rate high and one unfriendly would rate low.

Cheap energy has been regarded as an economic boon. But it's a problem. We are accustomed to thinking the cheaper the energy the better. But it doesn't work like that. Oil has been and still is amazingly powerful relative to its cost. But from its origins at extraction sites in the Ogoni territory in Nigeria to the U'wa lands of Columbia and Venezuela, where the damage to natural lands and native peoples is catastrophic, to the sprawl built with its help, to the reservoir of its CO_2 wastes in the atmosphere to accidentally dispersed oil in the oceans, from the smog caused by American attack aircraft taking off in Saudi Arabia and Iraq to the actual devastation of the bombs ripping buildings and people into clouds of air pollution where deployed — everywhere cheap energy comes from and everywhere it goes, it causes damage.

Cheap energy means we don't have to think through ecological design and building, we don't have to think through organic agriculture or the virtues of many very successful traditional low-energy ways of life. We can ignore the virtues of "access by

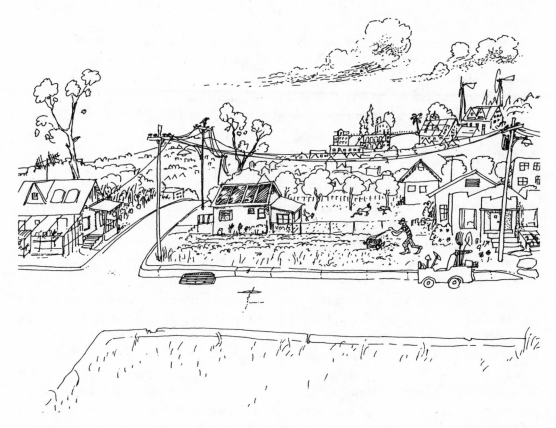

Fifteen years
or so later.

proximity" and simply ship in our "solutions" anywhere, paying as little as possible, thinking as little as possible. We've substituted savings in the cost of BTUs (British Thermal Units, a standard unit of energy) for clear thinking. Solar energy, thoughtful architectural design, carefully placed insulation, lifestyle adjustments to respect the seasonal and daily cycles, all look expensive compared with simply blasting in some more heat or air conditioning when fuel is cheap. If energy were expensive we would assuredly "discover" genuine efficiency on the larger scale. To help rebuild the city, then, we should consciously, as part of the strategy, increase energy prices by taxing fossil fuels while we still have the time, shifting resources toward renewable energy and the building of ecocities.

We should also be conscious of the simple weight of shipped goods and the basic physics that says it takes energy to bridge long distances. Frequently used and relatively heavy materials should come from relatively nearby sources. Using lumber in the United States coming from South East Asia and bottles of wine from Chile, for example, other things being even close to even, does much more damage ultimately than buying these products from close to home. Shipping lighter weight items from great distances and things purchased very infrequently and for special culturally enriching purposes is generally less of a problem.

Another thing we need to be aware of is this: we have to get over the voting threshold to ecocity building. In 1979, a solar-based town with many of the ecocity features described in this book was proposed for an old military base north of San Francisco. There was a clear strategy behind it, but politics defined the threshold for decision, commitment, investment, designing, planning, and building. What promoters of Marin Solar Village envisioned for the decommissioned Hamilton Air Force Base would have been both a community of solar homes and a manufacturing and solar services town, a whole mixed-use community large enough to justify reestablishing an abandoned rail link that ran through the area. The Air Force sold the base to the county for a dollar but the project lost in a countywide vote 49 to 51 percent. Jerry Brown was governor then and trying out all sorts of new ideas, and friends of ecocity work were in high places, but what should have been an enormously influential project at a crucial time just dissolved into oblivion. The crux of the whole thing was that northern Marin County thought of southern Marin County as snobby commuters to urban San Francisco who didn't appreciate the rural character of northern Marin, and southern Marin County people, who liked the Marin Solar Village idea, didn't reach out to the northerners because they normally just didn't have much to do with them. The whole thing was a pathetic morass of miscommunication on a different subject than the content of the proposal. Getting over these cultural blind spots to see the environmental and resources content of the proposal would have been crossing the threshold into new worlds.

Remember, only one out of ten people on this planet owns a car. That's something to be taken seriously in a world supposedly moving toward the goal of democracy and each person counting. It is in the great self-interest of the very large majority to adopt the ecocity economic strategy and get the appropriate land-use pattern established literally underneath the products and services of the green economy. On behalf of ecocities, many such thresholds of the sort that blocked Marin

Solar Village have to be crossed. The capitalist dream that we can all get rich contaminates reasonable voting for the public good, as millions of voters vote for privilege, hoping that one day they will be able to take advantage of it. On the lower rungs of the economic ladder, people who can't afford to own cars vote for support for cars, never even thinking about the possibility of ecological land-use changes. We need to begin voting for what's best for the great majority, realizing that we are part of it. When that majority includes the other species and the people of the deep future — the true Great Majority —

About forty years later.
These changes can be funded largely by "transfer of development rights," by which developers are encouraged to build larger ecological buildings in pedestrian/transit centers and required to help pay for purchase and removal of buildings — but only when there are willing sellers. This is a gradual approach, but if written into local ordinances, can cumulatively shift density and diversity to centers and create ecocities.

and we vote on its behalf, we will have a political/economic solution of profoundly creative and healthy power.

Third parties in the United States take votes away from one major party or the other. As we know so well after the 2000 US national elections, the third-party role is too often the role of the spoiler. The two major parties are considered the only viable ones, and they effectively steal the ideas generated by third parties when they appear to be on their way to becoming popular. This is a great way of taking away from creative people the opportunity of applying and refining their ideas in the crucible of practice and preventing them from having a voice in decision making. Thus the big parties are the real spoilers. The third parties seldom elect anyone and tend to fade out as their best and brightest ideas are stolen and the best and brightest people turn cynical.

An ecocity political strategy with real potential would seem to be closer to that of the German Greens of the 1980s, rethought in the ecocity context. "We're neither right nor left, we're ahead," they said. The strategic alliance with this way of thinking is between the environmentalist and the developer. Traditionally seen as archenemies, if they could get together, they could solve an enormous range of problems. The objective is to build a new infrastructure for humans, not cars, for health, not just for whatever we happen to be able to do on a whim with some technical ability backed by a few dollars. The present system promises profits to those who build and big profits to those who build big. What we need to do, therefore, is design a political/economic approach that rewards builders for building the right thing.

Environmentalists should understand what needs to be built and support the builders of those projects with public education and political backing, and the builders should support the environmentalists by helping pay for their work with the profits from the development that the environmentalists help make politically possible. Specifically, environmentalists and developers should work together to pass General Plans and zoning codes that make it possible for developers to make money building ecological features into their buildings and opening up landscapes for restoration of natural features like creeks and for community gardens, city parks, and the like. If the balance of wealth shifts too far toward the developer, those who see this as an inequity should pass higher taxes and use these moneys for public services — and to help restore nature and rebuild the city.

An ecocity environmentalist/developer alliance would have an interesting effect on the spoiler status of a green third party.

With voter support coming from both the traditional left/environmentalist and the right/business community, it would be far less clear which of the two major parties would have its ambitions spoiled by the candidate supported by those promoting ecocity policies. Third-party voters could vote their consciences without fear that doing so would shift the balance much between the two larger parties. Whether it would most hasten an ecological restructuring of society to build a third party largely around an ecological rebuilding program or to influence society enough that the major parties would champion ecocity policies doesn't really matter; we just need to get on with it. At the moment, though they are sometimes good at knowing what *not* to build, environmentalists are not much wiser than many developers when it comes to knowing what *to* build. The Green Party in Berkeley is promoting low-density development with all the enthusiasm of a loyal citizen of Atlanta or Phoenix. As it becomes ever more conspicuous just how destructive the car/sprawl/freeway/oil syndrome is and how sensible the ecocity alternative, perhaps everyone will be more willing to acknowledge the power and benefit of environmentalists and developers working together.

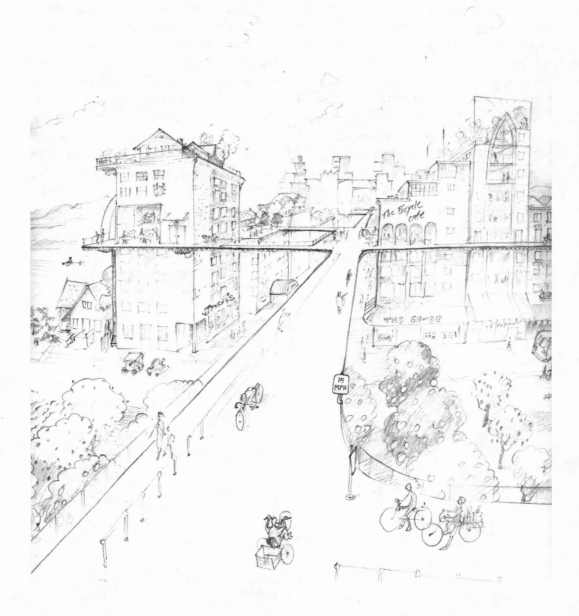

CHAPTER 9

Personal Odyssey

As far back as I can remember, I've had an urge to do something about the problems I see, to work with other people and to organize. Meanwhile I have also been an artist — sculpting and drawing — trying to create a more personal expression of what makes sense to me. It has been a difficult juggling act, but it has helped me discover and sometimes invent a few things regarding ecological city design and planning.

My father, Phil Register, is an architect and my mother, Jean, an enthusiastic supporter of the arts, sciences, and the garden. Many of my earliest memories are of either tagging along with my father as he surveyed the bunch grass, cholla, piñon, and juniper-dappled New Mexico sites of future buildings or digging in the garden, watching the bugs and birds, or drawing pictures of houses, plants, and animals. As a child I loved the countryside, the strongly punctuated seasons of the region around Santa Fe, the sunsets. I still see distant thunderclouds hurling purple lightning bolts ten or twenty miles across mountain ranges, across the face of the low orange sun, or between me and the crescent moon. I can't forget the lightning's metallic ozone smell on the breeze, the smell of rain on dust before it arrives. I remember the clean open sky, so rarefied and clear that the setting stars would hit the distant horizon and suddenly and silently disappear — blink. That almost surrealistic clarity and lack of pollution on a geophysical, even celestial, scale is gone now. Stars fade into the distant haze. They don't blink out. But perhaps we can bring the experience back in a few decades. If Jaime Lerner can improve the sunsets in Curitiba, maybe the rest of us can do the same elsewhere.

229

The year I was twenty-one was a big year for me in many ways. I left college to do my art work, moved from New Mexico to Venice, California, and started an organization called No War Toys, which was part of the peace movement and focused on the attitudes promoted by toys and entertainment. That was the year I began making sculpture for the sense of touch and the year I met the person — Paolo Soleri — who launched me into ecocity thinking. As chance would have it, the mother of a friend offered me a ride across Arizona and asked if I would mind if we visited a famous architect on our way. Of course I did not. It was her car, after all — and, as it turned out, my fascinating, if ever-after difficult, fortune. When we arrived in an area of Scottsdale, Arizona, called Paradise Valley, I didn't grasp what Soleri was up to right away. It was maybe 105 degrees, just after sunset, and Soleri, lean, wiry, and browner than the Arizona dirt, was scampering over a large mound of earth, cutting it with patterns in fine silt, using a trowel, working late to finish the job. Concrete would be poured over the mound tomorrow, I was told, and the shape, patterns, and colors of the fine clay would be transferred to the inside surface of the concrete. The concrete would set and the mound of earth under it would be dug away, and thus a half-dome shell would be created for the roof of a new building at his Cosanti Foundation.

That first evening Soleri talked with my driver about his search for land to build "the new city"; she wanted to help. New city? What new city? "Arcology," he called it, a combination of architecture and ecology. It would be the city to demonstrate his new ideas about ecologically fit architecture and whole-city design. I wasn't clear what that was, but I did pick up some of his papers and took them back to Venice.

They were stunning. It was hard work deciphering the language he was inventing as he went about exploring his idea of arcology and its place in the evolving universe. But I did make the effort and when I'd sorted out the terminology, the clarity of thinking and its implications for change struck me as one of those keys to the secrets of the universe — a revelation of how things work and what can be. Though I was swamped with work for No War Toys and busy making my tactile sculptures, I wanted to do what I could to help Soleri and better understand the issues he was working with. On the simplest level, he was talking about a single-structure city, like an Indian pueblo, but completely rethought and updated — a city that would occupy a small fraction of the land consumed by a conventional city or town, rising much higher, linked together physically as a real unity, conserving energy and natural and agricultural landscapes, and providing a container extremely good at firing the human imagination and making it real. On a deeper level, arcology promised nothing less than an instrument for the healthy evolution of life and consciousness deep into the future.

Over the next five years I kept in touch and wrote three or four feature articles on Soleri and arcology for the underground papers in Los Angeles and one for the *Los Angeles Times' West Magazine*. Meanwhile Soleri found and bought the land he had been looking for when I met him. Five years and one month after that meeting he started construction at Arcosanti, the first arcology of his Cosanti Foundation. A note on the name: Cosanti comes from Italian for *cosa*, meaning "thing," and *anti* meaning "against" or "before." So the Cosanti Foundation meant, for Paolo, the foundation for that which comes before things — or, half jokingly, the anti-materialism foundation, a non-profit educational and research foundation. The paradox lies in the fact that he believes the higher accomplishments of evolution up to and including the spirit emerge from the material universe evolving to ever higher states. One of his books is entitled *The Bridge between Matter and Spirit is Matter Becoming Spirit*. Arcosanti, the first arcology of the Cosanti Foundation, joins the two words "arcology" and "Cosanti".

I was among the twenty people that first day at Arcosanti, July 23, 1970, and with one other student of the budding ecological city raised the first structure: a shelter of two-by-fours and polyethylene film to protect boxes of nails and bags of cement from the rain. A cloudburst hit almost immediately, complete with lightning and thunder, and filled the sagging plastic roof with icy water. We scooped out a few glasses full and drank a toast to the project, and right on cue a rainbow appeared. We were off to a good start.

Berkeley: The Early Years

Three years later I had moved to the San Francisco Bay Area and decided to see if I could help spread the word about arcology. In a haphazard way I began making slide presentations for friends in my Berkeley apartment, at coffeehouses in San Francisco, in classes and at conferences at Stanford and Berkeley and other schools, for civic and fraternal groups, and so on. I sold bronze and ceramic windbells manufactured at Arcosanti and Cosanti to help defray costs and passed out brochures on workshops at Arcosanti after my talks. At the same time I was building a large tactile sculpture for San Francisco's Exploratorium and writing several more articles.

By 1975, I had met a number of others interested in getting organized and building an arcology or beginning the transformation of an existing city somewhere in or near the San Francisco Bay Area. We decided to start an organization. Our mission was to explore and educate about ecological city planning by building new towns or parts of an evolving ecological city or two, and by all other more conventional educational and policy means.

Most of our members were architects, engineers, contractors, carpenters, city planners, and technical tinkers enthusiastic about solar and wind energy. There were all sorts of other people, too, and we had two things in common: we thought that cities were great places but needed lots of work, and we all considered the idea of ecologically healthy cities magnificent, creative, and positive in difficult times. Some of us wanted to help save the world while others just wanted to build something better. I was interested in both, and everybody thought Soleri's ideas were a good point of departure for such explorations. They deserved serious experimental effort and experiential immersion. Most of us wanted to live in such a place and work intensively with one another to build it.

We called the organization Arcology Circle and set it up as a California educational, non-profit, tax-deductible organization. We chose the name "Circle" because we considered ourselves rigorously democratic. Each of our directors had to be a representative of an active committee: Land Search, Public Education and Outreach, Newsletter and Publications, Study Group (research and theory), Design and Model Building, Financial and Fundraising, and the like. We wanted no armchair philosophers, resume padders, or power-crazed bureaucrats telling the workers what to do. We were all workers. One of the things we did well was publicize Arcosanti and recruit people to lend their bodies to the cause by going to Soleri's experimental town for the workshops. The workshops are ongoing to this day.

By the end of the second year, however, it became all too clear that we were not going to be building a whole new city anytime soon. We'd gotten excellent publicity for the concept, but no one of means or connections was joining in. Our membership had leveled off at around 250, and the really active members numbered eight or nine. We planted street trees, got to know the neighbors, started working out a notion called an "integral neighborhood," spread the word through our newsletters, and began looking for a house we might buy to serve as office and home for those who wanted to live and work together. We went to Arcosanti twice for seminars on our aspiring arcology project (which wasn't shaping up) and our integral neighborhood project (which had real possibilities). Another year or two later we dropped the arcological new-town project, and Soleri was telling us that he simply couldn't endorse the piecemeal transformation of existing cities as arcology. "Arcology" meant single-structure city and we just were not building that.

We agreed that the term, by his definition, didn't really fit what we were doing but felt that we could explore the basic principles — three-dimensionality, relationship to ecology and evolution, mixed uses, and pedestrian environments — in existing cities. We were beginning to use the words "ecological city" and "ecocity" and saw what we were doing as a kind of urban ecology. And so we changed our name to Urban Ecology.

At that point, in 1980, we began to focus on our integral neighborhood project. The Farallones Institute, whose founding president was Sim Van der Ryn, had purchased and renovated a building in West Berkeley they called the Integral Urban House. The prime movers of that project were Bill and Helga Olkowski, who went on to become pioneers of integrated pest management. The Integral Urban House featured systems "integral" to one another. It had a passive solar greenhouse on its south side and solar hot water panels on the roof. Human waste was composted in a Clivus Multrum™ dry toilet, added to kitchen and yard compost, and used as fertilizer in the garden. Lots of food came out of the ground there — and out of the air, too, since they had bees that collected honey flying from neighborhood flower to neighborhood flower. Classes were held at the Integral House on dozens of subjects related to recycling, ecological building, and technology, and there were offices and residential spaces for students and staff. Our first Urban Ecology conference, "The City, the Garden, and the Future," a small one with about 45 people, was held there and featured Paolo Soleri, California State Department Director and founder of Trust for Public Land, Huey Johnson, *Ecotopia* author Ernest Callenbach, and biodynamic gardener and philosopher Alan Chadwick.

We in Urban Ecology were great admirers of the Integral Urban House. We proposed to take the next step with the Olkowskis and others by creating an "Integral Neighborhood." It happened that just one block south of the Integral House was a two-block area fiercely contested by the neighborhood and the city's Redevelopment Agency, which had been buying and tearing down houses for an industrial park. By the mid 1970s, the industrial park was highly unpopular, dead in the water, but what was to happen to the site hadn't been decided. About 80 percent of those two blocks were *tabula rasa* — nothing but weeds and sand.

With our friends in high places (Huey Johnson and Sim Van der Ryn) and a number of local allies, including the Farallones Institute, two churches, a plumbing company and several other businesses, several restorationists, and the recently founded Berkeley Architectural Heritage Association, we thought we had a very good chance. What we worked for was a cluster of apartments at the north end of the two blocks terracing up to six stories and flanked on the south with solar greenhouses and rooftop gardens. Active solar hot water was sketched into our drawings, and we included several electric and water pumping windmills for show and fun as well as utility. Six or seven ancient windmills with their water tanks still existed in the neighborhood then; only one remains as I write. We advocated restoration of a historic street with mixed uses — that was the historic pattern there.

On the theory that variety is the spice of life as well as the health of ecology, we planned to restore many of the old houses that still remained as well as move some other old ones onto the site and build new apartments. The scheme was to have open space made available by the clustering of the housing in apartments for both community and private gardens. We had drawn up an internal, rather European-looking pedestrian street system in the middle of the higher-density area complete with laundromat, corner food store, small movie theater, and shops on the ground floor. It would be a neighborhood center for that area of town. One version of the project was a poster drawn by Bill Mastin, an architect who has worked with me ever since. His integral neighborhood became a near-archetype appearing in four or five books and dozens of magazine articles.

But we learned a few nasty lessons about nonecological citizen planning. A number of low-income minority advocates we had originally thought of as our allies started circulating the thoroughly fabricated story that we were in bed with big developers planning a large shopping center. Fresh from the battle with the Redevelopment Agency over

"Hilly" height limits produce three-dimensional built landscape
Coupled with public access to rooftops, such height limits can create the city of high diversity and density with maximum views outward and variety of sun angles for the structures themselves. Several drawings follow and explore this experimental city design. Most of the energy efficiency, time efficiency and materials efficiency comes from the generally high order of compact diversity. The dashed lines indicate areas of not only particular general heights but also areas of very dense compact development, with buildings linked into an interlaced single structure.

the proposed industrial park, they had adopted the strategy to promote housing only. According to one of our friends who for a while was in planning sessions with our new "opposition," they said, "The people won't understand mixed uses anyway." Their strategy was to keep it simplistic, to keep it polarized, to make it a fight, and to win it. Keep the industrial park out and go "back" to housing. Never mind that the original uses there were not "housing only," but mixed uses with shops, small manufacturing, a lumber yard, fish market, candle factory, shoe factory, laundry, a kind of general store, and a bar that once served temporarily as a grade school by day — talk about mixed uses. We were the ones with a sense of time, looking to both the past and the future. Our opponents claimed to be working for the people — narrowly focused on *right now*.

They defeated us and built a pleasant enough but relatively nondescript one- and two-story all-housing project for low-to-moderate income people replete with what's advertised as "ample free parking." No significant progress, no city planning history, no story worth telling in it. Of course, an integral neighborhood is an equal-opportunity employer and housing provider. It can serve anyone very well, including low-income people, and, in addition, provide jobs that are so close you don't need a car, plus community gardens, energy conservation, and so on. Not only that, but our plan, since we proposed a taller set of solar apartments on one portion of the property, would have produced 50 percent more housing. It's worth noting that among our most energetic opponents against the integral neighborhood project were people claiming to represent "social justice," yet we would have delivered more housing than what was actually built — as well as many other good things.

Thus the next step up from the integral house to the integral city was never taken. It would have been an "in-city village," as one of our friends called it, an in-city ecovillage ten years ahead of the rural version promoted by the Global Ecovillage Network, but our "in-city village" actually would have used traditional village structure as its basic form, though none of us were consciously thinking about "traditional village structure" at the time. But with the loss of the integral neighborhood link, the chain from integral house to integral city was broken.

Failing that effort, we decided to settle into an existing neighborhood and see if we could transform it piecemeal. Several of us bought a house — 1939 Cedar Street in Berkeley — to serve as home and headquarters for our work. We designed and built a solar greenhouse facing the street on the south, shaped like a big bay window so that it would fit in with the architectural styles of the neighborhood. To build a greenhouse, we had to first pass an ordinance making it legal, since such greenhouses invariably protrude beyond the imaginary line called a "setback line" that prevents people from building out into their lawns and blocking the views of their neighbors. Our rationale for the ordinance was that the attached greenhouse constituted an energy-saving device, was mostly transparent glass so you could see through it, and helped with gardening, providing an early-season nursery for plants. City Council liked the idea and voted it in.

Once built, our greenhouse generated temperatures up to 107° F (41° C) on sunny winter days — we'd just open our windows to the greenhouse and the house was adequately warmed deep into the night. Two tomato plants in there grew up to the twelve-foot-high ceiling and back down to the ground and produced almost all our household's tomatoes for two years, most of which we had to harvest with a ladder. In the first of our depaving projects, we tore up the concrete planter strip between our sidewalk and curb, turned in some good compost, and planted fruit trees there and in the front yard. Of course we also set up recycling and composting.

We redesigned six blocks of an existing street, Milvia Street, terminating three doors from our house at one end and in the downtown at the other, and got it built by the city. We aspired to build a whole system of such streets and called our first project a "Slow Street." We had been impressed by Ivan Illich's

A closer look at the zones of highest density. These three-dimensional urban volumes look something like wedges, their steep slope angling toward the sun (right side of drawing).

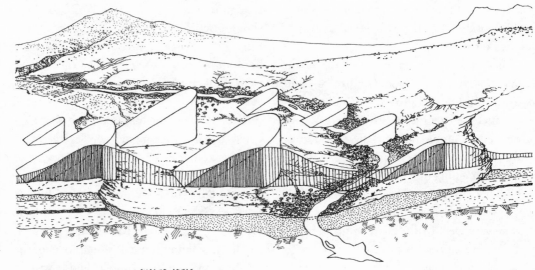

HIGHEST DENSITY AND MOST SHADED AREAS
(SLOPING WHITE BLOCK-LIKE FORMS)
HEIGHT LIMIT CROSS SECTION AND GEOLOGICAL
CROSS SECTION

small 1974 book *Energy and Equity*, in which he identified speed as the crucial element — if E.F. Schumacher said "Small is beautiful" (the title of one of his books), Illich added "slow" to the equation. With slow speed come safety, a quiet environment, low pollution and energy consumption, and so on. Speed is a primary, very basic consideration.

Our Slow Street design featured a street free of parking on one side, with the parking shifting from one side to the other about mid-block, causing the traffic to shift, too. Trees planted behind curbs that extend out a short distance into the street, obstructing views a bit, and low-profile speed bumps that were inconsequential at fifteen miles per hour also helped to slow traffic.

We also brought a bus to our street, Cedar Street (which runs perpendicular to Milvia), the only bus connecting the east and west sides of Berkeley through the middle of the twelve blocks of the northern half of the city. The bus filled up with lower income and minority people and took a fair number of cars off the street. Our opponents in the case of the bus were owner-neighbors who were securely established and had no interest in helping the bus riders or getting cars off the streets. They didn't like the noise and pollution of the buses. The noise and smoke bothered me too, but with high-density destinations at both ends of the line, evidence of a fair number of potential riders between, and a bus replacing five to thirty car trips on our busy street, our organization and our household felt that the balance tipped toward the buses. With the Slow Street, the bus line, the solar greenhouse, the fruit trees, and the composting system falling into place around and in our house and neighborhood, we had the feeling we were about 20 percent of the way to an Integral Neighborhood. It wasn't much, and there were no jobs, shops, or community gardens involved, but it was something.

One of our most enjoyable projects was a sort of street theater piece we called the "Vegetable Car." It was an attention-getting device for our Car Wars campaign, which hung out the dirty laundry on the car/sprawl/freeway/oil syndrome. We took a baby-blue, candy-apple, metal-flake 260-horse-power 1969 Pontiac GTO, cut its roof off with an acetylene torch, left its engine with the car recyclers, and turned it into a vegetable garden. It was a monument to the first automobile fatality in America, the death of one Henry H. Bliss killed by the best of all possible cars — an electric taxi — in New York City, on September 13, 1899. For three years it attracted photographers and international tourists to our house or one or another of the street and "energy fairs" that were popular at the beginning of the 1980s, with the memory of two oil crises fresh in the mind. What it meant exactly even I wasn't sure, though it was my project. Said one foreigner who had first seen the venerable vehicle in *Let's Go USA*, "What if everybody did that to their cars?"

Good question. Quieter neighborhoods, better transit, less pollution, more vegetables and flowers. Just dreaming, of course, but in a productive direction. It seemed to be working.

In 1988, we came up with the idea to mark the curbs around Berkeley with "creek critters" to let people know where their waterways flowed: 85 percent were under the streets and sidewalks, yards and buildings. I drew silhouettes of twelve animals associated with creeks — fish, crawdad, turtle, frog, snake, dragonfly, water strider, and others — made stencils, and, with 35 volunteers, hit the streets with spray cans of bright colored paint. City Council permission was needed. Otherwise the police would have stopped us at the first sound of spray cans in the graffiti-ridden streets. The stencils feature the name of the creek on the left, such as Strawberry, Schoolhouse, Derby and so on, with the creek critter cut out in the middle and the word "creek" on the right. They fit just right on curbs. One can do great things with a variety of colors, creating a contrasting background color first or spraying the image in one color, then offsetting the stencil for spraying a second color and thereby making a contrasting color fringe on the edge of the writing and creek critter. We marked 860 places on the curbs of the city, the whole creek system of Berkeley. Joggers, children, people going shopping, who never knew there were creeks underfoot, now knew.

One of our biggest successes was convening the First International Ecocity Conference in 1990, attended by 775 people, 153 of them speakers — not only some of the best-known, but many of the relatively obscure real pioneers in ecocity design and building, planning and activism. The conference was held in six separate buildings near Berkeley's Martin Luther King, Jr. Civic Center Park. The city itself was the conference center. The conferees were there between sessions out on the streets with everybody else in town, running into each other, smiling, waving, having lunch in the park, some sitting on the Peace Wall. We had big flags in bright colors on light poles welcoming our guests with the silhouettes of the creek animals we had used in marking the curbs representing the city's twelve creeks. Beautiful!

After the conference, however, there was a sea change in attitude inside Urban Ecology, and I found myself in a minority wanting to carry on with the actual buying and building of real physical projects. We had done very well with the house we had bought earlier, our headquarters for eight years, in which the organization was one of seven partners. The house had served as a strong anchor in a neighborhood where we were able to make substantial progress. But the board majority was not interested in that approach any more. They preferred the strictly educational approach. They wanted a strategy of supporting transit over cars and infill and transit village over sprawl, and of allying themselves with greenbelt advocates — all worthy as far as it went. Most of all they wanted to be identified with social justice issues. Their discussions about foundations being uninterested in ecological issues but very interested in social justice issues, which was why we should shift our focus, too, I found unnerving. First, ecological city design *is* a social justice issue. How else can one benefit lower-income people in a better way than to free them from having to pay thousands of dollars a year for a car and hundreds of gallons of gasoline just to survive in society? How are you going to build a car-free city or society where freedom from the injustices of the car is an option if you don't go directly for that goal? If it's an unpopular goal, try to make it popular. It has to be done. Second, there are plenty of organizations dealing with social justice issues and only one was tackling the ecological redesign of cities — ours. We were needed for that task.

To back off from that assignment and ask what everyone wants and then to try to deliver it rather than to say, "We've studied this for years and have some insights to solve these problems and here they are," seemed like selling out our contribution and the history that made it possible. The board seemed almost embarrassed by the early connection with Soleri and instead of explaining the evolution of our own thinking and our work to find and enhance the best of what went into arcology, they wanted to bring a shroud of silence down around the Arcosanti connection — and after I left, they did. As a dean of the Architecture

school once said to a young student while I listened (it happened to be at rock impresario Bill Graham's house at a party for Soleri), "If you want to get into Cal, don't mention you worked with Soleri." I saw folding before such styles and prejudice as cowardly. So I left.

Ecocity Builders

I wanted to continue doing the projects I thought were most important and participate much more directly in actually building things. I wanted to celebrate, not bury our history. Even if it didn't seem popular with foundations, community groups, and local politicians, explaining it would be educational — and education was our nonprofit status mandate. I felt we were dealing with a real science that had to be truthfully portrayed, like a mathematician saying two plus two equals four. If foundations and others in the community wanted to hear two plus two equals five, I wasn't going to agree.

By setting up a new organization, Ecocity Builders, I was able to go on dealing with issues on a very fundamental level. Ecocity Builders managed to maintain and strengthen an international network that led to four more ecocity conferences, one in Adelaide, Australia, one in Yoff, Senegal, one in Curitiba, Brazil, and one in Shenzhen, China. A sixth one — in Bangalore, India — is being planned as I write. These have been truly inspiring events, real adventures and opportunities for learning about ecological cities

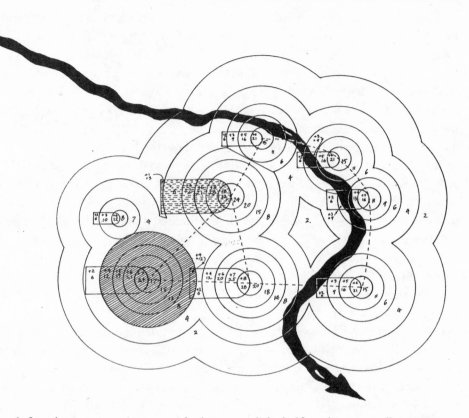

Mapping height and density. *Height limits are represented here with concentric circles and super dense areas with somewhat bullet-like shapes representing the slopes seen in the previous two drawings*

firsthand and for sharing experiences with many of the best innovators in the world.

At about the time I left Urban Ecology, I was drawing an ecocity zoning map for Berkeley. From developing the map and thinking through exactly where we thought some city infrastructure should be removed, as well as where transit centers should be strengthened, we evolved our depaving projects. We silkscreened T-shirts with the slogan (paraphrasing Joni Mitchell), "Save paradise — tear up a parking lot" and managed to

accomplish half a dozen small projects — depaving planter strips around the city, removing concrete from between sidewalk and curb, and planting a hundred donated fruit trees.

One of the largest depaving projects took place at University Avenue Homes, an old four-story hotel that had been converted to housing for the formerly homeless. Low-income and cooperative housing builder Susan Felix got permission from the city of Berkeley to remove about five parking spaces for a garden. After all, as she pointed out, of the 75 very low-income residents, only three owned cars and only two of them worked. A West Berkeley activist and mentor to several adolescent boys, David Eifler, wrote me about an encounter on the way to this depaving event. The boys with him suddenly darted across the street to look at something a friend seemed to be hiding in his shirt, and, after furtive glances and excited whispering, returned to his side. When he asked, "What's your friend got?" it turned out to be a gun. "Thanks for giving the kids something else to think about," he wrote. "They talk about this as their garden now." I noticed that they worked over that asphalt with sledge hammers and crowbars with real relish and then helped mix in the horse manure and finally plant and water the trees, vines, bushes, and seeds.

Our largest depaving project yet was a satisfying collaboration with Urban Creeks Council. Urban Creeks Council cofounder Carole Schemmerling did most of the negotiating with one private landowner, two cities, the University of California, the California Department of Water Resources, State Fish and Game, and the US Army Corps of Engineers — all agencies required to sign off on the creek project — to get permission to open a block of buried creek on Berkeley's border with Albany. Before we started working on this section of Codornices Creek, the waterway was in a concrete pipe under a little-used parking lot, a street that had been closed, a sidewalk, and a flat earth-filled area covered by almost nothing but fennel. Urban Creeks Council applied for and got $25,000 from the Department of Water Resources Board and hired a semi-volunteer bulldozer operator, who worked for half his usual pay. With the creek continuing to run through its culvert a few feet to the south, he carved out a rough cut about 80 percent the depth of the eventual creek canyon. Then for the next two years of Saturdays, with 375 volunteers, most of them only working a day or two but about a dozen working about once a month, we carved out a pleasant small valley and hilly area with picks, shovels, and wheelbarrows. Ecocity Builders' role has always been to coordinate the volunteers and to steward the land, both then and since.

This was a very low-budget project in a mainly warehouse-district neighborhood without landed gentry to protest or tell us with fifty conflicting voices what to do. The result was a great opportunity for innovation. Since funds were very limited, the landscape architect never specified what plants should go where. Our philosophy was to plant natives, but since we had a fairly large area, we decided to have an orchard as well. The University officials expressed no objections. Within a few months we had accumulated donations of twenty-five fruit trees of all sorts: apples, plums, apricots, figs, oranges, lemons, nectarines, peaches, cherries, persimmons,

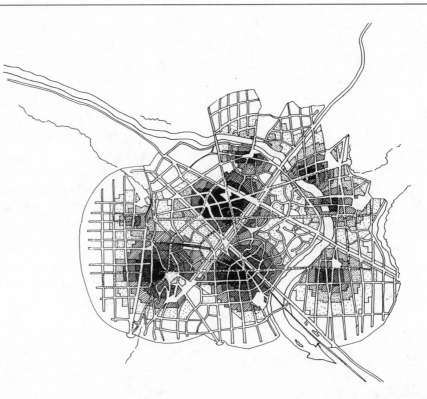

Asian pears. The place is a wonderful little paradise, home to a dozen types of butterflies, dozens of birds, steelhead trout, sticklebacks, crawdads, water snails, frogs, garter snakes, water striders, and lots of other insects, including sulfur-blue damselflies and vermilion dragonflies as brilliantly colored as tiny red flying neon lights. There are not many mammals yet: just pocket gophers, raccoons, and, unfortunately, stalking cats.

The place is amazing. We took cuttings from willow trees, cut them into stakes about an inch or two thick, sliced the trunk side of the branch at a sharp angle and the end toward the leaves square, and drove them with a hammer into the banks of the creek. Within days, roots were shooting out from the lower parts and branches popping out of the sides and beginning to reach for the sun. Now we have ten-year-old thirty-foot-high willows and alders that we planted. Watercress, cattails, horsetails, and bulrushes showed up on their own in our micro-wetlands and along the banks nasturtiums, wild grapes, mint, chicory, and plantain. Volunteers planted dozens of local native flowers, grasses, and bushes — ninebark, silver tassel, coffee berry, monkey flower, lupine, manzanita, and ceonothis — and trees: madrone, buckeye, bay laurel, California glory (fremontia), and two kinds of oak. You are all invited to come down on Sunday, midday, and help us water, weed, prune, harvest, care for the path, and occasionally plant more trees and bushes, vines, and flowers.

Assessments

In my forty years on this personal odyssey through ecocity territory, I've come to see the economics and politics of sprawl development something like this: Historically, towns have grown up around fairly high-density centers of work, commerce, and social life, with some people living in them but many more close by in lower-density housing. As towns have grown into cities, the close-in residents in low-density areas have resisted growth in their neighborhoods, forcing development farther and farther out. That next ring of low-density development, once established, forces development even farther out, and so on.

Why is sprawl being relayed outward virtually forever? At the start of our cause-and-effect chain, it's because people relatively near the centers are used to what they have and want to keep it that way. Some conscientious people, aware of the costs of sprawl, and some who actually like the company of other people in real urban neighborhoods near real city centers, see cars as a major problem and vote for greenbelts surrounding cities, theoretically restricting or slowing sprawl.

But just punching a hole through the greenbelt with a major transit line or a freeway extension or widening allows sprawl development to pop up another ten or twenty miles out. Slowing sprawl with greenbelts in many cases begets an even larger sprawl — this is called "leapfrog development." Due largely to the fears of suburbanites and their political influence, what infill that does go on is mid-range density — which is often located where it would be ecologically more sensible to unfill for restoration rather than infill for development. The whole pattern presently lacks a system, a vision, and a means both political and economic to reshape a healthy environment.

Where I lived in Berkeley, neighborhood activists and architectural conservatives now fight against new housing and especially higher-density housing in and near centers as well as in the lowest-density areas. They have made their emotional and financial investments and have become comfortable in their immediate environment. Higher density, they believe, would mean more shadows from taller buildings and many more cars on the street (though these conditions would improve radically with car-free buildings and car-free streets). The real problem is that there are well-established people working against growth even where it makes sense. Some of them deride

Green areas. *Stippling indicates green roofs, some yards — higher density areas have higher proportion of public access uses and somewhat less green. The highest points of each "hill" are indicated by stars. Striations indicate agriculture near the river in this dry landscape, like New Mexico's, and crosshatched lines represent natural riparian vegetation.*

opening creeks and expanding community gardens as silly, but never give reasons why. Just to condemn something makes people uncomfortable, gives it "bad press," and raises the fear that supports no change at all and the anger that closes minds.

Preservationists tend to share this anti-growth and anti-restorationist-anywhere point of view — but not always. Some want to freeze the whole town as it was at some point in time determined by a favored personal esthetic. Others, however, want to preserve just particularly worthy buildings and sometimes extraordinary clusters of buildings or districts. Some who really love architecture want to save

special buildings but also want to see what can be built that addresses present and future realities and uses the best, most creative designs the present generation can contribute to the next.

One important force in all of this is conscience. Sprawl really is an injustice against both other species and the struggling classes. Some people not only respond to that, they are aghast at how damaging sprawl is turning out to be, and they want to do something about it. They already vote for greenbelts and transit stations, as well as pay dues to the Sierra Club and other organizations. Among the disenfranchised — the students without housing, many of the elderly, the wheelchair-bound or blind, some types of professionals and artists who don't need large spaces — there are many who would be happy with car-free accommodations downtown near transit. Those who oppose higher density in these locations are quite simply denying housing to thousands of such people in city after city.

The no-growth-anywhere people often claim the moral high ground, saying that it is a grave injustice to force a change in the lives of people already living near or in the area of growing density. The reality is that the few already there are up against several times as many who often have only modest amounts of property and seldom any real estate in the community but may be local citizens nonetheless — or would be if they could find housing. The numbers game of democracy would favor providing housing for them, but money and established connections, not to mention deeds to real estate, work against them. Meanwhile it is seldom mentioned that the already established residents could move into the new development itself or sell their present real estate, often at a very big profit, and move to another pleasant neighborhood. Many of them have great flexibility but claim or actually believe they don't.

Then, too, circumstances change for all of us. Children grow up and move away, which means that both those children and their parents could be ready for a change, often a smaller place. Some of them do in fact end up interested in higher-density, pedestrian-environment living. What if nearby natural watercourses were available? What if they could get up into the buildings easily for local distant views? What if there were sunny public plazas and shady trees and pedestrian streets to stroll down without threat to life and limb from hurtling motor vehicles? We who would change the city toward an ecologically healthy one have a very full deck of cards. We need to be resourceful about playing them. Bringing this kind of argument to bear on the kind of resistance I experienced in Berkeley, which claimed to cherish the poor and downtrodden while opposing the housing that could actually allow some of them to live there, would help us begin building something that works both ecologically and socially.

Some of our local politicians have told me that theirs is the art of the possible, meaning that unless an idea is already popular, they will not represent it. They play the representative in this case, not the leader, much less the educator. But all the features of the ecocity exist — if scattered only thinly around the world — and, as the saying goes, whatever exists is possible. We have that foot in the door, too; we can tell the story of the Aztec toy with wheels. The reality is that politicians who don't help us identify those ecocity features and help assemble them are simply failing their duties as people who should professionally be concerned with change toward a healthy future. For forty years I've been doing just a little of what is "possible," making some progress. But being "visionary" and courageous is required of all of us if we want to bring our cities into balance with nature. My best offering: Try those four steps to an ecology of the economy.

One important force in all of this is conscience. Sprawl really is an injustice against both other species and the struggling classes.

We who would change the city toward an ecologically healthy one have a very full deck of cards. We need to be resourceful about playing them.

Tools to Fit the Task

Tools suited to the task of reshaping cities for a far healthier future than the one we are fabricating now — greenbelt laws, for example — have existed for a long time and will remain important far into the future. Some of them, such as the transfer of development rights (TDR), are being used effectively in many places but should be used much more widely and may require redesigning to work better and replicate themselves more quickly. In addition, completely new tools need to be designed to fill out a whole toolbox for ecocities. One of my own inventions, building on Ian McHarg's mapping system in *Design with Nature*, is the ecocity zoning overlay map. There are many more. When Jaime Lerner told the people of Curitiba that environmentally healthy policies and practices were important and that they, the people, were important, he helped create a culture of acceptance and support for very substantial urban transformations from the foundation in land uses on up. Given a culture of support in which people take problems and solutions of the sort addressed in this book seriously, these tools can be used to change the world profoundly. Some of them, in fact, can be used effectively by a small number of people right away, and this can build momentum toward more general public support. Then healthy cities and a vital biosphere become possible.

Ecocity Zoning: Mapping the Future

Many planners consider zoning a great invention that lends structure and order to city building. A vocal minority says, in contrast, that zoning has divided the city and precluded the natural development of land uses in complementary relationship with one another. The anti-zoning camp suggests that destructive segregation is intrinsic to all zoning. I don't agree. The problem, I think, is with the kind of zoning and the purposes it is designed to serve, and it can be largely solved by reshaping zoning itself — on the human measure. How, after all, will we talk about these complex things without a language, visualize them without images such as maps and graphically represented plans? Zoning provides these things, if not in an esthetically beautiful language, at least in words and images that can carry important meanings. Without words and pictures to represent human anatomy, it would be hard to understand and fix our flesh-and-blood physical equipment. So, too, for ecocities.

In support of zoning, it must be said that it does have a certain fairness in the sense that anyone who wants to play the real estate development game — or fight against it — knows generally what to expect. Zoning is simply a means of letting people know what they can build and where and what sorts of activities are allowed there. Many of the ecological and social disasters of ill-conceived and poorly applied zoning can be corrected simply by (1) planning for the city, town, and village walkable distances; (2) creating pleasant, inspiring pedestrian environments; (3) using not flat but three-dimensional thinking; (4) insisting on looking at whole-systems patterns; (5) long-term results. When these five major ideas are added to zoning for restoring

natural open spaces; (6) agricultural open space, you get ecocity zoning.

To create such ecocity zoning, first of all we have to acknowledge that the forces that gave us our present zoning are vested in the present system. They are personified, too. They include living, breathing people who are afraid of change in their neighborhoods, business people worried that customers might go away, people who just happen to like the way things are now, and people who know or care little about ecological collapse. Mostly, however, people just haven't heard about ecological planning, much less ecocity zoning. Many might see it as a good idea, including people who can make money on ecocity zoning — among them developers and business people in centers where density and activity increase.

There are people who will be interested in the greater cultural diversity possible with a shift of densities toward pedestrian/transit centers, and environmentalists these days understand that density and transit go well together. Of course they realize that restoration of nature is very important. The next step in their thinking is to realize that ecocity zoning makes their dreams possible as does nothing else. There are those who are unable to find housing near the town centers who want such housing — often desperately, as evidenced in a City Planning Commission meeting recently in Berkeley in which a University of California student said, "I'd be happy with a prison cell downtown if it were available." It is difficult to gather support from this diverse crowd to outnumber those afraid of the kind of changes represented by ecocity zoning, but it can — and must — be done. In any case, we have to start work on ecocity zoning by simply doing it ourselves as concerned citizens. If it has the value I think it has, it can then be held up to public scrutiny and found to be a powerful and positive tool.

The objective of an ecocity zoning map is to open up landscapes covered by car-dependent development and recover agricultural and natural landscapes while shifting density toward centers. The new density should be in buildings with the sort of ecological features described here in this book and other ones not yet invented. At the same time, the objective is to move toward a far more balanced set of land uses with most aspects of life provided for in a small area in the centers. This means, generally, mixed-use development and very little commuting. It means creating the physical structure of the city so that architecture, technologies, nature, and healthy lifeways can harmonize. It represents — it is — the first step in the four steps to an ecology of the economy (see Chapter 8) and shows how we can put a green infrastructure under a green economy. As a reminder, those four steps are map, list, incentives, and people.

With basic ideas from this book and your own knowledge of the place in which you live, you have enough to get started on an ecocity zoning map. If you consult aware ecologists

City of constructed hills in perspective. *For simplicity of expression, this city has only three such hills (compared to eight in the earlier illustrations). Towers can be seen to be compacted in the "dense" areas, but separated more widely for solar access deeper into the streets, on the sunny sides of the "hills."*

and concerned citizens *en route* and then revise, you will produce a good preliminary document. Then you can take a break for a week or two — visit a place that is mostly natural and similar to what your town's location looked like 100 or 1,000 years ago and wander around your town a little to see what attracts your attention. Then return to refine the map. If you want the map to look more professional or attractive, you can find a local geographer, cartographer or artist to help. Then you can put a date on it and start using it. It will never be finished and final; such is the way of all maps.

These are the seven steps essential to producing a viable ecocity zoning map:

1) *Produce a local natural history map.* Visit your library, local college, or historical society to locate the earliest available maps of your town and learn about its natural history — native plant and animal species, weather, climate, soils — and its cultural history. Features from these old maps may include creeks, original marshes, seasonal ponds, springs, shorelines, outcroppings of rock, ridgelines, major animal migratory routes, types of plant cover, areas of steep slope, sunny and shady slopes, archeological sites, old historic buildings, neighborhoods that may now be gone, and so on. Put this information on paper — call it Map #1. It prepares you to assess the priorities for restoration and development and where these

activities should take place. You may be going back thousands of years; the exercise will be fascinating. You might generate several maps: one of the natural environment, one of the early settlement and agriculture, one of the historic buildings and transit routes, and so on.

2) *Establish walkable centers.* On an up-to-date map of your town, which will be Map #2, locate the present city, town, and neighborhood centers and draw concentric circles indicating distances from these centers. These will look much like the concentric circles of a target. On about one-fifth to one-third of the land area of the town, in the zones closest to the centers, the density of development should be significantly greater than is the case presently. On about half to three-quarters of the land area of the town, in the zones farthest from the centers and most dependent upon automobiles, there should be much less density of development in the future and, ultimately, only natural or agricultural land uses. The lower the density of the whole town, the smaller should be the percentage in the increasing density area and the larger the percentage in the decreasing density area. Everywhere the mix of uses should become far more complex, even in the restoration areas on the future fringe; all sorts of diverse agriculture and networks and patches of nature corridors and zones can be established in time.

View from inside
the city looking
out over a terrace.

How many concentric circular zones — the bands of your "target" — you choose to draw and how wide they should be depends upon your own intuitions and experience. It also depends on the particular centers in question. Five to nine concentric circular band-like zones give enough definition to different areas to make it clear where more or less development should be happening. Using five zones, for instance, will mean highest density in the centers, second-highest density just outside that zone, a minimal-change area next, an area of reduced density outside that, and finally, farthest from the center, the areas of highest priority for "de-development," that is, for depaving and the removal of buildings, walls, streets, creek culverts, and other structures so that nature or agriculture can be reintroduced and their own regenerative forces be released.

The concentric zones around the centers should be generally larger for the larger centers and smaller for the smaller ones; their size will also depend on the total population of your town and the intensity of existing centers. Let me use Berkeley maps as an example here because this is a city I know well and have mapped in this manner. It is a town of 110,000 people that is, on average, approximately three-and-a-half miles wide, both west to east and north to south. The area around the downtown that should be up-zoned — that is, have its building heights and density increased — should be about three-quarters of a mile in diameter. The areas around the smaller neighborhood centers should be upzoned for only about two blocks from their centers, give or take about a half-block (four blocks in diameter). Middle-sized areas should be upzoned for about three or four blocks from their centers (or six to eight blocks in diameter). The eventual total area of land covered with development, both thin and intense, depends on these decisions, too. As a starting point, no less than half of the land area should be returned to nature or agriculture. This can be reassessed later. If you take the time to redraw the map several times, it will start making sense and become more self-evident.

3) *Adjust the circular zones and draw them in relation to nature corridors and agricultural areas.* Decide on the best locations for nature corridors and agricultural areas and draw them on a third map. You are now prepared to adjust the concentric circles to create nature corridors and to lay out areas for creek restoration and other special purposes. Some of the concentric circles around the downtown, major centers, and neighborhood centers are likely to overlap, cutting off potential connections between restoration areas. You will need to compress their edges so that the nature corridors can be established. The circular band-shaped zones will thus become somewhat flattened on the sides closest to other circles. The restoration areas between the circles with somewhat flattened sides, together with other natural and agricultural areas including creeks and ridgelines, indicate the location of the future nature corridors connecting future natural zones. Creeks become another kind of nature corridor that can penetrate right into the middle of a center. Creek setbacks — the distance between buildings, streets, walls, and other structures and the creek itself — should be wide in areas far from the centers and narrower in the centers, where land is of very high social and economic value. But creeks should *not* be buried.

4) *Show the limits of discontinuous boulevards and the location of railroad right-of-ways.* Since urban development naturally concentrates around transit hubs (future walkable centers or transit villages) and along transit corridors, draw in higher intensity development areas along boulevards that connect centers. But, somewhere in or near or probably just outside of the minimal change zone, make boulevard development discontinuous — that is, identify the entrance to the center right there. After major land use shifts in the course of many decades, the boulevard turns into a country road at this point, and there is the potential to do something interesting, even spectacular, here. Call these places "gates to the city" or, if defined by large structures, "ramparts." Not just architecture, but arches, sculptures, and big trees could mark these entrances, too.

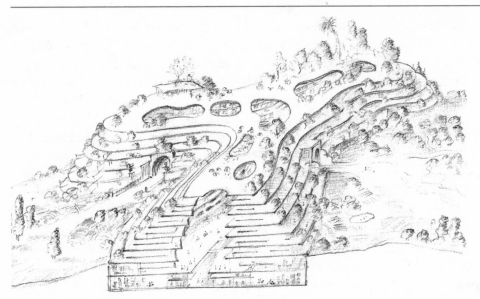

Hill towns of a special sort. *Gene Zellmer proposes towns with residence on outside terraces with views to nature and agriculture, inside streets sheltered by inside slope of the "hill" for public life.*

As the boulevard-become-country road enters the zone of highest priority for restoration or crosses any special nature corridor, it should rise up on a causeway-like structure or, preferably, plunge underground so that people and natural species can cross without undue disturbance of one another. Railroads should do this, and bicycle and foot paths, too. You can take advantage of hills, even very low ones, and valleys for burrowing under or rising over nature corridors.

Railroad right-of-ways should be featured on the map and if not active, saved for future rail lines or changed into bicycle paths. They should not be built on. Once sold off, this land becomes expensive and difficult to reassemble. Show all these features on Map #3. This becomes your ecocity zoning map.

5) *Prepare sample vertical cross sections.* To make clear the three-dimensionality of ecocity zoning, supplement Map #3 with drawings representing vertical slices through buildings and landscapes, which illustrate various arrangements of uses, ones on top of, as well as adjacent to, one another. Features such as rooftop cafés, bridges, elevators, and terracing, and the relationships of buildings to sun and views can be illustrated in this way to help explain the options for using the third dimension imaginatively and for ecological benefit. These images can be drawn in the margins of Map #3 or on a separate sheet.

6) *Provide keys for the maps in the usual way.*

7) *Add scenario maps.* To illustrate changes into the future, you might draw up several other maps representing different stages in the ecocity's development (Maps #4, #5, and so on.)

A refinement is needed here. You will notice the maps are "centers-oriented." This compares with general low-density "sprawl," which in the real world involves some zones of limited mixed use — hello 7-Eleven — but more prominently, CBD (Central Business District). That's the typical and not very healthy state of US cities. A third alternative general pattern is "corridors-oriented." Imagining these three basic patterns is useful for understanding in the most basic terms the nature of the land use issues we need to be dealing with.

The New Urbanists and people particularly focused on bus transit generally support building up density along corridors. Berkeley, my recent hometown, is loading up with corridors-oriented development as I write. Buses and streetcars stop every other block or so, so they tend to support corridors of development, and corridors of development, in turn, support buses and streetcars. Rail vehicles larger than streetcars stop less frequently and tend to work best with centers oriented land use patterns where larger numbers of people can get on and off at a single stop. Corridors are far better than sprawl, but when we think through ecocity zoning maps, the case has to be made that centers-oriented development is far better than corridors. First of all, New Urbanists and many people boosting buses who favor corridors frequently point out that shops on the ground floor with housing above up to four floors works well in Europe, so let's do it in the US, too. Problem! In Europe, the four-story corridor is typically backed by four-story housing extending perpendicularly to the corridor for several blocks, if not across the entire city. The US corridor, one building thick, backed by single family houses with front, side, and back yards, just doesn't have the density to provide customers for the corridor's ground floor businesses nor enough to make the bus system run without considerable subsidy. The answer: centers-oriented development with higher than four-story limits. Also, take the density back two, three, or more blocks behind the corridors in the area of the centers.

Another refinement: corridors of four-story development are dense enough (expensive enough in terms of investment) to effectively block opening the landscape for nature corridors and bicycle/pedestrian paths running along restored creeks. Therefore, here too, for the restoration of nature and its regenerative power, centers-oriented development is far superior to corridors.

Developing ecocity zoning maps is a challenge, but it is worth it. Doing it yourself will appear to take the initiative from the planners, competing citizen groups, and developers.

Once you've drawn a map, people will rage against your presumption in not having consulted them first (so they could stop you in your tracks). But if you don't draw one, it won't get done and no one will understand what you are talking about. It's damned if you do and damned if you don't, but worth the fight. Presenting two or three versions will help maintain flexibility of vision and invite better ideas. Someone has to exercise some imagination here and take some responsibility. When it comes to maps representing possible scenarios, many people will say they are unrealistic. They are not; they are simply long-range.

I am convinced that the entire ecocity map-making project has to be thought through publicly if it is ever to be adopted by the citizens of any city and serve as more than a fantasy exercise. Map #3 provides guidance. Maps #4, #5, and up, extrapolating into the future, help interpret that guidance. There all sorts of things can be featured, such as keyhole plazas, off-center parks and plazas providing urban views of nature, and quiet public spaces off main streets. "Lone-wolf buildings" — big buildings standing in restoration zones that don't make sense in the centers, scheme of things but have special economic or historic importance — should be saved. When they are relatively far from the centers, their uses may be changed to ones demanding little commuting. They may become very compact ecovillages or be remodeled to become part of very small arcologies. Transformed into factories that are incompatible with residential and social uses, lone-wolf buildings could appropriately stand separate from city centers, collecting workers daily with pleasant country bicycle rides where once there were one-story ranch houses, for example. Or a lone-wolf building's very existence could modify the ecocity zoning map. It could become the hub of a new neighborhood center, perhaps a small artist colony with a coffeehouse to which people could take the streetcar on a Saturday afternoon to watch the sunset and listen to poetry.

A usable ecocity zoning map in our hands provides broad outlines and a considerable number of details. It is essentially a "zoning overlay," as it does not represent the actual official zoning of a city. Its unofficialness is what makes it what I call "shadow zoning," an allusion to the "shadow ministers" of parties out of power in, for example, Australia's parliamentary system. Like shadow ministers, ecocity zoning maps stand ready to take over when there is a failure of confidence; in this case in relation to present zoning, perhaps in view of facts about peak oil, fear about climate change, and disgust with traffic jams. The rationale of ecocity zoning maps is impeccable, and it takes everyone in the Great Majority into consideration, including citizens of the future, animals, and plants. It does not have to wait for the future, either. It can be put to work immediately upon completion.

The ecocity zoning map is not as crisp, hard-edged, or directive as the actual zoning map, though it could be used to modify the existing one. Nor is it the "soft planning" of a regional metaphor like Frank and Deborah Popper's Buffalo Commons (see Chapter 3). It's somewhere in the middle, empowering the building in the physical world of something from the imagination.

The ecocity zoning map is an "overlay" in the sense that it can be imagined as superimposed on the existing map. It can then be used for influencing existing zoning and pushing its interpretation in ecologically healthy directions, encouraging more diversity and density in one place and restoration of natural habitat and agriculture in another while delineating ways to withdraw from automobile dominance everywhere.

The map can be used as a guide for activists. Many environmental organizations oppose, support, or comment on development

Table land 2. *The idea of building "artificial land in the sky" has been advocated but seldom executed in architecture. Would cost far less than a freeway system and would deliver access without motor transport, except for elevator — which would require far less energy than a freeway system.*

projects in their cities and counties. Ecological zoning maps clarify what should be supported and what should be opposed. As these maps are utilized, their legitimacy increases, and the chances of rebuilding cities for pedestrians instead of cars increases proportionally. Distributed to city council members, developers, and environmental groups, they let them know whether the map makers will support or oppose particular projects and why. It makes for a very fair game board.

Changing land uses in a major way may not yet be a traditional tactic for most environmentalists, transit boosters, creek fans, urban gardeners, energy conservers, bicyclists, and the disabled. But these activists will all discover that applying the ecocity zoning map provides the most powerful context for changes beneficial to their projects and positions that they are likely to ever find. The opportunity for the creation of a powerful coalition awaits their recognition of this potential. If they use the map and support its application, their work can become synergistically reinforcing, accelerating the effectiveness of all their various kinds of actions. This is one of the most powerful aspects of ecocity zoning: it clarifies what changes fit the whole, benefits all the groups striving for a healthier community, and avoids the pitfalls of placing a good development in a bad place.

Ecocity zoning maps help advance us beyond simplistic categorical thinking toward whole-systems thinking. For example, supporters of greenbelt initiatives who attempt to stop sprawl by placing land on the fringes in legally protected farms or nature areas call for infill development, that is, for filling in any vacant space available within existing cities. But if we are to restore nature in places where sprawl is now located, some vacant places should not be filled in. Buying a vacant lot for the restoration of nature or agriculture is far cheaper than buying a lot with a building on it for the same purpose and demolishing the building. Vacant land in or near existing centers should, in contrast, be filled in with appropriate development, in some cases with intensely utilized, big buildings. Ecocity zoning maps tell us where both open space preservation and infill development are best located and where unfill — removal of buildings, driveways, walls, culverts, etc. — should happen, too. Thus the slogan: "No infill without an equal and opposite unfill development!"

The ecocity zoning map can be a guide for developers and owner-builders as well as environmentalists and appropriate-technologists. Some developers would like to contribute to ecological health but don't know how. Ecological zoning maps can help them make decisions as to where particular projects may be helpful and, again, let them know in advance whether the map's supporters will be working for or against the approvals they are seeking from the city. The ecocity zoning map can also let everyone know how the city's zoning code needs to change if ecologically healthy and imaginative projects are to be built. Often the zoning and incentives are against such projects, but if enough people realize this, the zoning and incentives can be changed.

Thus the ecocity zoning map can also be a guide for policy makers and legislators. These maps begin to establish the framework for a new landscape of ecological laws and regulations — step three of our four steps to an ecology of the economy — as well as, eventually, for the actual physical city itself. They provide an idea around which imaginative legislators can design incentives, disincentives, changed tax structures and codes, and, some day — hopefully sooner than later — official ecocity zoning. The ecocity zoning map puts a land-use/infrastructure foundation under legislators' healthiest ambitions. If they want to be the builders of a civilization designed for the 21st century and beyond, ecocity zoning is indispensable.

At first, ecocity zoning maps will not be enforceable descriptions of how a city should be developed, but they start from what actually exists and therefore are partially implemented already. Even car cities, after all, do exhibit an almost natural expression of the basic pedestrian access-by-proximity principle in their cores where density and diversity are highest, and malls struggle to recreate the pedestrian magnetism that the automobile has close to annihilated by physical distance. These would-be pedestrian centers are engines of economic prosperity that can be tuned up for high economic and cultural performance.

It is very likely, then, that some town and city governments will eventually hire planning firms or knowledgeable local environmental organizations to draw up ecocity zoning maps. Early on, though, we will likely see such firms, organizations, or even teams of urban design students producing ecocity zoning maps without the assistance of governments, their work being paid for by organization membership dues, foundation grants, or the professionals, activists, or students themselves. So far, to the best of my knowledge, Ecocity Builders in Berkeley and Urban Ecology Australia in Adelaide are the only organizations to have produced such maps. Some people may produce ecocity zoning maps just for the fun of it. Maybe the video game SimCity could be redesigned for real relevance and applicability. If the maps are good and pass the test of reasonable local scrutiny, city councils may just endorse them as overlays to help guide zoning changes.

In the meantime, we will see small pieces of the puzzle fall into place in very different ways. State legislatures, for example, may write mixed-use car-free condos and apartments into their housing incentives and require that such developments be located in or near existing transit centers in order to qualify for certain state benefits. City governments may raise the height limit in one or several of their towns' future walkable centers without yet making a commitment to creek restoration or — much better — retain existing height limits but allow much taller buildings if the developers utilize ecological features in their buildings and purchase transferred development rights. A developer may decide not to develop at a particular location, even though the city zoning would allow it, because the ecocity zoning map indicates it should not be developed and people who understand the map will oppose the project. A downtown businessperson may decide to build a multi-story residential addition over his or her store because of the logic behind the ecocity zoning map, namely that the added population means more customers.

It will probably take a long time to reshape any city with an ecocity zoning map. Major changes in density shifts will take a long time if they proceed at the rate of normal replacement for aging infrastructure and would be expensive and constitute a societal investment if accelerated. However, I believe they need to be accelerated if we are to face the challenges of our times. In any case, improvement can be expected immediately, and we can begin moving resolutely step by step in the right direction. Remember Jaime Lerner's comment that major changes can be accomplished in just two years.

The ecocity zoning map is an offering — a kind of illustrated discussion paper — rather than the product of an all-inclusive public process. Simply calling a forum together and asking people how they would like to see their city changed will barely inch in this direction unless someone works resolutely and insistently to insert ecocity principles into the discussion. Ecocity mapping is complex and novel enough that it will have to come from people who have been thinking about it for some time. One cannot expect healthy results by asking a random sampling how to proceed with a medical operation. A surgeon is needed. The city's body is in need of ecocity doctors to get the urban anatomy back together after a terrible accident — a car accident. If there is respect for ecological city design knowledge, the citizen in the street and the ecocity expert can work together. After the pioneers have taken the risks to get the ball rolling, an open political process can amend and adopt it.

Probably the ultimate card up the sleeve of ecocity zoning mapmakers is that the map is based on important information that present zoning fails to consider. With an ecocity zoning map in hand, supplemented by descriptive explanations, you don't need to worry about whether anyone supports you initially. What you are saying makes sense. All ideas and built realities start somewhere as a tiny seed. In this case you have the logic of the human body's needs and dimensions and the logic of ecology on your side. You have good information about resources and ecology that conventional zoning has yet to deal with adequately if it has dealt with it at all, and your map is based on the spatial and ecological realities of your town. You can simply say, "I support this kind of project in this part of town and oppose this kind of project in this other part of town because these changes are needed to create a pedestrian, low energy, ecologically healthy city." You are in the world of development and city building like the intelligent consumer in the marketplace, and like that consumer you are in an extraordinarily powerful position. Just as the consumer armed with information on destructive companies and a list of green products can boycott or purchase new realities out of or into existence, the citizen equipped with an ecocity zoning map can change the physical structure of society. Starting in small but real ways immediately, by helping you support or withhold support from particular projects and from products and services offered there, the ecocity zoning map works.

A final important point about the ecocity zoning map: You don't have to wait for regional government; you can act now and act very effectively locally. Many thoughtful people promote the connection of land uses and transportation, encouraging higher density near transit and greenbelts. So far so good, but many of the best of them believe we can be only marginally effective until we create regional governments like Metro in Portland, Oregon — governments larger than the city and often embracing several cities and even counties. The idea is to gain the authority and power to rationalize transit, combine conflicting bus lines and commuter rail systems, coordinate schedules, devise greenbelts for whole regions, and select areas for future development. It's true that today's many separate municipal and county governments often create regional chaos in this regard, but it is not true that we can be effective only through regional government. In fact, the ecocity zoning strategy is safer because the intended specific results do not necessarily follow from setting up a regional government, which is almost as likely as a state government to support new freeways and acquiesce to pressure from sprawl developers and the more car-dependent drivers to continue their habits of highway building and car use.

Planning a recent trip from Oakland to Sacramento to see a friend and visit two government offices, I thought it would be fun to take the train. When I called the people I was planning to meet, I discovered that they were scattered all over the big flat town and I would not be able to visit all of them in the same day unless I went by car or spent the time to find a rental car when I got there. Sprawl at the other end of the intercity trip made the use of the train and local transit very difficult within what should have been a reasonable period of time. If Sacramento had been well along in the transitions proposed by an ecocity zoning map — finding its centers and shifting people to those centers so that its own transit system could work efficiently — I could have made the trip by train. What we do in the city we live in to make transit work with the land uses will enable people from far away to visit without bringing their cars. Thus if we act for ecocity zoning, we start to solve the regional problem locally — while reinforcing what a regional government should do when and if it is created.

Transfer of Development Rights

Transfer of Development Rights (TDR) is a real estate transaction tool established in zoning ordinances that makes it possible to buy and transfer the rights to develop from one piece of property to another. Most commonly, TDR is used to protect natural or open farmland from development or to save historic buildings. If the owners of real estate can sell their land for development, but there is good reason not to develop there, ordinances in some jurisdictions make it possible for developers to buy those rights and "sever" them from the deeds. The people selling the development rights get the money, but they and any future owners are prohibited from developing the property from then on. The developer who bought the rights, however, is allowed to shift those development rights elsewhere and build more than would otherwise be allowed by the local government. With the help of TDR, hundreds of thousands of acres of land and hundreds of buildings have been preserved in the United States that otherwise would have been developed in the case of open land, or demolished and then replaced with more development in the case of buildings.

Double TDRs are twice as good. They are a particular kind of TDR that removes the existing buildings, driveways, walls, culverts, or other such structures at the "sending site" (the location where the rights are purchased) as a condition of the developer being able to build more elsewhere at the "receiving site" (the location where the development rights are exercised and new development is built). South Lake Tahoe's TDR ordinance permitted and encouraged the removal of over 100 houses causing polluting runoff into the lake and the transfer of development rights elsewhere in the area. That is a lot of housing, but in Berkeley a single new apartment called the Gaia Building, eight stories high, houses twice as many people as those 100 plus South Lake Tahoe houses, and all on one sixth of a downtown city block. One hundred low-density properties in Berkeley translate into about twenty blocks of creek daylighting and ten community gardens, increasing in area the equivalent of two lots each in low-density areas. Density can do a lot in the right place and at the same time pay for the restoration of a great deal of open space, complete with creeks, ridgelines, farms, parks, and playing fields — whatever the community wants.

When we shift density away from car-dependent areas toward pedestrian/transit centers, it is important to create far more mixed-use development where the development takes place. The overall objective is to bring most people within walking or bicycling distance or a short transit ride of the places they need to be for a full range of their lives' important activities. For occasional pleasure trips, cultural and social involvements at greater distance, and relatively short commuting to jobs, they can use transit, but long-distance commuting is intrinsically a bad idea. Therefore development rights should be shifted so that if, say, there is mostly commerce, jobs, education, and so on near the centers but not much housing, more housing is created. That's often called "balanced development."

In addition, it's important to create the density in a way that is pleasurable and ecologically healthy. This book is replete with visions of terraces and rooftops buzzing with life. I am convinced that the meaning of a beautiful, fun, money-making larger building in the right place, with spectacular views to the local bioregion and associated natural features such as restored waterways, would be lost on no one. The fears that those projects arouse in distant neighborhoods would evaporate in the face of such successes. TDR can help build such projects and at the same time restore open space, thus benefiting the whole city.

The Double TDR functions because the developer is given a bonus in density when he or she pays for the transferred rights. One unit in a sending site might, for example, get the developer five in his or her new building — which is good for those who need housing — and, if built in the right place, greatly reduces commuting. It's important to hold standard height limits relatively low at the sites where the rights to build can be exercised, thus creating the incentive to trade upward to tall buildings as transferred rights are purchased and ecological building features are added. For example, a city might decide to limit buildings to five stories in its downtown but allow three more stories if TDRs are purchased in a particular quantity and another three stories if features such as bridges between buildings, terracing, public space on the sixth floor in a restaurant or café, promenade, or mini-park, or solar greenhouse are included. Each city would have its own formula depending on the climate, sun angles, history, and the hardiness of its population.

It should be emphasized that the Double TDR is a standard free-market exchange; it requires a willing seller and willing buyer. The idea is not condemnation, the compulsory purchase by eminent domain, or forced market value compensation. Instead, as a result of the design of the ordinances involved, the deal is attractive for all parties. If the new development is in the right place, of course, transit and bicycling work better, energy is conserved, and local businesses thrive. The ecocity zoning map is the key tool for directing where the development rights should come from and go. A sense of the proportions of restoration and development can be developed by thinking through the relationships indicated on the map.

To encourage the transfers that open up nature while building the city in the right places and with the right mix, restoration tax credits would help greatly. Developers would apply for credits by demonstrating that the project proposed is in the right area (in or close to a transit center), will add density and

diversity there, and will remove development to restore a creek, create a greenway, expand a community garden, or consolidate an interrupted railroad right-of-way. Any real estate within a hundred feet of the centerline of a creek in the outer zones of an ecocity zoning map, sixty feet in middle zones, and thirty feet in high-density inner zones, could be defined as an eligible sending site for development rights. Developers removing improvements there could total their expenses in purchasing the property, removing and recycling the improvements, and restoring the creek as restoration project expenses and be

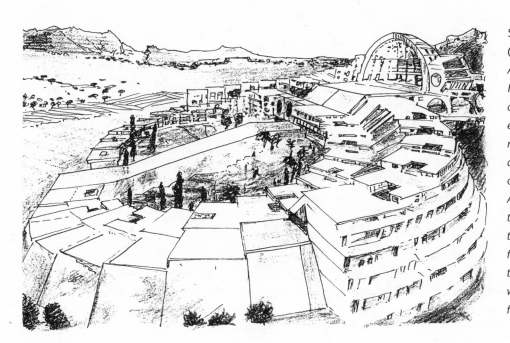

Soleri's "arcologies" (single-structure cities). *Architect philosopher Paolo Soleri has proposed architecture on the scale of ecological cities tuned to nature. In this illustration and the next we see drawings of two of his many models. Above, a proposed research town in a cool desert region that receives most of its heat from solar greenhouses on the sunny side, with the whole town accessible by foot, bicycle and elevator.*

awarded tax credit certificates by the state and or federal government. Developers with large tax bills might decide to use the tax credits to reduce them. Others might sell the tax credit certificates to a company that wanted them right away. Anyone who bought the tax credits would be helping the restoration project. Because the restoration would now be assured, the developer could be awarded the building permits for the restoration site work and approvals for the added height, density, and fancy features of ecological development at the project construction site. This is essentially how preservation tax credits work for development projects protecting historic architecture. It's time that waterways, hills, trees, and soils were considered as valuable as historic architecture and as worthy of restoration.

Here, then, with Double TDR and restoration tax credits, we have a means for significantly reshaping cities. In many cities there are constant complaints that there is not enough open space, not enough parkland, and not enough money to buy more land for them and pay for maintenance. But with Double TDR we have a mechanism for creating open spaces and parks while providing new housing and developing a larger tax base from the new higher-density development.

An important detail: the wider the setbacks for the creeks, the less maintenance work and expense per acre. This is because creeks then have enough room to meander back and forth a bit, causing nobody any problem, just eroding one way for a few years, then the other, and back. The normal level of erosion is a good thing since that means sifting out sand here, mud there, pebbles somewhere else, creating many different shifting environments for the eggs of different insects and fish and creating other beneficial micro-environments in and next to the creek. The whole idea is to let the streams be as natural as possible. If the creek corridor is wide enough, there can be a buffer zone of minimally managed landscape with perhaps a bicycle/pedestrian path or urban orchard of the sort we created at our restoration project on Codornice Creek. The creek itself and its banks can be almost completely wild. TDR and restoration tax credits make it feasible to purchase and maintain enough land to restore major natural and civic open space. In Berkeley, at the number of lineal feet per year daylighting is progressing, it will take over 5,000 years to open up the buried creek system. This is not an exaggeration or a joke. Simply divide the number of feet opened since 1982 in three small projects by the number of years elapsed, and you'll get the number of feet opened per year. Then divide that number into the number of feet still remaining locked in underground culverts, and it comes out to over 5,000 years. Real estate tools like ecocity zoning maps and Double TDR can speed that rate up many times over.

Rick Pruetz, who wrote a book on TDR called *Saved by Development*[1], says that a revolving fund is helpful for rolling back sprawl development through Double TDR. Any nonprofit or municipality can create such a fund to buy land and sell development rights so that the rights to develop can be shifted to other parts of town. They can call it a Double TDR Bank. Funds can be gathered from contributions from individuals, foundations, businesses, governments, or any combination of these, and the nest egg turns into land and buildings at the time of purchase. The seller gets the money from the bank. The building or buildings are then removed and the building materials recycled. Then nature, agriculture, or some other open space for other purposes is restored. At a later time when another developer buys the development rights for use elsewhere, the fund in the Double TDR bank is recapitalized with the developer's money and the land can be maintained or deeded over to the city, a land trust, community group, or some

Another Soleri "arcology." *Drawing of a model of a hypothetical town for about 20,000 built on an artificial lake on a natural river.*

other steward. This puts the bank in a position to buy more real estate for further transformation of the urban structure and further restoration. If major foundations or big donors to civic or environmental causes catch on to the potential, the fund might grow quickly and some truly magnificent projects might transpire.

We could think of this revolving fund as a "should-be open space acquisition fund." It would buy real estate where buildings are in the wrong place with regard to automobile dependence, floods, efficient urban structure, railroad right-of-ways ill-advisedly built upon, and so on. The municipal, state, or federal government could set up or contribute to such a fund and eventually come out ahead. They would save money by not having to build as many highways, and the city government would make more money in taxes from the new development. The developer who buys the transferable development rights will make more money, too, by being able to build more. The private individuals needing workplaces and housing will get just that, and in a place served well by transit and full of cultural benefits. And, in the most general terms, people and nature will thrive in an urban environment that is ever healthier and more vital.

The Ecological General Plan

Cities get their broad directives for zoning from a general or master plan. This document expresses the intended tenor of relations between citizens and their built and natural environments and is one of the most important instruments in shaping cities and their functioning for decades at a time. The zoning ordinance, which has the force of law, is based on the General Plan and is supposed to be consistent with it, though politics does produce inconsistencies as people and their leaders change their minds over the years. In its broadest definition, the General Plan is a framework for public decision making. It is made up of different elements: land use, transportation, public safety, open space, citizen participation, and so on. General plans and the zoning codes based upon them lay out a vision for the city and policy directives for actualizing that vision. The zoning ordinance gets precise about the details, down to specifying dollar fines for violations of the ordinance.

Citizens wanting to build and maintain an ecologically healthy city would do well to make sure that vision is explicitly stated in the introductory paragraphs of their city's General Plan. Then the document should follow through by spelling out policies and actions throughout. General plans draw general guidelines but can also get quite specific about what should go on in particular locations. One could argue that General Plans calling for the health and safety of its people would provide a policy context for ecological ordinances so that an ecologically healthy city can be built. But unfortunately, much contained in

General Plans, as in the zoning codes, too, gets in the way, such as calling for low height limits near transit which makes it impossible for transit to function efficiently, or calling for large quantities of parking for new buildings that guarantee a glut of cars.

In Berkeley, for example, there are "eco" elements in the General Plan, but some of the most important pieces are missing. Berkeley has good recycling and a Styrofoam ban supported by the General Plan and ordinances. Other policies that are at least somewhat encouraged in the city's General Plan include moderate but significant support for bicycle parking, bicycle paths, and bike lanes marked on streets; green building policies for nontoxic, energy-conserving, and recycled building materials; encouragement for at least one block of pedestrian street (though no such street exists at this writing); encouragement for traffic calming on residential streets such as the Slow Street; encouragement for creek daylighting and community gardening. But the degree of support in all these cases could be much stronger, and a long list of different and even more important policies would be required to qualify the plan for something heading towards an ecocity general plan.

In the summer of 2000, Ecocity Builders Program Director Kirstin Miller and Berkeley Planning Staff member Andrew Thomas suggested that concerned community members write a list of crucial policies we thought could nudge the General Plan toward becoming a real Ecocity General Plan. We would call them collectively the "Ecocity Amendment" and try to convince City Council to adopt them into the General Plan. We decided that there were four policies that had a fairly good chance to be adopted: encouraging centers-oriented development, establishing TDR, supporting an ecological demonstration project such as Ecocity Builders' Heart of the City Project (which we will look at in some detail later), and laying out some funding mechanisms for these policies. We could argue the policies' positive contributions and gather support.

To say Kirstin and I were methodical and perseverant is an understatement. One year later and after up to as many as four meetings with some of the boards of directors, we had assembled 103 organizations. We had bicycle and transit organizations, the Berkeley High School Ecology Club, University Coop Housing and the Student Union at UCB. We had the Berkeley Ecology Center and local Sierra Club chapter, two creek groups, gardening and park organizations, several architectural offices, the CoHousing Company, some businesses including two developers, and many other organizations signed on in support of the policies. Then, after meeting with council members individually, we went before City Council with thirty supporters in the audience to face the Berkeley Architectural Heritage Association (BAHA) and the Council of Neighborhood

Associations, an organization that tries to stop virtually all development everywhere in the city.

To put it charitably, our opponents, numbering less than half our supporters that night in Council Chambers, used distortion, fear, and anger. Council gave us a mixed but mainly very unhelpful bag, encouraging with weak language the ecological demonstration project and rejecting the other three policies. TDR lost with no specific debate on its potential or merits. That was the strong one, the policy that cold have made substantial progress in the city, from the land use foundation on up, toward transit efficiency, creek restoration and biodiversity enrichment, pedestrian convenience, more housing for people needing it, energy conservation, reduction of CO_2 output, lessening of automobile traffic and hazards, and more. Again, as in the case of car-free housing two years earlier, it was the City Council "progressives" voting no. Why? To maintain a left-wing stance against developers? To support a conservative no-change agenda? Threats?

I got a call from the General Manager of Chez Panise, the famous California/French cuisine restaurant that had signed on for supporting our Amendment. One of the officers of BAHA had called the restaurant and warned them they would be in trouble if they didn't remove their name from our list of supporters. "Do you think we'll get in trouble for signing that petition?" the man-ager asked me. "It's a free country," I said. "I don't see how you could." They stuck with us but it didn't convince City Council. I doubt that ultimately the threats really mattered; instead, the council members in question are among the leaders in keeping everyone comfortable with as close to no change as possible. Whether they believe it or not, and despite their claims to progressive sentiments, they lost a major opportunity in city design history and helped maintain the cultural denial of the growing environmental debacle. They also squelched a social justice policy of real strength, belying their espoused position on that issue as well. It was a tragic lost opportunity. Since then, Patrick Kennedy, the developer of the Gaia Building, has completed five more residential buildings in Berkeley. Four years earlier, he said, he would have been happy to buy TDR for an extra floor or two. If City Council had passed the TDR policy that night, we would have had twelve to twenty-four properties purchased for opening creeks — in other words, a lot. And I am convinced people would have been happy with the results. Several homeowners with houses over crumbling old creek culverts and no way to sell their endangered houses in a normal real estate market have wanted to sell for some time. City Council cancelled that option for them that night as well as dropped the ball for leadership in ecological city design and planning.

We need to be very clear about bad process, even bad "democratic" process. Berkeley has a reputation for giving its citizens repeated and substantive opportunity for participation in government, so much so I used to joke that City Hall needed a plaque over the door saying, "Process is Our Most Important Product." But when elected officials endorse what they know full well to be contrary to their own supposed values (in this case environmental and social justice values) and go against as many organizations as Ecocity Builders had assembled wanting to try out a set of ordinances that provide more, not fewer, free choices and options to their citizens, it should be known they are doing a disservice to democratic practice itself. One should not endorse people using misinformation, fear, and anger over those who go directly to the missions and values of a community's best organizations. In the meetings with the organizations that supported the Ecocity Amendment, real deliberation took place. Before City Council, the sum of that effort was clearly stated and yet, knowing the difference, the elected officials went with the approach that used disinformation, emotional extremes, fear, and anger.

Listening to the story later, an environmental activist asked why, with that sort of support in the community, we had not gone back to the community and fought it back before Council again. Answer: exhaustion. The community was exhausted with a contentious planning process and didn't want to revisit it after that final vote. On our own, a small organization making a gigantic effort like that, exhausted our resources — and that means a lot too. In addition, we had no confidence that the council members would be any more reasonable the second time around.

Regarding General Plans in all cities, we need to methodically shift from automobile-dictated development patterns to pedestrian-, bicycle-, and transit-oriented land uses and development — and to say so explicitly in the General Plan. There should be a policy to methodically reduce parking — a good model is Copenhagen, which is cutting back about two percent per year — while encouraging bikes and transit and, especially, while taking care to shift land uses toward balanced development. Any new parking built should be a temporary replacement for parking lost due to other changes in the city infrastructure, and it should be easily convertible to other uses — in short, convertible parking. Low ceilings and sloped floors must be avoided so that other uses can be easily accommodated in remodeling, for example for housing, shops, or, as in the urban permaculture example from Berlin, day care and nursery schools. Of course, any ecocity General Plan worthy the name would have to adopt an ecocity zoning map; a very major step would be the establishment of an Office of Ecological Development (which we will look at shortly) as part of an "International Ecological Rebuilding Program".

With an ecological General Plan in place, a city would have the written mandate to shape policies to manipulate the city's land use infrastructure and create the physical reality of an ecocity over, I'd estimate, two to five decades. Not for the impatient, but substantial benefits would start accumulating with the pursuit of such ecocity policies, as Jaime Lerner says, within two years.

Roll Back Sprawl

You've heard of "slum clearance." What we need is "sprawl clearance." Tools to roll back sprawl development exist. With strong interest on the part of legislators, we can strengthen them considerably, craft a few more, and make it profitable to implement and replicate them. With Jaime Lerner-like appeals to the people, we can create a culture of acceptance with its own imagination to shape the many unique places in this country, and in all countries. Millions of people lament the loss of better times and the better towns that went along with them. Here's a way to get them back and at the same time build better cores for our cities in ways that actually address the future. The strategy of a Roll Back Sprawl campaign is simply to identify means to remove sprawl and shift development toward evolving pedestrian/transit centers. Double TDR, supportive zoning, and city government commitments to purchasing and removing car-dependent real estate, foundation and investor support for the transition — all these

can be facilitated and accelerated by such a campaign. It's first order of business should simply be to let people know that such changes are possible, that tools exist, and that we already know they can work well.

I spent several months in 1999 researching not just the possibilities for a campaign against sprawl but also means to reverse its spread, to roll back sprawl toward pedestrian/transit centers. I found that of the several larger environmental organizations I talked with, none wanted to join such a campaign unless other major organizations or foundations got on board first. The Sierra Club national office, which runs a campaign called "Challenge to Sprawl," was satisfied with action far short of working systematically to remove sprawl development and was unclear on the concept of urban ecological whole systems. Most Sierra Club members seemed to think they needed their cars to get out into the wilderness. Said one Sierra Club leader, "Our members say, 'sprawl very bad. Cars? Pretty good!'"

Once we could use trains, horses, and bikes to get out into nature — and if we design the right way, we still can. Sounds fantastic from inside the blinders of today's auto world, doesn't it? The Sierra Club's campaign against sprawl supports infill development along corridors up to about four stories but is fearful of talking about higher-density centers, convinced that higher density than that is politically unpopular. I question their assumption, since millions of people in the

United States work and/or live higher up than four stories — hundreds of millions worldwide. There is the theoretical problem with filling up corridors with four-story development too. As mentioned earlier, it hardens the arteries and makes restoring natural zones and corridors that would cross such streets much more difficult to obtain. It perpetuates the pattern of low-density cities surrounding little islands of "park" and works against the centers-oriented pattern of cities as pedestrian islands in nature.

As we spread the word about ecocity design and planning and continue to refine tools for rolling back sprawl, the day may soon come when a Roll Back Sprawl campaign will make so much sense as to be easily organized. I am convinced that it could be among our most important tools for creating ecocities and building a healthy future. With the sort of demonstration projects I've been describing, people could begin to put two and two together — building right in the first place and also removing wrong. If the possibilities offered by reshaping our cities with these restoration/development tools can capture the attention of creative people and tweak their sense of the possible, an explosion of good projects could ripple, then roll in waves across the continents. The land under millions of acres of asphalt yearns to breathe free, and real community longs for expression — a Roll Back Sprawl campaign is the means to achieving both.

The International Ecological Rebuilding Program

We have now looked at several new tools designed specifically for ecocity building, but where is the institutional support for all of this? Where is the scheme, plan, or program in which the ecocity zoning map would work and we would design Double TDR, pass them as zoning, and apply them through everyday administrative practices rather than trying to improve, by making less pollution, the very infrastructure causing the problem? Perhaps ecological rebuilding could come about chaotically — a little here, a little there, in a pattern not too different from today's groping forward and backward and around in circles — but I doubt it. In a crisis like the one enveloping the biosphere today, it would be helpful if there were a concerted effort to build as if we thought building had something to do with a crisis like the one enveloping the biosphere today. If we set the goal of bringing society into balance with nature — and set out to develop a methodology for achieving that goal — we would have a context in which the transformation would have a far better chance of success. But we have not yet made such an effort.

Why don't the governments of the world have ecological development departments dedicated to a vision of an ecocity civilization unfolding? Are they not supposed to be working for the common good? Haven't there been enough discussions in

In a crisis like the one enveloping the biosphere today, it would be helpful if there were a concerted effort to build as if we thought building had something to do with a crisis like the one enveloping the biosphere today.

the environmental protection agencies of the world, enough environmental conferences, for people to have caught on to the necessity of a major rebuilding? There are serious international efforts to cut CO_2 emissions, but where is the work going on to create a treaty on ecocity development and restoration that would solve the problem at the level of its causes? The governments of New Zealand and the Netherlands are leading the way with their own national "green plans," but they are not focusing on the built habitat as centrally as they should be — they are not being quite so presumptuous as to call for a genuine rebuilding of our Western technological civilization. The US Green Building Council and the authors of the LEED (Leadership in Energy and Environmental Design) standards and certification process support better buildings but have no LEED standards for better whole communities and thus bestow high marks on buildings dependent on hundreds of thousands of gallons of gasoline every year so that people, in their cars can even get to the buildings. If there were a scheme for rebuilding civilization, it would sort out contradictions like that.

If there were a scheme for rebuilding civilization, its name would be something like the International Ecological Rebuilding Program. Al Gore had a similar idea in 1991 when he was writing his book *Earth in the Balance*. He developed the idea in some detail, addressing

the need for a major reduction of pollution, a restoration of nature where possible, and an organized effort to promote technologies that conserve resources. "Human civilization is now so complex and diverse, so sprawling and massive," he writes, "that it is difficult to see how we can respond in a coordinated, collective way to the global environmental crisis. But circumstances are forcing just such a response; if we cannot embrace the preservation of the earth as our new organizing principle, the very survival of our civilization will be in doubt."[2]

Gore proclaims that there are "no precedents for the kind of global response now required" but does point to the Marshall Plan, which organized much of the rebuilding of Europe after World War II. He credits that plan with enormous success and proposes naming a new initiative after it, a Global Marshall Plan that would have five major goals: (1) population stabilization, (2) the development and sharing of appropriate technologies, (3) new global "eco-nomics," meaning ecological economics, (4) a new generation of treaties and agreements to accomplish ecologically healthy ends, and (5) a new global environmental consensus.

The chapter on "Developing and Sharing Appropriate Technologies" is as close to confronting the built civilization as Gore will bring us. Under the subheading "Building Technology," he calls for passive solar design and greater energy efficiency in buildings. In other places he speaks of the benefits of decentralized electricity generation and expresses some surprise and delight that wind electric energy is economically viable and promising for larger scale applications in many locations. He speaks of "emphasizing attractive and efficient forms of mass transportation."[4] He even makes one of the most stunning statements against automobiles I have ever seen: "We now know that their cumulative impact on the global environment is posing a mortal threat to the security of every nation that is more deadly than that of any military enemy we are ever again likely to confront."[5] (Where was he when we needed him during his eight years as head of the White House Office on Environmental Policy?)

Gore's Global Marshall Plan, however, says almost nothing of that created object in which most of us live, that invention for maximizing exchange and minimizing transportation that Jane Jacobs describes as the chief engine of industrial and cultural production and consumption, that thing that can be designed and physically rearranged to reduce demand radically and therefore add to energy efficiency like nothing else: the built community rearranged as ecocity, ecotown, or ecovillage. The city simply does not appear, much less serve as the foundation of the plan — even though it could and should.

More recently, in 2003, Lester Brown, founder of Worldwatch Institute and Earth Policy Institute, wrote a book called *Plan B*.[6] (Plan A is the conquer, exploit, and control approach while hoping for the technological fixes for those pesky environmental problems, commonly used by government). In his second edition of the book Plan B 2.0, Brown has a whole chapter on "Designing sustainable cities". In addition, he provides much of the kind of detailed information about the condition of the planet's resources that anyone interested in a systematic approach to reshaping our civilization needs to know. He knows that we need such a plan and that it needs to be pursued with the resolve of fighting a war for our defense and survival. Writing the book, he provided the germ of an idea that could coalesce the real thing. Now we need to get that idea out to people everywhere.

In 1991, I tried my own hand at an outline for an International Ecological Rebuilding Program and took it with me to the Second International Ecocity Conference in Adelaide, Australia, the next year. There it was amended and adopted. Two later versions were adopted at subsequent International Ecocity Conferences, in Yoff, Senegal and Shenzhen, China. The following paraphrases parts of the various:

1. *We must declare an emergency in human and environmental affairs and create programs specifically for ecological rebuilding in every country and in the United Nations. The emer-*

gency is not temporary. We are entering a period of permanent emergency, and we will cling to the edge of this precipice until we fall off or solve the problem.

2. *Energy policies must be linked to ecological development.* We need to recognize that energy powers *something*, and mostly it is the city, town, and village — the built human habitat. The ecologically healthy structure of the city is the foundation for energy conservation and should be item number one in any energy strategy.

3. *Because living systems cannot function well when they are effectively cut up into isolated chunks, we should establish Departments of Ecological Development on the national, state, and local level.* The United States and other countries have environmental protection agencies empowered to enforce environmental regulations for the prevention and amelioration of pollution, but these agencies don't build. There are, however, housing agencies and other departments that do build, using their own construction corps or directing grants, loans, and contracts to builders of machines, buildings, infrastructure, and products of all sorts. We need governmental departments or agencies that coordinate ecological objectives with actual construction. It is important to see that building right in the first place is at the root of environmental protection, and the Department of Ecological Development would be charged with just that. Under it there would be research wings such as the National Renewable Energy Laboratory and an Ecocity Research Institute that would assist projects from the small integral neighborhood scale up to whole new-town projects like Arcosanti and major ecological urban demonstration projects in cities of any size.

Departments of Ecological Development should initiate aggressive spending programs to develop renewable energy technologies and ecological community building as two coordinated facets of the same overall effort. They should transform defense programs and companies into builders of elements of ecocities and associated technologies and products and reward pioneering companies in these fields with profitable contracts. They should make federal, state, and local moneys available to ecocity projects as loans, grants, and research and development contracts. They could provide assistance and oversight to other governmental branches as well, so that in relation to ecological building and ecological policy in general the left hand would know what the right was doing. They could even build their own experimental projects. Whereas environmental protection agencies function appropriately on the federal and state level, there should be Departments of Ecological Development on the municipal level as well.

4. *We need ecological rezoning, complete with ecocity zoning maps.* We need programs to roll back sprawl and restore wildlife habitat and farmlands, withdrawing from tracts as large as the proposed Buffalo Commons and as small as narrow creek and wildlife corridors in the cities.

5. *We need economic restructuring, i.e., phased, steadily increasing taxes on pollution and energy waste.* A land tax could be designed to shift society toward ecocity development patterns. Taxes per square foot of developed usable floor space should be descending toward the centers, while toward the fringes, in automobile-dependent areas, taxes should be rising (except for natural habitat and agricultural land, which should pay no taxes in the city at all). Such taxing can work as powerfully as outright zoning change, and so can the restoration tax credits described earlier.

6. *We need not only to develop foot, bicycle, and public transportation, we also need to put transportation into the land use context.* Politicians and everyday citizens can use imaginative leadership and planning to allocate city, state, and national funds to nontransportation modes of access, a practice that was started in the US with the Intermodal Surface Transportation Efficiency Act. Building diversity at close proximity is the most effective route to the same end as efficient transportation: access. Therefore we need Departments of Access and Transportation on the federal and state levels that could still deal with conventional transportation strategies, but would emphasize providing access through ecological urban and architectural design and planning of the city layout.

7. *Automobile subsidies must be ended.* We can start with a steadily increasing gasoline tax and a tax on second cars, then add a tax on all cars, then increase taxes on all of them, and finally charge drivers for the smog damage to crops (money to be transferred to farmers) and to people with lung cancer and emphysema (money to the victims). Insurance companies could pay these victims and pass the cost along in higher automobile insurance rates. As a pedestrian advocate in my neighborhood suggests, we could require drivers to pay pedestrians for time wasted at traffic lights — hours every month — by redistributing part of the car taxes as tax rebates to non-drivers.

8. *We need to develop strong educational and economic incentive programs for the ecological rebuilding effort.* No one should be abandoned in the transition. Retraining workers and retooling industry following the four steps to an ecology of the economy to produce and operate ecocities is the plan.

The Ecocity Organization
The ecocity organization is a rather everyday type of association of people working together, chipping in dues, running fund rais-

ers, doing mailings, hosting events, promoting what they feel improves life, and so on, but it has an extraordinary mission: it's an organization designed specifically for exploring the theory of the ecocity as well as for experimenting, learning, teaching, and building ecocities. I know of very few organizations that are explicitly just that: Ecocity Builders, Urban Ecology Australia, Urban Ecology China, Ecocity Cleveland, and the Cosanti Foundation. If we included ecovillages, then the Global Ecovillage Network out of Denmark would also qualify.

There are many organizations that protect one aspect of the environment or another and a fair number that provide expertise on energy conservation and recycling to community groups. There are professional associations like the Congress for the New Urbanism that have theories on urban design and work to promote their ideas tending in an ecocity direction while benefiting their architect and planner members. There are public transport and greenbelt advocates, bicycle clubs and bicycle-promoting organizations, strictly anti-car organizations, "road ripping" and dam removing organizations, anti-oil industry groups, wilderness protectors and river and creek restorationists, and community gardening associations, and permaculture groups whose design principles are closely related to ecocitology's. There are academic institutions like Jeff Kenworthy and Peter Newman's Institute for Science and Technology Policy at Murdoch University in Perth, Australia, that study the structure and functioning of cities and advocate for pedestrians, bicycles, and urban transit anatomy over automobile land use infrastructure. There are city governments like Vancouver's and Curitiba's that are writing and executing policy while building features that help them convert their cities in an ecocity direction. They run in-house ecocity organizations, such as the Planning Department in Vancouver and Curitiba's IPPUC.

But we need a clear, specific focus on basic principles — a scientific approach that is not yet a popular preoccupation, but that is simply looking for the truth about the relationship of the physical community to ecology and evolution. We also need to involve millions of people, and therefore we need organizations in every city. We need organizations that try to put all the pieces together. Rusong Wang, host of the Fifth International Ecocity Conference held in 2002 in Shenzhen, China, and president of Urban Ecology China, has proposed an International Ecocity Society that would promulgate ecocities and consult on ecocity development around the world. To join in the real action in Vancouver or Curitiba, you'll have to be hired onto the government team. But you can also join one of the ecocity non-profits (usually called non-governmental organizations, or NGOs in international circles) or start your own organization for similar purposes. With your supportive thought, work, time, and money these organizations could do more than practically any other conceivable tool to transform our cities.

Arcology Circle, as it was turning into Urban Ecology around 1980, was probably the first real ecocity organization. It was not just taking on the theory and practice of the three-dimensional pedestrian city as was its predecessor, the Cosanti Foundation, but applying the ideas to existing cities. After all those years, all those minuscule budgets wrung from a few hardy and faithful souls and a small number of unusual foundations willing to take a little risk on a new idea, it is evident that our work is a genuine struggle. Some people congratulate us on doing exactly what has to be done but decline to join or help because we will do it anyway.

I've thought long and hard about why so few become involved in ecocity organizations and why most foundations decline to help us while telling us we are doing great pioneering work on one of the most important issues going. Now I think I know the answer. We point the finger at ourselves, and only a few are strong enough to face that truth in us. It's one thing to blame distant corporations, globalization, the loggers and industrial farmers, the greedy shareholders, power-hungry executives, vote-grasping politicians, and those other folks who drive their cars too much. It's quite another thing to see that we may all have to change — "We have met the enemy and he is us" — and not only that, but build something that has never been built before. The ecocity organization requires three rare things of its members: a willingness to confront our complicity, a great deal of creative imagination, and hope in the face of depressing facts about biodiversity collapse and climate change. A very small band of supporters from a diversity of perspectives — barely enough to keep things going — is all we have had for more than thirty years. But the point of greatest resistance, in typical paradigm shift theory, is also the place where we may well have the real breakthrough. Nobody said this would be easy.

CHAPTER 11

What the Fast-Breaking News May Mean

FOR THE BIG THINGS, time ticks very slowly. "Old as the mountains." I am writing this at the start of the Chinese New Year. We are launching the year 4703, aka 2005. Emperor Huang -ti introduced the first cycle of the zodiac in ancient China about the year Westerners call 2600 BC.

This means more to me now than at the time I'd finished my first edition of this book (2002) because I've been to China several times since then. Big place, long history. Yet things are changing quickly in China. I watched over three short years as roads stretched out and buildings popped up with amazing rapidity. Earth Policy Institute founder Lester Brown says the big change for the planet might not wait for the effects of collapsing oil availability but come rather suddenly with China crossing over a certain divide from producing more food than it con-

sumes to consuming more than it produces, almost immediately changing the world food and economics calculus. China's grain harvest peaked at 392 million tons in 1998. By 2003, the harvest was down by 70 million tons. For comparison, that exceeds the entire grain harvest of Canada. Nonetheless, in 2003 China was still self-sufficient in wheat. But in 2004, China imported 8 million tons of wheat. Says Lester Brown, China's "purchase of 8 million tons of wheat to import in 2004 could signal the beginning of a shift from a world economy dominated by surpluses to one dominated by scarcity. Overnight, China has become the world's largest wheat importer."[1] Could 2005 or 2006 be the threshold of the era of "peak everything"? That is, of the average peak of the total human production of oil, food, wood, fish, etc. Just a thought. I don't have the figures.

I mentioned my 1971 interview of solar energy scientist Aden Meinel in the preface to this edition. He had warned that we would need to invest a considerable amount of our cheap fossil fuel energy in renewable energy technologies or else the option of a civilization powered by renewable energy at anything close to our "World Civilization's" level of cultural product would evaporate before our eyes. Now I think again of China, this time of the ancient symbol often called the "yin-yang" symbol. It represents one of those big things, one of those ancient, maybe permanent principles. It's the symbol that looks something like two water droplets or two tadpoles swimming around one another, one black, one white, together forming a circle. The black shape has a white dot in its middle; the white shape has a black dot. In other words, in the opposites there is always something of the other. The symbol exists only with the two shapes together, or not at all. Very interesting.

To me it describes those eternal dimensional pairs, each of which is nothing without the other. It represents a pair of dimensions of our reality, like time and space, or energy and matter. More to the point, I suddenly saw the black droplet shape as the built, physical aspect of our civilization and the white shape as the renewable energy flows we need to tap into — or else the thought of sustainability of life on Earth is strictly delusional. The renewable energy system needs a built infrastructure of our civilization to fit. Simple as that! No fit, won't work. The physical stuff and the energy that animates it are "integral" to one another. To be integral to one another, the infrastructure system and the energy system have to be designed for one another, and essentially designed together, designed at the same time. The two parts of the yin-yang symbol, in this case, are the ecocity and renewable energy.

There is a big problem here. As mentioned earlier, no such renewable energy investment of the sort required by Aden Meinel's warning has been made. Thirty-five years later, there has been even less of an investment in the ecologically tuned physical infrastructure, the ecocity.

To put that into the context of the history of Western cultural civilization and its history of ideas gives us an interesting perspective. In 1933, philosopher Alfred North Whitehead said that general ideas are often so much accepted without question that they are not even discussed in the culture that holds them. Like the idea on Easter Island that trees will always just be there no matter what Easter Islanders do; like the idea in America that cars will always be there no matter what Americans do to oil reserves. Whitehead surprised me when he revealed one such absolutely basic idea in Western Civilization that I had not thought of in this way thusly: "Throughout the Hellenic and Hellenistic Roman civilizations — those civilizations which we term 'classical' — it was universally assumed that a large slave population was required to perform services which were unworthy to engage the activities of a fully civilized man. In other words in that epoch a civilized community could not be self-sustaining."[2] He pointed out the great contradiction between Greek ideas of democracy and the institution of slavery, which was virtually not questioned at all, despite the Greek's famously wide-ranging discussion of everything philosophical.

This goes way back. Remember Egypt, Babylon, Rome, China, the Mayan, Aztec, and Incan civilizations. But early in the 19th century, England had the moral fortitude and imagination, but mostly its place as the first nation on the scene of the Industrial Revolution, to eliminate slavery. In 1808, it abolished the British slave trade. In 1833, it bought and set free all the remaining slaves in the British Empire at enormous expense to the country's treasury. Energy slaves, some suggest — that is, machines powered by fossil

Same general arcology approach for three climates. *Clusters of "arcological" structures for mild, hot and cold climates of my own design follow. The first, the mild climate version, lets sun fall into interiors while shutting out breezes on cold days with low sun angles.*

fuel — were taking the place of human slaves. Instead of the leverage of solar energy going through food to human muscle power, it has become solar energy going through fossil fuels to machine power. Those have been the real energetics of civilization; they build cities and make them and their various systems tick. Today, buying a gallon of gasoline (at, say, $2 per gallon), a typical American appropriates the energy of something like 60 hours of human labor for approximately 10 minutes of his or her own labor (at, say, $12 an hour). That ratio of 1 to 360 units of labor expended to labor acquired through gasoline is so gigantic as to approximate what was once dreamed of as the kind of "free energy" people might get from the mythic and elusive perpetual motion machine.

In other words, complex society has been profoundly dependent upon an enormous influx of energy to the human enterprise, whether through human slaves (very cheap labor) or machines powered by energy (very cheap labor). What happens when peak oil, if not peak everything, hits very soon? When machine slaves, powered by mechanical energy other than muscles, are cut way back, will we be constrained to choose between reintroducing human slavery and giving up complex material civilization? Very hard to say.

Peak oil, as mentioned here before, is particularly significant because oil was in massive supply and we developed a massive use of it as if it was infinite, but, alas, it is not. We can't

No, energy is not going to be easy to come by in the long term. And yes, the largest part of the solution is the extremely low-energy city, town, and village teamed up with renewable energy sources.

use the usual replacement strategy: find something else. Everything else is far weaker in energy delivered for energy invested — "net energy" — and far more environmentally damaging than the already damaging petroleum has been (coal, oil shale, tar sands, and nuclear). Or, more benign energy sources have not been adequately invested in (wind and solar). Hydroelectric power? The dams are slowly filling with silt and sand, some faster than others. Where I grew up in New Mexico, even in the 1960s there were several dams near home filled completely with dirt, covered with scrub, grass, and brush. I've walked across their bizarre out-of-place flat lands in valley bottoms. Theirs is the destiny of many dams, some with water flowing over the dirt and over the edge, with zero storage and electric generating capacity.

Finally we can begin to grasp the dimensions of the problem when we see that complex civilizations have always had (one is tempted to say, needed) enormous supplies of cheap labor, from slaves in the past to the cheap energy machines of today. No, energy is not going to be easy to come by in the long term. And yes, the largest part of the solution is the extremely low-energy city, town, and village teamed up with renewable energy sources. Think dimensional pairs, think yin-yang.

9/11 and 12/12: Dreams and Nightmares

In the preface to this book I said that the situation has changed greatly in the last three years, forcing something more than a simple updating in these pages. In addition to the kind of incremental changes in society, technology, economy and ecology we might expect at any time in history, there have also been enormous qualitative changes that indicate that we have passed over some threshold and — this is the bad news — into a different and rapidly degenerating situation.

Some say it all began with 9/11. But I'd say we have been plummeting backward for four years as of this writing (2005), unwinding decades of civility and whatever there has been of genuine progress in human affairs. I'd say an even more important date than 9/11 was 12/12 of 2000, the day the United States Supreme Court decided there wasn't enough time left at 10:00 pm — two hours — for the state of Florida, according to its own laws, to recount the ballots in the Gore-Bush election for President of the United States. The Supreme Court had stalled its decision for two

Hot climate version.
Large shade structures of the sort that are appearing in some of China's hot climate larger modern buildings, are prominently featured here. Cooling breezes blow horizontally underneath.

days and twenty-two hours while time ran out on the recount. Then the justices cited Florida state law, not their own action in making the law of Florida and the intent of that law void, for making it impossible to execute a recount. Thus the recount could not proceed, they — not the voters of the United States — said, and George W. Bush would be the next President of the United States.

Was the Age of Aquarius stillborn on 12/12/2000? "It was an appalling decision," said Bill Clinton in his autobiography *My Life*: "A narrow conservative majority ... had now stripped Florida of a clear state function: the right to recount votes the way it always had .... *Bush v. Gore* will go down in history as one of the worst decisions the Supreme Court ever made ...."[3] That's not even considering the disenfranchisement of thousands of lower-income and racial-minority voters in Florida obviously ready to vote for Al Gore who were harassed by police roadblock, made to stand in line until after the polls closed, left off of precinct roles, and otherwise prevented from voting at all. Gore won the popular vote by over half a million, not even counting those sabotaged and missing votes, but lost the Presidency by one vote at the Supreme Court. It was a *coup d'état*, not a bloody one, but one doing inestimable violence to democracy and justice nonetheless.

Obviously I'm a dreamer or I wouldn't be writing a book like this with the sort of drawings you see in these pages. But mine are dreams based on some solid thinking I see others are assiduously avoiding — at the peril of us all. I believe I have good information, and I am not in denial. My dreams are not waiting for miracles like refrigerator-sized clean nuclear power plants, but dreams that could easily be built with today's technology. I don't think it was a dream impression to think that humanity might have been inching forward toward an ecologically informed and humanitarian future. The ideal of a compassionate, creative society was actually making a little headway. Europeans had had enough of war and — amazing to me — were pulling together well enough to even mint and print a common currency, the euro. The World Court was gaining legitimacy with increased use. Co-operation, not one of humanity's strong cards, was nonetheless progressing in fits and starts. Environmental treaties were getting better and being respected by more people and governments every year, their legitimacy gradually growing. Information on resources and ecology was almost racing forward, and — window to the infinite — the most spectacular images were coming back to Earth from the fringes of time and the universe via the Hubble Space Telescope. I wasn't exactly expecting the Age of Aquarius to burst forth with Al Gore as President, but the Kyoto Protocols would have been signed and we would have continued to see sanity and grace gaining an ever better footing internationally. Or so I'd dared to hope based on some solid evidence.

But nine months before 9/11, the monster began gestating. Regression in civil rights of Americans at home, the posturing of arrogant unilateralism in foreign policy, and the advocacy of outright violence to gain wealth and control of resources, especially oil, began growing rapidly in the new administration. We all know the list, which is updated below. And by the way, about half of us in the United States *like* that list. Beginning before 9/11 and increasingly so thereafter, the Bush administration made policies and pronouncements to:

- disenfranchise voters through obstruction and whatever other means necessary
- abrogate past treaties and refuse to sign the Kyoto Protocols ("needs more study")
- dismantle environmental protections
- prepare for bringing back nuclear power plants and accelerating drawdown of fossil fuel resources
- assault protections for forests and endangered species
- empower religious intolerance while

Cold climate version in the snow.
Large openings to the south are featured here and partially enclosed with permanent glass windows and wind screens. Some of the glass is removable for the warm season.

bringing religion into government rather than maintaining the separation of church and state

- promote fear through gross overemphasis of violent threat and violent counter threat
- neglect policy initiatives that could reduce tensions caused by US exploitation
- glorify war and violence ("shock and awe," as compared to blowing people's bodies to pieces by the tens of thousands, the non-glorious reality)
- lie about reasons for going to war in Iraq — relentlessly
- bribe other nations to join in the American/Iraq War with tens of billions of dollars
- ignore past allies in decisions of war and international security
- vilify America's long-time friends who didn't approve of the invasion of Iraq
- break the tradition of international civility by declaring the intent of killing a head of state (as despicable as Saddam Hussein was)
- obscure the fact that Saddam Hussein had largely been elevated by CIA support and was partially a creation of people in the Bush Administration
- break international law by invading a country that had no weapons of

mass destruction or connections to al Qaeda and was no credible threat to the United States

- privatize the military at enormous profits to war-supporting companies and rebuilding contractors and at enormous cost to the American tax payer
- hold without charge and torture prisoners of war
- support the development of a new generation of smaller-yield "usable" nuclear weapons
- reserve the option for first-strike nuclear attacks against anyone deemed an enemy by the Administration
- plan the unilateral militarization of weapons in space ("In Your Face From Outer Space")
- transfer wealth and services from the lower-income people to the wealthiest
- drive the country into massive debt
- neglect the development of renewable energy

I'm not convinced the Democrats would be better than the Republicans in regard to helping build ecocities, but the above is plenty to sound the alarm that things are in a critical state of degeneration and possibly early stages of collapse. Besides, the Democrat/Republican divide seems almost beside the point since John Kerry ran for a "better war" and most Democrats in government supported most of the actions just listed.

How does one get a handle on building ecocities in the face of all that? One doesn't. New strategies, other institutions, and alert, caring, and courageous people of vision are needed. People desiring to build a peaceful and ecologically healthy civilization should expect little if any help, not even recognition, from that quarter. In fact, they — we, all of us who have somehow wandered into these pages — should be alert to the possibility that dirty tricks are being used routinely and that we could be targeted at any time, like anyone else. It's almost egalitarian how victimized the vast majority of us are lined up to be, even those who voted for George W. Bush who don't happen to be part of the great American Empire's aristocracy.

I have my own history and perspective in regard to 9/11, nothing new about how it happened or how the investigation into the events went or about US policy since, all of which is murky and intentionally so. Instead I was involved in what might be (or might have been) a meaningful response at that location in relation to building a healthy civilization. It goes something like this.

Sue Labouvie is a graphic designer I met at the Fourth International Ecocity Conference in Curitiba, Brazil. A bicoastal woman, she has a studio in New York City and a small house in Northern California.

Four months after the attack on the World Trade Center she asked me if I might have some ideas about what to build to replace the destroyed buildings. Because she had been doing pro bono graphics work for a group of architects in New York City called New York New Visions, a group of 21 architecture, planning, and design organizations encouraging good design responses to the catastrophe, the organizers said they would be happy for me to present. Though I wasn't a practicing architect, they would welcome some novel ideas from the environmental perspective from a Californian. They would be having a meeting shortly in which about fifteen architects would be showing models representing their ideas for replacement structures at Ground Zero.

I drew a number of pictures for the site on the airplane to New York and visited Ground Zero with Sue and her business partner, Max Heim. We explored the neighborhood, the first time for me since I was there in 1977. At that time I took the elevator to the top of the World Trade Center — the roof of one of the towers was open then. I took a picture of a small airplane that day in 1977 that was about 50 stories below me flying close in to the tower and over the Hudson River. I was impressed with how small the airplane looked below, later barely showing up on the photograph. Now it was all gone, two of the world's tallest five or six buildings. It still seemed there was a kind of stunned gentleness in the eyes of people on the quiet streets. Busy streets but strangely quiet and personal compared to what I'd remembered. On the way home to Sue's studio I bought some clay at an art store, and the next day made a model that I thought exemplified what should be built.

The day after that we walked to the gathering at the office of Fox & Fowle Architects hosted by Bruce Fowle. A large model of lower Manhattan was laid out on a table. Participating architects had made smaller models to fit the open space at Ground Zero in the larger model. My model was a little too big because, at a scale that would fit, I couldn't get the kind of detail I felt I needed to show my ideas. No problem. I placed my model on the table just to the west of downtown, in what would be the Hudson River, trusting those present could imagine it transposed slightly to the east. One young architect couldn't help herself and said, "Clay?!" with an involuntary laugh. I explained it was a great medium when you didn't have much time. I could change things quickly and see how they looked. I also didn't have woodworking tools with me.

One set of shapes after another filled part of the open space in the large model, avoiding the footprints of the original twin towers of the World Trade Center, leaving them open for a memorial. All models shown, other than mine, represented about 11 million square feet of office space. They were business as usual, in triangular and rectangular solids, looking pretty much like you'd expect in the modern style. The twisted and sliced forms of the architect who would win the competition for the site, Daniel Libeskind, were many months in the future. These forms, on the day I participated, were much more conventional, though one looked like a set of giant lanky sculpture legs.

When my turn came I started off saying we should first of all realize the role of city planning, design, and architecture in history at this time, and especially in relation to Lower Manhattan and what the attack on the World Trade Center might mean in the larger world picture. I suggested that Manhattan had a great deal to offer the world at a time of dwindling oil and growing extinctions, at a time of ecological crisis world-wide. I said that an enormous cloud hangs over this enterprise of designing replacement buildings at that site and that we need to see the horrifying events of 9/11 as a warning that things are not well in what is happening globally and at that location and in regard to universal issues like climate change, species collapse, and peak oil. Therefore we should look to the best New York and the United States has to offer at the World Trade Center site as an alternative to violence, exploitation, and business as usual. The best trade is human communications and understanding through mutually beneficial exchange. But we have to be careful it doesn't transform, as it often does, into exploitation and even war. (I already noticed a few eyes glazing over and some other people getting uncomfortable. Bruce Fowle and three or four others seemed interested, however.)

I noted that the density of New York, the stunning diversity of activity, and the well-used public transit system mean very few people have cars and the whole city runs on about half the energy per person as the average American city. In days of a coming and permanent energy crisis, that is very important to notice, I said. In addition, the people who live and work in Lower Manhattan represent the whole world of people, almost every race, nationality and ethnicity, and accept one another fairly well. Finally, perhaps the best and most auspicious symbol of the United States every created stands in clear view of Ground Zero, if you are up about 20 stories in the air and look over and between nearby buildings, namely the Statue of Liberty. All these things suggest a particular sort of design.

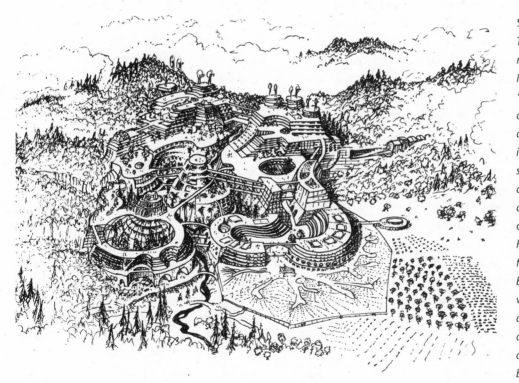

Semi-arcology.
This structure is also related to the "artificial land in the sky" concept of "table land." A large developer or government agency builds the basic infrastructure of slabs separated vertically by about three stories. Further development as residences, offices, shops, etc. — the highly mixed use city — follows with people building a kind of super version of "interior decorating." This town for about 20,000 is completely accessible on several levels by foot and bicycle.

Run with the head start New York has over the other cities, I suggested: density, mixed uses, and an international community. Build for that, and in a way cognizant of the environmental situation the whole planet is in at the beginning of the 21st century. Build a fully functioning international community. Build lots of business and commercial space as planned, but also housing and other activities. It can still be mostly commercial since my proposed replacement project would link to the residential buildings nearby in a highly conscious pedestrian way, including pedestrian bridges. In lower Manhattan more than 70 buildings had been converted from office to housing uses over the last two decades, I had heard. This trend could continue with the goal of maximum walkability for the area.

Specifically I suggested a large plaza or park floating about 20 stories in the sky over a very dense structure with pedestrian streets passing right through in large hallways. Skylights would drop beams of sunshine into the interior streets at various times of day. The main elevated public level with the plaza would have ample space for a memorial to the victims of the catastrophe, but also be dedicated to a peaceful future. Taller towers of varying heights would divide the open space and surround the main plaza in the keyhole plaza pattern described before, buildings framing the view. Happenstance has it that the site is positioned geographically so that the view to the Statue of Liberty, symbolizing welcome and acceptance of people from all over the world, is due south so that the whole structure would be canted toward the view and be solar passive designed. The linked and united buildings of this version of the replacement project, apropos Manhattan, would be very tall, perhaps almost as tall as the original towers but probably not quite that tall since, unlike the original towers, they would see themselves as harmonizing with Lower Manhattan, maybe stretching it higher, but not dominating it.

As readers of this book might expect, the whole project would constitute an enormous Heart of the City project, an urban fractal symbolizing and embodying a fully functioning, highly mixed-use community.

When I'd finished, the architects looked at me a little blankly, in silence. After an awkward moment with no questions at all — every presentation before and after got five to ten minutes of scrutiny — they went on to more presentations. When we were breaking at lunchtime, Bruce Fowle politely suggested I might like to talk with a couple New York "green" architects he knew.

What might actually get built? A set of commercial buildings re-expressing the commercial, rather than human, community of the world. Instead of welcoming people from around the world, it'd be saying, "We stand taller than you. Our 'Freedom Tower' is taller than any other building on Earth. We are free — whatever you happen to be." In fact, the whole enterprise is confused in many particulars. When the basic design was going on, from about the time I went to New York (I went twice more, attempting some influence) and for two years thereafter, people insisting on "sustainability" and "green" features did have some influence. And so, the tower from about the half way level up to 75 percent of its full height is actually an enormous skinny, linear, vertical greenhouse, which sort of counts as a building, but not in the sense of it being habitable or a working space for people, though I like the idea and have drawn rooftop greenhouses (claiming no influence there, by the way) on lots of my imaginary buildings. The "Freedom Tower" is, to my taste, overly symbolic and its vertical greenhouse inaccessible to people except in a rather remote viewing sort of way. With a very narrow needle going up the last quarter of its height and rising there only so that the tower can claim to be the tallest "building" in the world, and with the coincidence of the building's height of 1,776 feet and the date on the Western calendar of the US Declaration of Independence, something strictly nationalist and not oriented toward the best of world trade at all, I wonder what the message is here. Being "the tallest building in the world" and being named for the stated, if distorted, justification for invading Iraq — providing the people there "freedom" rather than wanting access to their oil — it presents more of a dare for future terrorism than a means of cooling the heat of exploitation that fuels terrorism. If someone makes such a dare at Ground Zero and such an obvious nationalistic game of it, is that someone being immature or what? Who wants an ego-bound hypodermic needle to represent freedom when just a few miles south a strong and nurturing woman stands in her sandals holding the light of liberty where ocean, land, and sky come together?

In the replacement project now planned, Libeskind has sliced through the buildings in a conspicuous effort to be dramatically expressive. I hate to say it, but it reminds me all too much of buildings hacked down by scimitars. I don't know if I'm being a cranky, sensitive peacenik in this, but on a deep psychological level it seems to me the planes of varying steepness that slice through the forms of the towers from one diagonal corner to the other speak unconsciously of the very sort of violence that leveled the original buildings. As a work of art trying to make people think about a gigantic disaster at the beginning of the 21st century, this is OK, but as a building to live with dominating the skyline, something more hopeful and looking to the future is called for. Even something just plain neutral would be better. I wouldn't want to be taking a romantic moonlit walk in view of a skyline like that. Overall, the replacement project as it is currently unfolding is conceptually muddled. It is not relevant to the needs of the times and the future. It is not a case of people's thinking rising to the occasion. It's a shame to have another lost opportunity, and one on such a grand scale as this.

The New American Empire

Soon we will look at climate change and peak oil — after a few words regarding the New American Empire, which is causing the largest percentage of the first problem and where most of that oil is going, leading to the second problem. To launch into this line of thinking and to take a strictly economic view of it, why do about half of us Americans actually like the list five pages back, the list representing the erosion of our rights and the escalation of international belligerence on our behalf by the United States government? Because we, like our oil executive government leaders, but in a very different way, are rich because of their exploitation. We all know it takes less

Tropical city with canvas shade. *A lower-cost version of the shading roofs of China mentioned three drawing preceding, this hot region city uses shade and breezes to cool.*

effort to survive and thrive in the US than in other countries, even if many of us are relatively low-income, because on some level we all understand the obvious, namely that our "power," meaning our use of violent force, and our wealth, meaning our ability to buy off key people in all parts of the world, guarantee compliance with excellent business deals that keep a flood of wealth pouring into the country. Some of us may be fairly poor — we don't get fabulously wealthy like the captains of our industry and banking — but imagine how much poorer we would be if the United States government, our giant corporations, and our global banking institutions like the World Bank and the International Monetary Fund (IMF) were not funneling the money in our direction. Better let them continue playing the game, just as the democratic Greeks permitted slavery and exploitation of their colonies to

Inside the city with canvas shades.
A market place cooled by shading with cloth. Trees could also help cast pleasant shadows into such urban interiors.

stand and grow. Ronald Reagan forthrightly promoted it all as "trickle-down economics". The rich get fabulously wealthy as the empire expands, and when they spend and invest their money, some of it percolates down to the rest of us. It actually does, too, just enough that around a majority think it's a good deal.

John Perkins, in *Confessions of an Economic Hit Man*, lays it out as a former insider and as clearly as you are likely to hear it. "My job," he writes, was "to encourage world leaders to become part of a vast network that promotes US commercial interests. In the end, those leaders become ensnared in a web of debt that ensures their loyalty. We can draw on them whenever we desire — to satisfy our political, economic, or military needs. In turn, they bolster their political positions by bringing industrial parks, power plants, and airports to their people. The owners of US engineering /construction companies become fabulously wealthy."[4] But *which* people are "their people?" Mainly the compliant elite. A bottom line for the ostensibly generous "aid" programs — loans, not gifts — to developing nations is this: in 1960, the income the wealthiest one-fifth of the people in the world compared to the one-fifth at the financial bottom was 30 to 1. In 1995, the ratio was 74 to 1.[5] Aid to the wealthy.

Cities and the energy flows that build and sustain them play a big role in the game. Sprawling urban infrastructures make people who live in such cities generally unwitting players in the game as they become deeply dependent upon the entire delivery infrastructure from oil wells to shipping lanes to defense of that delivery infrastructure. The economic hit men, says John Perkins, represent large engineering firms like Kellog, Brown and Root, Halliburton, Bechtel, and his own firm in the 1970s, MAIN, out of Boston. These "EHMs," as he refers to himself and his colleagues, exaggerate benefits of infrastructure projects with complex economic development planning studies claiming inflated likely rewards. They work to convince national leaders to borrow money and go deeply into debt to buy oil facilities, power plants, hydroelectric dams, and the highways, ports, and airfields to get there, plus whole cities to house workers pulled off the land and the managers from what Perkins calls the "corporatocracy." (I like calling it the oiligarchy.) The World Bank or the IMF loans enormous sums to the national leaders who play the game and who can profit personally. They have to agree to business deals immensely profitable to the American companies, oil companies in Ecuador, for example, receiving 75 percent of the proceeds, and 85 percent in the case of Iran under the Shah, says Perkins. If the developing country leaders don't want to play along, they often die violently like the democratically elected Torrijos of Panama, Roldos of Ecuador, and Allende of Chile or are overthrown and driven into house arrest like Massadegh of Iran or exiled like Arbenz of Guatemala. Perkins says, should the economic hit men fail to convince a head of state to be ensnared in debts and forced loyalty to the American Empire, the CIA-sanctioned or -led "jackals" move in. And if the jackals fail, the military invades — for the benefit of the invaded country, as it is always portrayed.

What is the dream of the Empire's leaders? To feel the power and experience the heights of wealth, of controlling the largest empire in history? Why do millions buy into it? Fear of other people like that on the "other side?" I'm guessing, but I think their assumption is that humanity has always been and always will be stuck in a universe that is essentially violent. *Star Wars* is a movie set in a distant past, far, far away ... or in a distant future? Its assumption is eternal violence. It doesn't matter whether it's past or future, for in this nightmare, which many romanticize as a fun adventure, humanity is destined to eternal war and violence. Some millions, probably billions of people buy into this dream that we can't change our violent world and violent souls, despite the fact that there have been a very wide variety of cultures from generally peaceful to violent ones and some very successful philosophers and strategists of near-nonviolence like Christ, Gandhi, and Mandela. In the real world, not the world of *Star Wars* and John Wayne westerns, the excitement and drama end with the shocking gore and pain of the real thing and the confusion of the crippled veteran with ghastly wounds that will never heal. Children wouldn't play war games if they had to spend the rest of the year stuck on a bedpan with their "killed" playmates banned permanently.

It's a rotten dream, the glorification of violence, and the opposite of democracy and justice, compassion and creativity. We need to be clear on that and stop glorifying violence. Learning to live fully and in harmony with ecology and vital evolution forever is another dream altogether. Gandhi said, "Nonviolence is the greatest force at the disposal of Mankind. It is the supreme law. By it alone can mankind be saved." I say, we have to build for it.

And now we find ourselves in a situation where the New American Empire is trying to conquer the very last of the oil — right when China and India are rapidly growing their own massive automobile dependence and are also wanting in on the oil. Where might it all end?

Climate Change

After participating in the Towards Car-Free Cities 4 conference in Berlin — a delightful event filled with enthusiastic young bicyclists and other people hotly antagonistic to cars — I traveled to Lyon, France, to see if I could find possible hosts for an International Ecocity Conference. I was hoping to set in motion some contacts for an event that would have as a major theme the links between city design and global climate change. That part of the trip didn't pan out, but I also had a hankering to visit Mont Blanc, the tallest mountain in the Alps. The previous summer, in 2003, the region had been the epicenter of one of the worst heat waves in history, killing 35,000 people in Western Europe. The top of Mont Blanc is a great dome of snow and ice that accumulates and spreads toward steep slopes and cliffs, then plunges into deep clefts in the mountain, past spectacular spires of stone, and drops into the pine-draped canyons far below as great green-white glaciers.

According to my atlas, the mountain is 15,771 feet high, but I learned that over the last ten years it has lost 8 feet in height as the snow and ice turned directly to vapor — sublimated, as they say — faster than new precipitation accumulated. It's been a hot ten years. My friend Alain Bourgeat, a nuclear waste disposal scientist (there is no safe way, he says, but he tries) drove us to Chamonix, and from there we climbed as far as open trails permitted, getting as close to the nearby glacier as possible. Hundreds of other trails had been closed due to the melting of ice and the associated destabilization of the rocks above, for thousands of years glued together by ice, now melted away, with the rocks loose and, some, teetering on their balance points high above.

I learned a few things. Despite what I thought would be a major blow to the tourist industry, one of Alain's best friends, who owns a small hotel in Chamonix where we enjoyed a beer or two looking way up the valley to the retreating glacier, said that they were as busy as usual. The tourists just do something else. They hike other trails or go across to the other side of the canyon from the peak, take other lifts, shop, dine, and socialize in Chamonix, like we were doing at the time. Pictures on the wall of Alain's friend's hotel showed the glacier 30 years ago, breathing down his backyard's neck. He may like it better the way it is now. And what I learned was that people might adjust without much worry. In fact, Alain waxed cynical, saying that back in the 1980s there was a considerable interest in solar energy, but now nobody was interested in such energy alternatives. Everyone was just interested in him- or herself. He was discouraged enough to be contagious.

Back in California, as I was cruising web sites on the subject, a few facts popped out quickly about a mountain top in Hawaii almost as high as Mont Blanc, Mona Loa, where CO_2 measurements reinforce the idea that something large and disturbing is going on with global climate change. For example, at the beginning of the 1990s, CO_2 in the atmosphere was increasing at one part per million (ppm) every year, which doesn't sound like much but adds up. The concern is that since 2001 it has increased at about 3 ppm, which was a surprise to scientists. Some said the tripling of the rate of increase could be an anomalous situation and the rate of increase would settle down to the already disturbing steady levels seen since the beginning of the industrial era. At that time, about 250 years ago, CO_2 began creeping up from 280 ppm and has, in 2004, passed 380 ppm. What could be especially disturbing about this information is the fact that warmer oceans can dissolve less CO_2 than cooler oceans. So too for soils. Many scientists feel that this may indicate a positive feedback loop has started to accelerate global warming in an unanticipated speed-up.

This is not a book on climate change, however, so I will not belabor the details. It's a book about city structure being probably the chief cause of the problem — and about ways to solve most of that problem with redesign. But I think it is important to get an almost visceral sense of what's happening to our dear planet. Therefore I will mention just a few items and then move on.

Reports have been coming back from Alaska and northern Canada that the Eskimos, or Inuit, as they are also known, are experiencing global warming coming on at a much higher rate than in areas farther south. The region they live and hunt in is now around 6 degrees Fahrenheit warmer than it was 20 years ago. Their experience is of ice

Curitiba implantations – next four illustrations. *Curitiba's five arms of high density development served by exclusive bus arterials in Brazil, top image seen perpendicular to the arterial, could be the site for Richard Levine's "implantations" on the scale of small arcologies. Bottom picture: one such implantations along one of Curitiba's high density arms – 16 stories*

thinning to the point of danger, with hunters occasionally falling through. The ice melts two months earlier and freezes two months later than in 1990. That's not very long ago! One 85-year-old hunter named Inusiq Nasalik said to a reporter for the *International Herald Tribune*, "Animals are changing and I cannot tell you why." He related, said the writer of the article, "that caribou are skinny, and so are ringed seals, whose fur has become thin and patchy. The arctic char that swim in local streams are full of scratches, apparently from sharp rocks in waters that are becoming more shallow because of shrinking glaciers .... These animals are showing abnormally hard livers .... The fat in Beluga whales is changing color. Hunters across the eastern Canadian Arctic are reporting that an increasing number of polar bears look emaciated, probably because their hunting season has been shortened by the shrinking ice cover."[6]

Waterspouts whirl and rise in the vapors over fjords, something completely new and downright creepy in Eskimo experience. Forests of short trees in nearby regions, formerly standing straight on thin soil over hard ice, are now pitching back and forth like pick-up sticks on mud. They have earned the name "drunken forest." Overall, Artic Ocean ice is 40 percent thinner than it was in the 1970s and covers 20 percent less of the sea, meaning its volume has been reduced by 52 percent.

Most readers will also know about the oceans rising, with small island nations like Tuvalu in the west-central Pacific, an atoll of 10,000 inhabitants, flooding more frequently in higher tides and storm surges and having to eventually be evacuated forever. You will have heard of the bleaching and dying of, so far, approximately 25 percent of the world's corals reefs due mostly to warming of the sea. So much for life in the land of palms and paradise, or under the shimmering northern lights.

Hanx Blix, the United Nations weapons inspector who reported that there was no evidence of weapons of mass destruction in Iraq and who fled Iraq days before the United States invaded, is also concerned about climate change. He responded to an interview question like this: "On big issues like war in Iraq, but in many other issues, [the US government] simply must be multilateral. There's no other way around. You have the instances like the global warming convention, the Kyoto Protocol, when the US went its own way. I regret it. To me the question of the environment is more ominous than that of peace and war. We will have regional conflicts and use of force, but world conflicts I do not believe will happen any longer. But the environment, that is a creeping danger. I'm more worried about global warming than I am of any major military conflict."[7]

Probably the ultimate apocalyptic worry is something called a "runaway greenhouse effect," a type of "positive feedback" condition: warmer water and soil absorb less CO_2

than cooler water and soil. In fact, CO_2 dissolved in water and soil, as the water and soil warm, out-gasses into the atmosphere, adding to the natural and human-made (mainly automobile/sprawl-generated) CO_2. Thus more warmth means more heat holding CO_2 entering the atmosphere from ocean and soil, which means more warmth, which means more CO_2 in the atmosphere, and around and around in the positive feedback loop which, on Venus, turned into a runaway greenhouse effect so far gone that even during Venus's long two month nights, the rocks glow a faint red from their heat.

Could humans trigger such a feedback system? *Are we triggering one?* The atmospheric system is very complex and relates intimately to water, land, and biota, especially plants. But by stripping so much of the natural forest cover of the world and reducing the resilience of biodiversity through increasing species extinctions, any such runaway greenhouse effect on our human-damaged planet would lack much of the buffering action provided by Gaia in the past in her three billion years of evolution. It would be far more prudent to look at the larger patterns in what we are doing and take such large-scale action as prescribed by ecocity building.

It's a little depressing to look out upon such big facts and across the largest things people create on the face of the planet — sprawling cities covered in smog spreading out through most of the whole planet's atmosphere — and still find "environmentalists," who will refuse to even listen to the

Larger implantation – 35 stories.

above two paragraphs before cutting off the conversation saying, "Stop being so gloom and doom. You'll never get anywhere that way." They love their comfortable lives and cars that much, it seems.

Peak Oil

Let me first summarize some of the main features of the "Peak Oil" story before I look for a moment at the new book called *Collapse: How Societies Choose to Fail or Succeed*, by Jared Diamond.

In 1954, petroleum geologist M. King Hubbard got a firm round of derision from other oil geologists and oil company executives for predicting oil production would peak in the United States in 1971. If you don't know the story already, you can guess. That's exactly what happened. In 1954, the US was the first Saudi Arabia of oil and had approximately half of all the 55 million cars in the world. (Today there are more than ten times that many cars in the world — about 600 million — and about three times as many in the US.) It was as hard for the established powers and opinion makers, riding the euphoria of that brand of material success, to believe the supply could ever start falling any time in the foreseeable future as it is for people like me with my international lecture circuit to imagine that one day a flight to Asia might cost $30,000 and a few years later be absolutely unavailable. Back to sail ships — which sounds relaxing after all,

if you can find the time — and avoid the pirates. That's the kind of thinking — pirates boarding your ship, steeling your Rolex, and feeding you to the sharks for entertainment — that starts happening when one contemplates peak oil.

Shocking to me when I first read it, the United States peak for discovery of domestic oil fields was 1931. Forty years ahead of peak production, and a long time ago.

Consensus among the geologists and energy analysts at the Association for the Study of Peak Oil (ASPO) on the date of peak oil ranges from 2006 to around 2010. These are, again, oil experts who are not on oil company payrolls, though many used to be. Oil companies and "optimistic" business analysts estimate generally around 2020. As mentioned in an earlier chapter, to maintain investor confidence, oil companies tend to exaggerate reserves and, probably more to the point, portray getting more oil to be a technical problem that can be readily solved. But as depth of drilling increases, wells are placed in deep water or icy regions and ever more complex and difficult techniques are needed to get at the more inaccessible material. The material becomes contaminated by the techniques themselves, and the energy won in the process diminishes ever more relative to the energy that goes into the process. In the business they call this EROEI, energy returned on energy invested, the difference between the two being "net energy." In 1916, the energy returned compared to the energy invested by the US oil industry was 28 to 1. By 1985, the ratio had dropped to approximately 2 to 1.[8]

Collapse

Jared Diamond's book *Collapse: How Societies Choose to Fail or Succeed* arrived in the bookstores just as I started writing my *Ecocities* update. It's a powerful exposition of societies facing changes and prevailing, hanging on with adjustments or collapsing and disappearing utterly. Unforgettable touchstones remain with the reader afterward. Every bit as stunning as the Easter Islanders cutting down every last tree is this: The occasional reasonable and courageous steps taken by a few societies to survive. Take Tikopia, for example, a small island in the West Pacific. It is so small, just shy of two square miles, that directions are given as "inward" and "outward" meaning inward toward the center of the island and outward toward the sea, as in, "Your outward cheek has a fly on it." We chuckle, but then hear this: they had high spiritual ceremonies to celebrate zero population growth. Zero Population Growth (ZPG) was the name of an organization set up in the "developed" world following the publication of Paul Ehrlich's pathbreaking book *The Population Bomb*,[9] a very sophisticated and

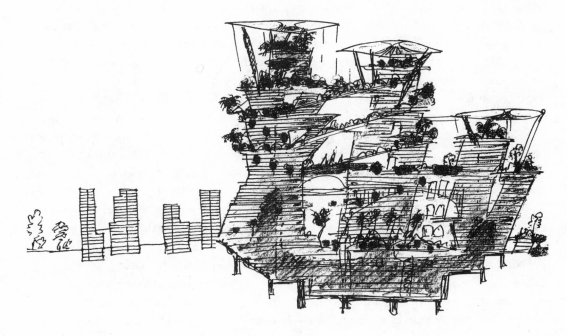

Larger yet implantation – 50 stories.

effective organization in the heart of the modern beast (in 2002, ZPG was renamed Population Connection). But zero population growth is the best English translation of the annual ritual on Tikopia to remind its people to keep the population from overreaching the carrying capacity of the small island.

When the Tikopians arrived almost 3,000 years ago, according to archeological evidence, there were several more species of bird, fish, and shellfish living there. They brought pigs with them, too, with the deep Polynesian tradition of relating to them as a staple food of life and ever-present front, back, and side yard companion. Over the centuries, the natural species were reduced in number and the surviving shellfish species grew smaller as Tikopians did selective harvesting in favor of the larger animalss and left the small ones to survive and reproduce. But eventually, sometime around AD 1600, Tikopians learned and stabilized their relationship with their resource base and natural biotic community. They stopped the slide toward island species extinctions. They made very explicit their several means of birth control and integrated them into their cultural daily life and special ceremonies. They made a momentous decision, too, reports Diamond, and slaughtered every single last pig for the broad damage they were doing to the island. Harvesting the sun's energy via the plants, then the pigs, they realized, was far less efficient on an island susceptible to famine than to eat the products of the plants directly. About ten times less efficient. Plus the pigs damaged the remaining native species. Getting rid of these animals and finding a balance that recognized their world's carrying capacity shows an intellectual sophistication that would be the equivalent of Americans slaughtering their road hogs, every last one.

Though "essential reading," as the book blurb writers are prone to say, for anyone concerned about the directions of contemporary society or worried by environmental trends, Jared Diamond joins the long list of perceptive people and crystal-clear writers who have not noticed the 800-pound gorilla in their own living rooms. Or rather, more personally, the 800-pound gorilla in the bioregion where he lives: Los Angeles, California, famous as the world's original capital of cars and sprawl and the inventor of smog. He wanders off to the far corners of the world — the South and West Pacific, New Guinea, Australia, China, Japan, Greenland, Iran, Rwanda, Russia, Haiti, Guatemala, and even Montana and New Mexico — and back again to the thin, suffocating coating of concrete, asphalt, roofing tar, grass, and air borne particulates that is his city's physical structure. Yet he never says that its very design or the shape and organization of its anatomy have anything to do with the calamities we are facing. Nor does he notice, with all the successful adaptations cultures devise in forestry and agricultural practices and policies, stewardship of the

Shenzhen average big building

Shenzhen arcology?

The 20 to 30 story buildings to the left of the 70 story structure above are typical building profiles in Shenzhen, China, where the Fifth International Ecocity Conference was held. There are many bridge buildings and very large shade structures in Shenzhen, including the world's largest roof, the size of twelve football fields, several stories above and shading the new City Hall and main conference center. Here we imagine an interlaced in-city city for 250,000 completely accessible by foot and elevator. If development rights were transferred from elsewhere in the region, the structure could be built while financing considerable natural and agricultural preservation and restoration. Construction costs would be relatively high, though car and bus infrastructure requirements would be zero instead of extremely expensive. If maintained for centuries, the economies of amortization would be stunning — and ecologically very healthy.

ocean harvest, and so on, that this scattered urban structure is a gigantic maladaptation in its own right that needs fundamental change in *our* society.

Some of the clearest thinking in regard to the collapse of societies I have read lately comes in a discussion of Diamond's new book by Richard Heinberg, author of the before-mentioned *The Party's Over: Oil, War and the Fate of Industrial Societies* and *Powerdown: Options for a Post-Carbon World*.

Heinberg believes, as I do, that Western culture has not invested in renewable energy and at a time in the very near future will face an inevitable enormous and once-in-the-lifetime-of-the-planet problem defined by collapsing cheap energy availability and continuing growth in demand. I think we may have a few more options than he does, but basically he is saying collapse is coming, and soon. Diamond opens the subject, says Heinberg in his monthly newsletter called *Museletter*,[10] but could say much more: "He offers a message of the type we have come to expect: Humanity is undermining its ecological viability, but there are things we can do to turn the tide. Indeed, Diamond predictably devotes the last section of his last chapter to 'reasons for hope,' leaving the reader with evidence for thinking that collapse will not occur in our own instance after all." But, says Heinberg, there is a question "tugging at our minds and with more urgency every day: *What if it's already too late?*"

Heinberg points to reasons "for concluding that Diamond has in fact made an extremely timid case for the likelihood of global industrial collapse" including these: First, "Diamond does not even hint at the phenomenon of the imminent global oil production peak." Let me just stop for a moment to remind ourselves of what is in Heinberg's mind: massive energy use, no alternative to oil of remotely similar low price, conveniently portable form, and highly concentrated energy. This resource also produces clothes, furnishings, building materials, cities, vehicles, and even foods and medicines. You could say with only a little exaggeration it is around, under, sheltering, transporting, feeding, and medicating us all at the same time. A lot! Also, Diamond says that we will need "coura-

geous, successful long term planning." I hope I don't sound too discourteous but I couldn't help a chuckle myself when I saw that, considering recent trends in American national and international politics. Heinberg writes:

> Averting collapse would require changes that must be championed and partly implemented by political leaders: unprecedented levels of national and international cooperation would be needed in order to allocate essential resources in order to avert deadly competition for them as they become scarce, and our economic and monetary systems would have to be reformed despite pressure from the entrenched interests of wealthy elites. Yet the American political regime ...as evidently become terminally dysfunctional, and is now the province of a group of extremist ideologues who apparently have virtually no interest in international cooperation or economic reform. This is a fact widely recognized outside the US, and by many sober observers within the country.
>
> ...the American media have been so cowed and co-opted by the dominant party that most of the citizenry is blissfully unaware of its plight and is thus extremely unlikely to vigorously oppose the current trends. Diamond shows some limited awareness of this truly horrifying state of affairs .... Yet he refuses to draw the obvious conclusion: the most powerful of the

world's current leaders are every bit as irrational as the befuddled kings and chiefs who brought the Maya and Easter Islanders to their ruin .... The question is no longer that of avoiding collapse, but rather of making the best of it.[11]

Best Options in Dark Times

So let's talk about hopefulness. Recently, I went to a lecture by Lester Brown, and he offered some. But first he elaborated on one of his own favorite peaks, peak grain. Following the theme in his recent book *Outgrowing the Earth*, he described the importance of grains in world food supply. Not only are they staples consumed in very large volume world-wide, but they are relatively easily stored, and so they become important for weathering food shortages, especially in countries without great wealth and wide options.

Just a little more sobriety before the hope: "In each of the first four years of this new century, world grain production has fallen short of consumption. The shortfalls in 2002 and 2003, the largest on record, and the smaller ones in 2000 and 2001 were covered by drawing down stocks. These four consecutive shortfalls in the world grain harvest have dropped stocks to their lowest level in 30 years." Brown continues:

The best weather in a decade raised the 2004 grain harvest by 124 million tons .... For the first time in five years, production matched consumption, but only barely. Even with this exceptional harvest,

the world was still unable to rebuild depleted grain stocks. The immediate question is, Will the 2005 harvest be sufficient to meet growing world demand, or will it again fall short? If the latter, then world grain stocks will drop to their lowest level ever — and the world will be in uncharted territory on the food front. The risk is that another large shortfall could drive prices off the top of the chart, leading to widespread political instability in low-income countries that import part of their grain.[12]

Peak grain in 2005, followed by peak oil shortly after that? Certainly population and consumption are not slated to peak anytime close to that date. What begins, then, which is also uncharted territory, is the rapid divergence between demand and supply.

On the hopeful side, Brown has always provided the best data — an absolute necessity as a starting point — and clear thinking for the use of many good technologies. He's been promoting the "better" car lately, but with a little ecocity thinking that wrinkle might get ironed out and my main complaint about his work would be history. At the talk he suggested, in addition, that unexpected, pleasant surprises can appear, real breakthroughs. For example, he pointed out that the Berlin Wall suddenly came down and shortly thereafter the Damoclean threat of nuclear war suddenly and quietly evaporated.

That sullen, looming, and ever-present image from my childhood, when I hid under desks and sirens screamed, the mushroom cloud, just disappeared. Somehow, without me really knowing why, everything about life seemed just a little more relaxed. Another wonderful surprise: suddenly, rather unexpectedly, you could actually go to restaurants and fly in airplanes without gagging on people's rancid, carcinogenic smoke. More "breakthrough" observations would include the fact that in the US in the last couple of decades finally the horrible edge is off racism and sexism. However, classism is alive and well, and rich and poor are diverging toward every more hostile and far separate places.

Thinking a little more systematically about this, one realizes that there were certainly thousands if not millions of people working for a long time toward relief from communist state fascism, the tyranny of smoke and smoking, and the scourges of racism and sexism. We may not have been reading the signs right and been surprised, but it is not as if there was not a lot of work going on in preparation for the dreamed-of breakthroughs. Then when they came, people forgot almost reflexively that they had helped create them. Very strange! For example, people have ripped into Paul Ehrlich for his ranting prediction in his 1968 book *The Population Bomb* that famine would come if they didn't start cutting down on high population numbers. Then even he seemed to fail to give himself credit when a combination of agricultural methods and better education and family planning policies, partially created by the concern he and Zero Population Growth helped generate, actually did start slowing down population growth. Suddenly a chorus broke forth saying he had failed miserably because the famines didn't come as soon or weren't so bad as he'd predicted. Yet he had worked for just that success. Another little-noted good turn of events accruing to no one's credit was the follwing: Gorbachev's attempt to reform Communism to make it more peaceful and open and the resulting evaporation of not just the Soviet state but the world's major threat of nuclear war.

Generally, even though there didn't seem to be a major breakthrough, there were steady advances in international cooperation, democracy between nations, international law, cultural exchange, international tourism and "getting to know you" civility — up to the tipping point of 12/12/02 when all that began unraveling. For the first time in my life, after all my travels, soon after Bush' *coup d'état* I ran into genuine, personally directed anti-Americanism in a foreign country.

Lester Brown's work is especially strong with respect to agricultural solutions. But by suggesting ways to increase water use efficiency, to shift from one crop to another, to move as quickly as possible to wind and solar energy, he gives the impression, as does Diamond, "that collapse will not occur in our own instance after all." Yet when he describes how Brazil can coax more soybeans and cattle out of its landscape, he lets drop a statement such as this one: "This pressure to clear more land means the worst fears of environmentalists may be realized. The prospect of losing so much of the earth's remaining biodiversity is scary, to say the least."[13] He says the long distances from the interior agricultural areas — more than a 1,000 miles in many cases — to cities on the coast and ports for the beginning of the long haul to foreign markets "can be costly. In a world where oil prices are likely to be rising …"[14] Likely?! Brown's desire to keep alive the sense that we can do a lot to improve things is mine too, but I think we have to look at the fact that "peak everything" is on the way and assess what it might mean that synergistic systems happen. "Synergy" describes the interaction of two or more agents or forces where the whole is much more than simply the sum of its parts. How can we say at this point in history that we are not about to hit the wall of a negative synergy that is planetary and permanent in scale and character? It's hitting us from climate change, biodiversity collapse, and the peak of several resources, including a key one called oil. We need to admit that, wake up, and do something.

So what have we really to be hopeful about? Heinberg says we can "powerdown" and offers what one can do to ease the transition. First, he and Diamond both point out that collapse doesn't necessarily happen overnight, though once you start exploring the literature, you find panic under every unturned stone: people worry that billions will die in resource wars and anarchy. But when Richard Heinberg lists the things people

actually can do to powerdown with grace, they look, compared to the task, a little vague and faint hearted: (1) protest war, (2) protect ecosystems, (3) defend traditional and indigenous cultures, (4) adopt lifestyles of "voluntary simplicity," (5) relocalize the economy by buying locally produced things and services preferentially and adopting local currencies, (6) build intentional communities, and (7) regain forgotten handcraft skills. Though I think this is a good start and find it harmonious with ecocities, it appears to me far from enough to give us a really humane "powerdown." And if it can't be humane, he needs to add (8) stockpile a large supply of dried and canned food and water, and (9) stock up on guns and ammo and face up to the possibility that his "last man standing" situation might not be part of the strategy embraced but part of the undesirable necessity that comes knocking at the door.

The movie *The End of Suburbia*[15] is making rounds as I write this second edition of *Ecocities*. In fact, I've shown it to a number of small audiences myself. Its main theme is that suburbia will not survive long after peak oil hits. The film ends with some hopeful words on behalf of the New Urbanists that seem starkly shy of the mark: images of slightly more compact two- and three-story housing and mixed-use traditional American towns that look like they were build in the 1940s, small cozy downtowns with cars still parked out front illustrating the usual New Urbanist theory that the street energizes the people if there are cars zipping up and down. The small step in the right direction from abject sprawl this represents hardly sounds like big guns against global economic and ecological collapse.

What can I add? I think that building small intentional communities and growing your own vegetables is fine, but if we descend that far, if those are supposed to be realistic levels of organization to meet the problem, we have already lost. I don't see how such small-scale levels of operation could maintain peace. True, the larger-scale level has a bad habit of war itself, made much worse recently under the bad leadership of the United States. But tiny-scale organizations, I think, would be less likely to produce a bucolic rootedness in the soil than something kindred to what is going on now in Somalia and Sudan.

The redesigned built community is precisely what has a chance to cushion the blow, and maybe even, after a few generations, help bring us back to some of the options of complex technological society, but with a stable and slowly regenerating environment.

I think it is possible we could "powerdown" in a way that maintains systems of enough complexity to continue meeting most of society's needs, with some of the dieback resulting from disasters, but with purposeful, peaceful efforts for slowly reducing the population playing a large role, too. That seems to me to mean government institutions continuously learning their lessons and becoming cognizant of the bioregional realities of resources and life systems, being, in that sense, "relocalized". It means businesses of a scale to deliver, if not the glut of superfluous things we produce now, still a very wide variety including essential tools so that the valuable specializations remain viable. Metropolises breaking up into scattered cities and towns of a range of sizes all interlaced with agriculture and restored natural areas, requiring far less energy than the cities we build now, could maintain systems at a decent level of delivery of necessity and maintain systems of order and defense. These units would have to be large enough to preserve essential props without which our current gigantic population could only topple into chaos. And defense would have to be a strategy amounting mainly to the kind of cooperation and mutual respect and economic justice in regard to which we are currently losing ground, not gaining.

To me, then, the redesigned built community is precisely what has a chance to cushion the blow, and maybe even, after a few generations, help bring us back to some of the options of complex technological society, but with a stable and slowly regenerating environment. Since the investment in renewable energy and ecocity infrastructure has not been made and the foundation has not been built, I have a feeling the dreams I had for ecological cities in the 1970s are a little more flamboyant than will be possible to realize until we survive the collapse and emerge a whole lot wiser than we are now. (Jane Jacobs, however, brings us the haunting idea that dark ages erase memory inexorably, which would include the memory of this book.) If we rise from the seeds of an ecocity civilization during the powerdown or after the collapse, building a complex society in a healthy environment — a relative term since both society and environment will be so badly damaged by then — would also be a very slow process. I hope there will still be hummingbirds. But perhaps we would by then have learned the tough lessons of the 20th and

21st centuries and be slowly building up both the renewable energy system and the built community to fit. We will have to design things for a very low EROEI compared to what we grew accustomed to in the age of oil.

For better or worse, then, all I have to suggest is only what's in this book. But if we have a "Plan B," and should we some how find wise leadership and "courageous, successful long-term planning" emerge despite all, maybe even because of the shock of it all, we might just find a positive synergy in time to effectively counter much of the negative synergy and make some real miracles happen. There have been preparations building just below the skin of public awareness for a long time. Maybe we will have one of those positive surprises yet. If we do, ecocities will be right there at the foundation.

And in any case, it is always too late for our dreamed-of best, but never too late for making things much better. So, though the hour is late, don't get depressed by people saying it's too late — other than to allow them to help wake you up. Then get moving.

CHAPTER 12

Toward Strategies for Success

WHATEVER STRATEGIES WE ARE USING TO build our society aren't working very well. We need ones that work significantly better, strategies that can build ecologically healthy cities, towns, and villages that further creative and compassionate human evolution. First, though, partly by way of summary, here is some general guidance for people who want to change things:

1) *Follow the builder's sequence*: start with the land use foundations of the city and then build up the details — the ecocity features — upon that foundation.

2) *Pursue the four steps to an ecology of the economy*: the ecocity zoning map, the list of technologies, businesses, and jobs for building and maintaining the ecocity, the incentives, and the people to animate it and get it built.

3) *Move steadily and as vigorously as possible away from automobile sprawl and toward the pedestrian compact city.* The crucial step is to begin removing low-density development in a systematic way that can be replicated. Add to the pedestrian compact city appropriate technology and biodiversity and you have the ecocity. Wherever people have been satisfied with small steps, failing to see them as early steps in a long journey, progress has soon been overwhelmed by cars and sprawl. If we start with one car-free building, we have to move on toward more of them. One car-free street needs to beget another faster than car streets increase in number, length, and area. Being proud of your city's one solar greenhouse, one block of opened creek, one park with a fruit tree is being proud of going nowhere fast. Symbolism has its place, but mainstreaming ecologically healthy functioning is what we need. Remember the Aztec toy animals with wheels.

311

The Ecocity Zoning Overlay Mapping System.
First step; the way it used to be. The following drawing and maps represent a system for reshaping cities by informing official zoning codes and maps where best to place new pedestrian / transit centers of higher diversity and density of activity and people, and where to remove more automobile or other damaging development. First learn what was here before humans altered the environment radically. This illustration pictures a site on the Berkeley shoreline circa 1750 during the period of minimal disruption to biodiversity from Native Americans and previous to settlement by the whites.

4) *Take small steps as needed but move on to bigger things when you can*: plant fruit trees on the curbside, help with thorough recycling, or collect signatures on an open-space initiative; then arrange work and home close together, give up driving, become committed to an ecocity organization. You don't have to wait for enlightened regional government to have a profound effect. The Ecocity Zoning Map shows us that so much room has been set aside for cars that density shifts within the city can liberate hundreds of acres of land even in small cities. Just use the map to see where to open up the landscape and where to reposition the displaced uses.

5) *Use whatever tools you feel comfortable with and no others; make commitments you can keep and those only.* People who overextend themselves for causes and community work often burn out and then quit, resentful because of their exhaustion and the perceived lack of appreciation on the part of others. Substantial change is difficult to accomplish,

and every individual is in an excellent position to see his or her own efforts and in a poor position to see the full efforts of others. Of course we all, from our limited perspective, know that we work harder than others recognize, so to avoid burnout and thus damage to what we believe in, we need to pace ourselves. Many people volunteer to do more than they can deliver, and when they fall short, they are sometimes so embarrassed that they never

come back. Pace your promises as well as your efforts. We need you with us for the long haul.

6) *Be flexible about ecocity changes.* Take the time to consider the connections that make up a holistic view. Ecology is whole-systems thinking, and it's complex enough that understanding usually comes slowly. The passage of time gives us the opportunity to think about benefits and costs. If your house would be best

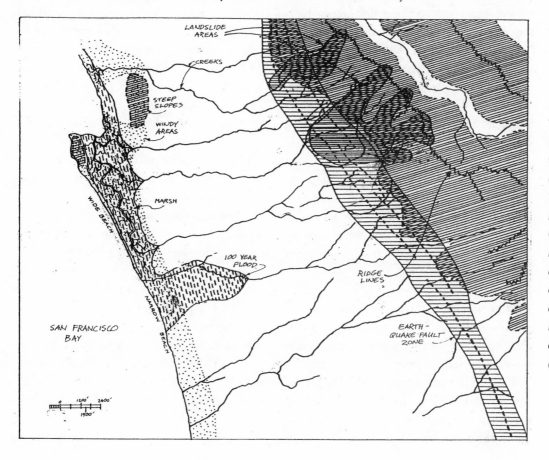

The map representing the natural environment. *This one was based on Berkeley, California with its many creeks, ridges and steep slopes on the east (right) and San Francisco Bay shore, beach and wetlands (on the left).*

removed for a creek restoration, look at the positive side of selling and moving closer to a center and having your old environment become a wonderful place for wildlife to return to and for children to learn in. This effort to build ecologically healthy cities is new. We are learning on the way. We will make mistakes — but not as many as we would if we were not making the effort. Therefore practice flexibility, forgiveness, and perseverance — and expect improvement, at least in regard to the built environment.

There are a number of actions that any of us can take, either alone or in a group. In the "alone" category are these: voting and spreading the word simply by talking with friends and acquaintances; learning more — using the books referenced here and in other volumes on related subjects to follow your interests, using keywords on the Internet; attending ecocity conferences; conserving, recycling, avoiding toxics; buying, boycotting, and investing accordingly; helping restoration projects by dropping in on their organized activities.

In the "group" category are these possible actions: joining and financially supporting an ecocity-building organization; supporting good transit and delivery services, especially by pedal power; adopting proximity policies, such as hiring and renting locally; supporting ecocity design for conservation, recycling, avoidance of toxics, building soils, biodiversity, and car-free buildings; recruiting people

to participate in the above through education and personal effort; and moving to have the International Ecological Rebuilding Program adopted and implemented by governments or supporting such a move. Maybe Al Gore and Lester Brown will help.

It's important to choose the levels of commitment that best suit you. Know your starting level and work into a higher commitment level if it is in your nature. The first level is joining a group and learning more, volunteering in a modest way. The second level is making lifestyle changes to fit the pattern. The third level is becoming more deeply involved in major projects, perhaps helping to develop or use the tools for ecocity building described in this book. The fourth level is becoming professionally involved and an activist for ecocity policies and projects at the same time — as an architect, planner, builder, journalist, or educator.

Though professionals are supposed to serve their clients (architects and carpenters serving developers and homeowners, for example) or elected officials (city planning and management staff members serving city councils), they influence outcomes by bringing professional insights to their employers. They can always remind everyone concerned of what they think is best for the community, and ecocity ideas and actions are good for the community and environment. If the compromise that would be required is just too much, professionals can in many cases refuse to take the job. That action is often dramatic and has real educational potential.

The fifth level is working for an ecocity-building organization, lending your skills to helping it succeed, forming one with other people, or becoming a genuine ecocity developer gathering investment capital, hiring ecocity designers and building contractors, and actually creating important parts of the evolving ecocity.

Back to Basics:
Example Strategies for Curitiba

When it comes to ecocity successes, it's hard to beat Jaime Lerner. Though his triumphs in Curitiba are many, two loom over all the others. First, by resolving the land-use issues at the foundation of the physical city, he made possible the successes of transit and left room for the restoration of waterways, the expansion of public open space, tree planting, and very efficient recycling. At the same time, he set up a system so efficient that money was saved to provide Curitiba's remarkable human services. He knew how to prioritize. Starting with the land-use/transportation layout, he got the builder's sequence right. Second, he recognized that his fellow citizens were absolutely essential in achieving ecocity design goals, and

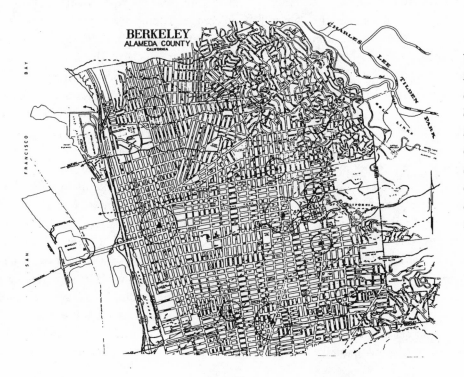

The existing city.
Start the new mapping by using an up-to-date map, finding the centers of most liveliness and locating these as centers of high diversity. The downtown and major urban centers will get more emphasis, higher height limits etc. than the smaller neighborhood centers.

he told them so. Step by step, from that basis up, he proved that various pieces of the ecocity worked well, and the citizens' trust grew.

Lerner didn't postpone anything. "The starting is very important. If you wait until you have all the answers," he said, "you will never start."[1] Thus he dispensed with excuses and launched into major changes for Curitiba almost the minute he became mayor. His administration quickly created institutional support for the new ideas and, by actually building things, he guaranteed that the community would experience and remember ecocity solutions deep into the future. "Being enthusiastic is not enough," he argues. "They [the people and especially the children] have to have knowledge."[2] And so his administration founded the Institute for Urban Research and Planning of Curitiba (IPPUC) and the Open University for the Environment. Lerner and all the imaginative people working with him got it right.

But, as good as it gets with regard to ecocity building, all is not well in Curitiba. In an interview at the Fourth International Ecocity Conference with the journalist and Ecocity Builders Program Director Kirstin Miller, Lerner began to hint that the time for a re-evaluation was at hand: "You have to always have the way to combine, to propose a better offer, a better alternative. From time to time, you can lose two points in the game, but as soon as you realize this — and this is very important — always maintain a high quality of public transport, so the people use it in their normal itineraries." Then he seemed to equivocate somewhat: "I'm not against the car. I'm against using the car too much. That's the problem. You shouldn't need the car for normal itineraries."

Later he added, "You have to take risks, because you cannot have all the answers. But fortunately, planning is a process that you can always correct. So, that's why you don't have to wait. You don't have to be perfect first."[3] But the car is being used more and more there, even for "normal itineraries," as we learned dodging them to save our lives while visiting Curitiba for the Fourth International Ecocity Conference. Twenty-seven blocks of pedestrian streets and five major dedicated routes for buses, but much of the rest is another story. Said our conference videographer Tim Alley, who had made a film on Curitiba just four years earlier, revisiting an Italian restaurant in April of 2000 was a shock: it had been in a quiet and very pleasant part of town, but in the intervening years the area had become clogged with nervous, fuming traffic. The conference convener, Clovis Ultramari, complained to me about the traffic, too.

What had happened, I think, is this: As the city's five high-density arms reached farther out into the countryside, diverging ever more from one another, the area of land between the arms grew much faster than the area of the arms themselves. At some point in the city's recent history, instead of the open space between the arms remaining open space, low-density and much more automobile-dependent development between those arms began growing proportionally faster than the transit arms themselves. Some large areas of open space have been preserved, but evermore low-density development is happening as well. The twenty-seven blocks of pedestrian streets have grown slowly compared to the hundreds of blocks of streets for cars, which are also serviced by busses of the sort that I rode with few other passengers aboard.

At the same time, with the dedicated route bus system working so well from the early 1970s through the early 1990s, the city prospered. Lower-income people in particular could save money on the low-cost transit, and when they could afford it, many started buying cars. Some of them moved to the low-density areas, and others looked for high-density housing with parking structures built in. Alfredo Vincente de Castro Trindade, the city's environmental planning manager, told me that the cars had suddenly appeared in one stunning month in 1994: "Hard though it is to believe, the Brazilian currency, the *real*, was re-evaluated and suddenly people who were making three hundred dollars per month were making eight hundred dollars in purchasing power. So, naturally, after riding the bus for years, many people decided to buy cars.

"In one month car ownership went up here by an amazing 20 percent and it has continued to go up but not at that extreme rate since then." Most disturbing to me, three new automobile factories have been built in Curitiba in the past few years: Renault, Peugot, and Chrysler. Volvo has been in town making both cars and buses for decades. For Curitiba to make such a large commitment to cars is something like a health food company going into tobacco in a big way. Why not bicycle and streetcar factories and equipment for solar power manufacturing rather than cars? Why not organic farming equipment and healthy fiber products, even everyday competitive electronic products rather than cars? "When you are talking about five thousand high-paying jobs at a time," Trindade responded, "a car factory can really help the economy. The voters want it."[4]

Back to the basics again: flat sprawl works with cars (discounting, as they say in military circles, collateral damage) and linear higher-density corridors work well with transit, three-dimensional high-density centers, if articulated with open space, work well for pedestrians. Transit is certainly better for the planet than cars but it is still far more energy-consuming, polluting, and pushy than people on foot. Well-designed pedestrian cities are by far the best way to go. Curitiba put most of its eggs in the transit-corridors basket; its highest-density centers and pedestrian streets were a kind of beautiful civic flourish. But with the historic integrity of the parks, many buildings, and

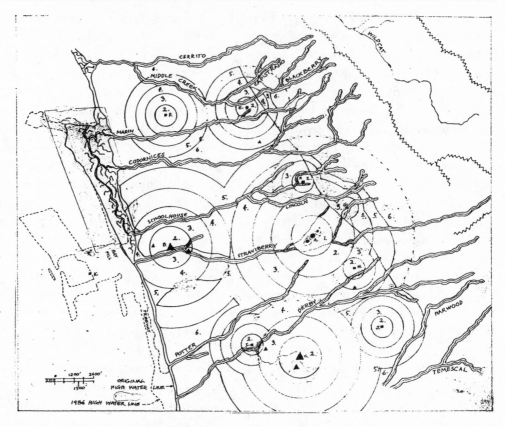

Centers - major and minor.
The map representing the diversity/density of centers based on walkable distances and featuring the potentially restorable waterways, wetlands, ridgelines (with great views) and other natural features.

plazas, with Curitiba's cheerful patterns in the paving underfoot, with the actual vital social and economic function of the spaces for people in highly mixed-use areas, it was more than just a promotional "civic flourish." The content of socially lively and economically vital functioning was always there in these pedestrian centers. However, two apparent assumptions are problematic in Curitiba: that cars really aren't too bad if you resist the temptation to use them overmuch

and that pedestrian centers are special in the sense of "rare" as well as special in the sense of "important."

The great thing about pedestrian centers is that they are functionally the solution to environmental problems. They should be valued so highly that they become common. Every city center, even every neighborhood center, should have a car-free area as soon as possible. Curitiba's great pedestrian streets and parks should be the model for most of that city's future development. In fact, whole car-free cities are possible in infinite variety. It is truly bizarre that the planet has been capable of creating and maintaining only one, but Venice proves that hundreds of them are possible, each with its own spectacular features, cozy recesses and "grand canals," in endless variety.

In my trips to Curitiba I fell in love with this picture: tall buildings — some modern, some colonial or Art Nouveau or almost baroque — rising up behind the big epiphyte-covered trees and palms of the parks and, viewed from the Telecommunications Tower, a green and rolling landscape with craggy distant coastal mountains capped with streaming clouds in the east. An idea for Curitiba in particular, but applicable anywhere else, came from that. Why not, along the transit corridors and just a block or two to one side, create

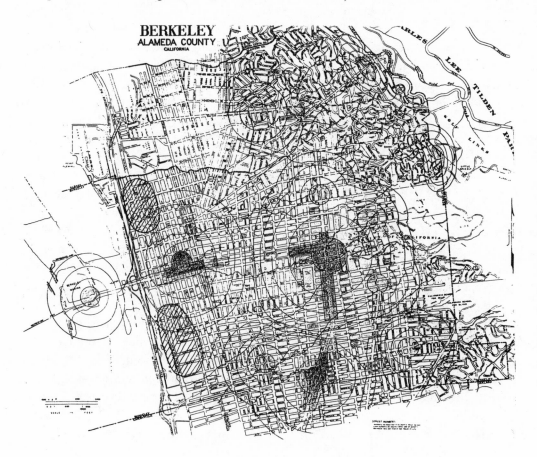

The Ecocity Zoning Overlay Map.
This map adjusts the round circles representing density of development into the future so that areas of highest priority for natural restoration can constitute future nature corridors, and labels the demarked zones for clear reference and planning.

centers that become new pedestrian town centers — encouraging taller buildings there and, rising up in their midst, elevated parks and plazas, say, five or ten stories in the air, with views between the buildings toward the coastal mountains or along one of the city's rivers. These would be keyhole plazas and parks lifted high into the views. They would seem to float in the sky, or give the impression of being near the top of hills. Underneath them would be warehousing and other uses that don't have much need of natural lighting. There are six stories of parking under the San Francisco plaza called Union Square. Union Square is a roof garden on top of a six-story building placed in a six-story hole in the ground. Take the same idea, raise it up, and make it for warehousing something other than cars. Curitiba could do it.

As new streets are built like ramps between buildings, leading "up hill" to the new parks, storefronts that used to be at street level will be lit by skylights and artificial light under the level of new sloping pedestrian streets. At the same time, valuable new commercial window and floor space will be created on the higher floors as the street rises toward the park. The new sloping street might be partially a staircase, something like the Spanish Steps in Rome, with landings and terraces at the edges of the buildings defining a new set of vital areas in front of doors and display windows. Essentially we would be trading valuable ground-level storefronts for new ones on the sloping street. There would also be new space in the taller new buildings, much of it enhanced in value by the busy pedestrian street with its sculpted terraces and the elevated park with its great views.

In this context, transfers of development rights could function three-dimensionally: trade outward and upward from the darker zones to the new sunny spaces in the sloping canyons and around the rooftop parks. These could be called 3D TDRs. The total development value of the new "hill" with park and taller buildings would be extraordinary and would make possible not just the storage space at arm's reach, but high-density pedestrian environments and the newly created public parks and plazas. With the help of TDRs of the sort that shift density horizontally from car-dependent areas to pedestrian/transit centers, natural and agricultural lands could, as a result of more dense building, be restored, and in sufficient quantity to actually make a dent on sprawl. As Jaime Lerner says, "fortunately, planning is a process that you can always correct."[5] Given Curitiba's capability for ecological imagination and leadership, it might be among the first cities to begin removing large areas of low-density development, not to build just a few pedestrian centers but to move a whole city toward pedestrian/ecological design.

I suggested to Maria do Rocio Quandt, information officer for IPPUC, that Curitiba might want to create high-density pedestrian areas with elevated public parks in a few centers. In fact, she reported, IPPUC was beginning to explore expanding some of the community centers on the transit corridors called "Citizenship Streets" into higher-density real town centers in a similar way, though they were not planning strictly pedestrian zones as part of them. "Citizenship Streets" are actually not streets but community facilities for health, sports, education, and economic assistance services. Elevating parks and plazas for views as an expanded version of such a place and adding full village functions would be a new idea for their toolbox. Curitiba and other cities should consider a strategy that says, "We will build pedestrian infrastructure at a proportionally faster rate than we build for the car." Then see what happens. If combined with sprawl clearance, paradise, I predict.

Additional pedestrian streets could be located largely around new public squares and involve not just the squares but also a block or two surrounding them in new centers. In the short term, this would not be too disruptive of the automobile infrastructure being steadily replaced. Adding in a closed street or two at an old square every so often, requiring the designers of new developments everywhere to provide 50 percent more pedestrian blocks than automobile blocks, and encouraging innovation in the third dimension would produce some very imaginative solutions.

These innovations could take Curitiba to its next level of ecocity leadership, from being primarily a transit city featuring a good complement of pedestrian areas to being primarily a pedestrian city with assistance from transit — and this just as it was sliding off into automobile dependence and degradation and the world was approaching peak oil.

Another piece of a strategy for Curitiba would be to identify some small neighborhood centers as satellites of the larger Citizenship Street centers. These smaller centers would be situated between the five high-density arms of the city and agricultural ecovillages would be located there. Double TDRs could reinforce those small centers through a staged removal of automobile-dependent development around them and between the five transit corridor arms. These new pedestrian ecovillages would be focal places for another kind of new pedestrian environment that was structurally very similar to the traditional village and was functioning in the manner proposed by Joan Bokaer of EcoVillage at Ithaca, namely providing agricultural services to the city.

Jon Jerde and Ken Yeang

The new Citizenship Street centers could become places for commercial pedestrian center projects similar to those of the architect Jon Jerde. Canal City in Hakata, Japan, created around an artificial branch of a natural local waterway, is the largest private construction project in Japanese history, according to Jerde's enormous picture book *You Are Here*.[6]

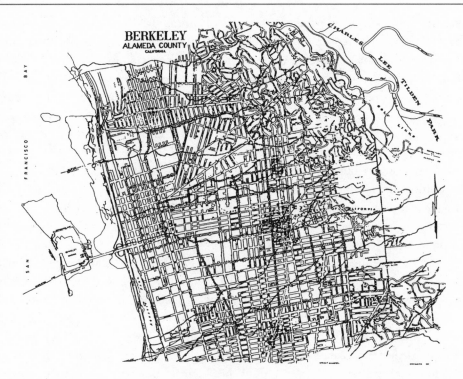

Applying the map. *Over several generations development — and de-development — using the ecocity overlay maps gives us the city of extremely low energy and land consumption. This sequence uses Berkeley as a model for the following scenario in maps. This first projection takes us 30 to 40 years after completing and publishing the map.*

Jerde's gargantuan enclosed Mall of America in Minneapolis, Minnesota, with its 1.2 miles of skylights and 4.5 miles of looping walkway through the interior, might at first glance be puzzling in a book on ecocities — consider the parking lot! But a look at what's function-ally going on there suggests profound ecological possibilities. These projects are intensely pedestrian and very three-dimen-sional. That they are unabashedly, aggressively commercial makes them not so different from the farmers' markets and central city markets throughout history. The Grand Bazaar in Istanbul is a spectacular ancient version, if only on a single level. Such markets have been and still are vital economic and social founda-tions for cities, crass though they may seem to those of refined aristocratic or escapist envi-ronmentalist tastes. They work.

Where they require a veritable geological-scale parking desert to function, giant pedestrian malls like Jerde's Fashion Island in Newport Beach, California, constitute a con-tradiction and in fact an assault on nature. But where, as in his Horton Plaza in San Diego, California, they are part of the urban fabric and enhance the diversity of human life in a center, they are part of the wave of healthy possible futures. Tens of thousands of people live close enough to walk, bike, or bus to Horton Plaza, with its pleasurable environment of pedestrian streets and multilevel structures with many bridges creating the kind of adult jungle gym I've mentioned in these pages. Much of Jerde's work can serve as precedent and idea source for three-dimensional pedestrian areas at all population levels, and as a kind of node of DNA in the living protoplasm of the city, containing enough "genetic" information to precipitate change throughout the whole city and in its progeny cities into the future. There is powerful urban change potential here.

Jon Jerde says that he was profoundly influenced by Paolo Soleri. Jerde arrived at Soleri's workshop in Paradise Valley in 1959 and was amazed by his drawings. He asked if he might stay a couple of days and, he says, "ended up sleeping under Paolo's drafting table for three nights, poring over his drawings and notebooks until 2:00 A.M."[7] Until very recently, however, he was designing a crucial component of the basic three-dimensional structure Soleri had proposed as an arcology while letting the existing urban infrastructure take care of access to it. Now he's thinking through ways he might be able to connect his work with Soleri's.

If Curitiba is not too lost in the recent automobile-induced trance that its transit-assisted prosperity has helped produce, it could be a prime candidate for Jerde-type commercial centers finessed toward the biodiverse and the technologically "appropriate." Such centers could be integrally connected to the Citizenship Street core areas, or the central business district itself could be reshaped with three-dimensionality that creates the taller buildings and elevated parks. Underground space there would provide commercial and private storage for the businesses and residents in the downtown. Jerde's bridges could link many old and new buildings alike. Mid-block passageways already exist in considerable number in Curitiba. There could be more.

Ken Yeang of Kuala Lumpur, Malaysia, has recently been designing "bioclimatic" skyscrapers — typically towers with movable windscreens and sunshades, some of which circulate around the outside of the tower on fixed tracks, regulated according to daily conditions, to let in or keep out breezes and sunshine. One of his smaller buildings — three stories high — is shaded under a hovering sunscreen that looks something like a pair of pointed pillows or flying saucers tethered by steel cables between the ground and high-tech crane towers. These things cast cool shadows in a hot climate, letting the breezes slide on pleasantly by, saving the building enormous sums in air conditioning. When I met Paolo Soleri in 1965, he spoke about urban "garments" and immediately introduced me to a micro-version. Stretched over his semi-outdoor drawing table was a modest

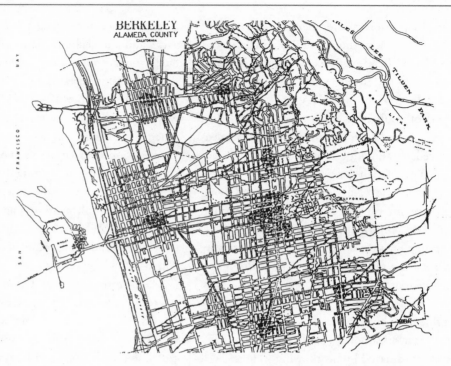

BERKELEY
ALAMEDA COUNTY
CALIFORNIA

More development in centers, less farther away from centers.
In 40 to 80 years the Marina peninsula has become a distinct island, wetlands with their biodiversity have returned, development has reinforced centers with enlarged and connected buildings in pedestrian areas and open landscapes have increased areas of parks, food production, recycling areas, sports areas and such natural features as creeks and ridgelines. Transit and bicycling works much better, saving energy and reducing pollution and CO$_2$, due to centers-oriented development.

canvas awning on poles, perhaps twenty feet long. Jon Jerde slept there.

In that sweltering climate, Soleri would occasionally take a hot hose from under the Arizona sun and sprinkle the canvas awning over his drafting table, the water getting cooler as it snaked out from underground and sprinkled over the cloth. The evaporation and shade cooled the drafting room, the moist canvas giving off a pleasant fresh air. On a larger scale, marketplaces around the hot regions of the world shade the stands and stalls, vegetables, fish, customers, and dusty traders under canvases similar to Soleri's. Ken Yeang blows them up to shade whole build-

ings. For cities, I advocate this idea writ large: billowing horizontal clouds of cloth over whole sections of cities in the hottest climates, with thin steel cables and towers carrying them up into the daytime skies and electric winches reeling them in when the sun settles near the horizon so we can see the stars at night. The cloth could be tapered in thickness, reinforced toward the edges of higher stress or cut into checkerboard or strip patterns to let dappled shade and sunshine fall into the activity below. I've always marveled at the great billows of dark netting at construction sites draped over scaffolding ten, twenty, fifty stories high to catch and deflect falling construction debris, protecting people in the streets below. Federico Fellini celebrated the windy billowing and rippling of large cloth shrouds in the movie *8 1/2*.

It may seem excessive, expensive, experimental to a fault to suggest creating such immense city garments, but they could save copious amounts of energy producing comfortable environments at a cost that would be nothing compared with that of hundreds of air conditioners, much less the cost of a new freeway interchange. In what way, in fact, are innovations such as urban garments and in-city basement warehousing, large-building solar greenhouses, and publicly accessible high places "impractical"? A few thousand years ago our grandparents a few dozen times removed, slaves or not much better, were rowing big ships through the Mediterranean or up and down the Tigris and Euphrates. Then someone had the idea of catching the wind by lashing cloth into giant sheets, and a whole new age of utility and exploration erupted into reality: the sail ship. Maybe it was a woman who, hanging out the laundry one day, said to her husband, the captain who directed the drum beat for the rowing crew, that if the clothesline pole were made bigger and attached to the boat and a big piece of cloth were provided, the rowers could take it easy for a while and practice for their next war or read some cuneiform. Who knows? But at that point, raising fabric to the gods of the wind created a new world. Moreover, given its

absolutely incredible historic leverage, the creative innovation is the cheapest investment on this planet.

Yeang's buildings are situated in hot climates. I'm not aware of what he might do in terms of capturing and storing heat in cooler climes, but similar thinking with regard to taller buildings in these areas could be promising, too. This is the "climatic" part of his architecture. The "bio" features of his skyscrapers include continuous gardens circulating around and up the building from one terrace to another, sometimes connecting with "skycourts" (three- or four-story notches, atriums, or arboretums) cut into the twenty-, thirty- or forty-story structures. The continuous ascending terraces are like three-dimensional nature corridors, with insects and lizards, birds and tree frogs. The skycourts become small neighborhood parks of a sort, but at the edge you might look up or down at ten or thirty more stories of building, and then to the sky or the ground. They are like the celestial garden terraces in the Andes, à la Machu Picchu, but carved into modern buildings. These vertical greenbelts and parks work in concert with the climate-tempering structures to help refresh and cool the air and support the biology and appropriate technology. All these become integral to one another.

In Curitiba or other cities willing and able to experiment with larger projects, Jerde's commercial centers and new clusters

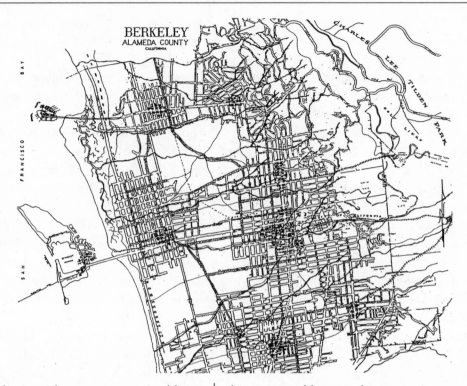

Development of centers and de-development outside of centers continues. In 60 to 100 years reshaping yields enormous gains in transportation and energy efficiency while restoration of agricultural and natural areas progresses. Now, no one has to own a car to connect with the full range of their life's activities.

of buildings and open spaces could use Yeang's ideas to temper the extremes of climate. Yeang says that his buildings are more expensive to build in the first place, but consume far less energy and in a few years are far less expensive overall, even with construction, maintenance, and climate moderation costs included. My suspicion is that covering public spaces with canvas awnings would save enormous energy and financial investment. If the strategy were used to block unpleasant winds between buildings with transparent movable windscreens operated by the municipality as cities currently maintain streets,

this, too, would create pleasant environments while saving energy. In the pedestrian city of the future, moveable devices for moderating temperatures would probably run on less energy than the street signal lights required in a car city of the same population, saving the city millions of dollars.

Pulling these elements together, from Curitiba's land uses and the refinements of architects like Jerde and Yeang, is a complex undertaking, to say the least. But is it very much more complex than cities already are? Or is it any larger an investment than much of the infrastructure we are already used to, such as freeways, automobile factories, petroleum cracking plants, subway systems, and the like? Even today's large buildings are formidably complex, with their elevator systems, grand lobbies, machinery for completely controlled climate maintenance, and electronic hook-ups. Would it be that much more difficult to include ecological design? More functions would be closer together in the ecocity, but substituting the complexity of the pedestrian city for the complications and contradictions of the automobile city would end up freeing up money and time for the highest levels of complexity — in the form of cultural activity.

Strategies for Built-out Cities

Ideas for Curitiba can apply well to any growing city in an open area, but a similar arrangement of pedestrian land uses and infrastructure can be created for any existing "built-out" town, whether it is growing or even shrinking, by transferring density within the city limits. In all cases the strategy for creating an ecocity is a steady and methodical moving away from automobile toward human infrastructure — and also away from "auto" as in "mechanical" and "reactive" and toward the reasoning and intuitive capacities of the human mind.

The "built-out" city, hemmed in by other cities, greenbelts, or unfillable waters, is typical in our large metropolitan areas these days, but if it is build-out, it is very seldom "built-up" or arranged on the principle of access by proximity. It lacks what Michael David Lipkan calls "proximity power."[8] Instead, it utilizes gasoline power.

Dealing with the built-out city is the real action for those who believe ecocities are important. First, if cities are sprawled out two-dimensionally and thus built out already, shrinking back to smaller footprints is called for immediately. Also, if organic agriculture, permaculture, and intensely managed, ecologically informed bioregional farming of any style make sense, then the population shift should be generally away from further urbanization and largely toward the building of towns and villages associated with this sort of agriculture, along with forest management, eco-tourism, and study. Perhaps the most important role for new or transforming cities will be as experimental ecocities testing various structures, arrangements, technologies, and citizen reactions.

One approach, as we have seen in the case of Curitiba, is to start anywhere you can in a particular city with car-free streets and to systematically expand from there, building pedestrian infrastructure more quickly than auto infrastructure. Simply with the idea and political will your town can go directly to a car-free street or district or car-free apartments or condos or both at once. A building with an interior garage can be remodeled to replace the garage with better uses as units shift out of the typical lease agreement and over to a car-free

The strategy for creating an ecocity is a steady and methodical moving away from automobile toward human infrastructure.

Differentiation and diversity of activity reaches high levels. *Variety of human, "working landscapes" and natural areas reaches high levels in this representation of a time 100 to 150 years in the future. The dashed line, above, represents the underground Bay Area Rapid Transit line existing from the 1960s. Dotted lines represent vehicle, bicycle and foot tunnels under open land allowing natural habitat continuity. Thin lines are surface foot and bicycle paths. Car-free has been attained and from here forward we are moving toward highest levels of cultural and ecological harmony.*

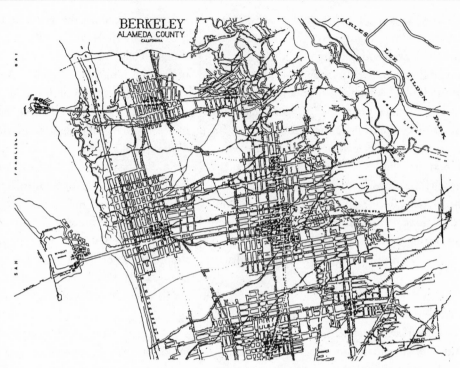

agreement, thus making more housing or a wider variety of services available to the residents of the area — the car-free conversion mentioned earlier. Another approach is to simply build the better higher-density building with ecological features in a good location for transit and a good location for other urban services and features, TDR or not.

The Gaia Building constructed by Berkeley developer Patrick Kennedy is a good example. Kennedy has built five residential buildings in downtown Berkeley, each with shops at street level. Neighborhood activists and architectural conservatives have opposed him, arguing for low height limits, and he has countered by rounding up and bringing to hearings all the low-income, disabled, minority people and environmentalists that he has benefited so that he can win his next approval and build his next building. He was amenable to including ecological design features in the Gaia Building and adopted several from Ecocity Builders, including terracing and rooftop flowers, bushes, trees, trellises, vines, and accessibility for renters, office workers, and guests. I sketched out the solar-oriented rooftop features, and he incorporated them into his design. The plans then sailed through

City Council, and the building, with the first genuinely alive roof in Berkeley's history, was historic upon completion.

The Heart of the City Project

For twenty-two years I worked in Berkeley with others of similar mind trying to transform the city into an ecocity, adding one good project to another and supporting everything that might help. I didn't have a real system or an awareness that a particular sequence of efforts might further the cause. I had no real overall strategy. The theory was simply that good changes would add up and the process would accelerate. By 1997, however, I began to face the fact that very little was adding up. I was beginning to comprehend the larger forces working against ecocity efforts locally and, it seemed, almost everywhere else. I also noticed that there wasn't much coherence in the details we ecocity advocates were trying to weave together in our various projects. People walking across town would see an open creek here or a solar greenhouse there, some fruit trees by the street or the Vegetable Car with its garden in front of our house but connect none of these to the idea that some kind of ecological change was afoot. The ecocity features weren't together physically, so putting them together in the mind was a stretch. If we could literally put them together and meaningfully connect them in one place, a special place visited by many people — if these features collectively looked like an ecocity — we

might finally be able to communicate something about the whole integral pattern of parts. We'd have, in other words, one of Paul Downton's "urban fractals" mentioned earlier, an "integral" piece of the city where all the pieces came together.

By 1997, we had some very particular ideas for downtown Berkeley in mind. Ecocity Builders proposed to restore Strawberry Creek downtown, create a public plaza and pedestrian street or two, and add considerable new housing there, right next to the main transit station, in a set of taller-than-usual buildings with ecocity features. If, in addition, these features — creek, plaza, housing, transit — were conspicuously linked together with busy footbridges, if they were spectacular in their regalia of solar greenhouse glass, covered with people on the rooftops under colorful umbrellas, and obviously a celebration of nature, then maybe people would begin to grasp the ecocity concept. In fact, it would be hard not to notice it, not to be intrigued and inspired by it. The project would be profitable for local businesses, a destination for tourists curious about the future, and something to photograph to show to friends back home and buy postcards of to send off around the world. It would be a "postcard breakthrough." With a single glance at an image as small as a postcard, the observer would say, "Ah-hah! So this is what ecocities are all about!" It would be a downtown

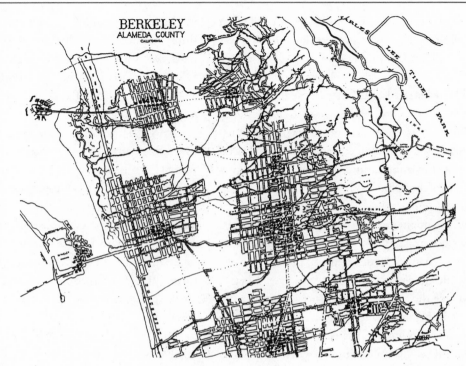

BERKELEY
ALAMEDA COUNTY
CALIFORNIA

Distinct and well-balanced
human and natural areas
achieved.

*Low-density uniformity with a
downtown, the condition in
the early 2000s, becomes
discrete towns with their own
character and a small city
(formerly downtown Berkeley)
surrounded by open space
rather than the automobile
dominated, oil dependent
infrastructure of the past.*

integral neighborhood, a big one, with nature and regional transit. It would communicate the story and affect the entire city. Like a strand of DNA, it could replicate mixed-use, bio-rich urban tissue throughout the anatomy of the city and spawn ecocity projects elsewhere by example.

Most cities have quite a few functions already established in their downtowns: jobs, housing, shops, transit, entertainment, food. Typically, though, as in Berkeley, housing is very limited, and ecological architectural features are close to non-existent. If you simply supply the miss-

ing pieces, however, presto! You have what we are calling a "Heart of the City" project.

Downtown Berkeley has a perfect place for this, as do most cities. The Bay Area Rapid Transit (BART) Station is located on the corner of Shattuck Avenue and Center Street just a block from the University of California campus. Eight thousand people walk here every school day from the station to campus and back. Thousands of other people see this corner as the center of town. Along this route to campus there is a one-story Bank of America building that would look completely at home in the most car-dominated areas of suburbia. There's a parking lot for the bank and next to that a printing press building owned by the university, displaying a blank wall to passing pedestrians. The university, the bank, and the city of Berkeley all seem to see this urban space as a bit of an urban design embarrassment.

We suggested building housing, opening the creek, creating a public plaza and pedestrian street, connecting buildings with bridges, putting in public rooftop cafés and gardens, and firing up the project with the power of the sun in solar greenhouses cascading down from the heights to the lower floors. There was enough room for all this detail on this site because it covers more than a whole city block. Many found the idea intriguing, and forty small businesses and nonprofit organizations joined a coalition to explore the idea further. A charrette, a workshop for designing an architectural project, was held where half a dozen architects contributed drawings of possible variations on the theme. We organized two small international conferences with speakers from six countries focused on city center plazas and works of water restoration and celebration (see illustrations of various versions of Heart

Corridors vs. pedestrian/transit centers.

The top map represents New Urbanist and other advocates recommendations to add one row of four or five story buildings on transit corridors. Where this creates modest density and adds efficiency to transit, not much. It also tends to lock into place the lower density development around the corridors and prevent opening creeks and reclaiming open space for food growing and other uses by creating a barrier of expensive-to-remove medium density development where creeks and other natural or agricultural areas could be restored in the future far from the centers. The much better solution is stronger centers that are from a couple to several blocks across — not just one row of buildings — that can host real density and diversity and become stronger centers for transit, especially streetcar and regional rail systems.

Present Berkeley Zoning Map — Built out — same color code as Ecocity Zoning Overlay Map of the City projects for this location, p. 142, p. 147, p. 149, p. 151, p. 153, and p. 156).

Realizing that we were facing height limits too low to allow a Heart of the City Project with relevance for an urban future, we began working on amending the Berkeley General Plan to allow taller buildings if they included ecocity features or helped finance the opening of creeks and the expansion of community gardens and public parks. Years before, we had begun to think that we needed major land use changes to accommodate creek restoration in the city. Now we realized from another angle how important modification of the city's land use ordinances would be if we were to be able to create ecologically healthy rearrangements of the city. We simply had to change zoning for our Heart of the City Project to be possible.

We knew that the General Plan needed to provide means to remove development from areas farthest from the transit centers and shift density toward those centers. The Heart of the City Project area should, then, be a place where more density should be built, while buildings in car-dependent areas should be removed. Thus it was natural to imagine Heart of the City projects as receiving sites for Double TDR, and so we developed language specifically calling for "ecological demonstration projects" like the Heart of the City Project and recommended that the land use element of the General Plan include a policy encouraging the use of TDR. As mentioned earlier,

weak support did get into the General Plan for an "ecological demonstration project" of a definition very much like our description of the Heart of the City project, though the TDR policy failed. Still, we continued working on it.

Harrison Fraker, who had been involved in the Phalen Village project in St. Paul, Minnesota, and had moved to Berkeley to become the Dean of Environmental Design at the University of California, suggested that our project was lacking something. What he had in mind was liberating the buried Strawberry Creek from downtown all the way to San Francisco Bay and building a cycling, walking, and jogging path near the creek that would connect with the San Francisco Bay Trail. This Strawberry Creek Greenway could also connect with an existing greenway named after the local Ohlone Indians that comes partially into town from the north and could be extended to Strawberry Creek and on to Oakland to the south. An excellent idea. A heart needs a vascular system and here it was, both water-powered and human-powered, linking the heart of the city to the San Francisco Bay and the whole region.

At the time of writing, the University of California and the City of Berkeley are reviewing recommendations provided by numerous groups in the community, which include all the essential elements of the Heart of the City Project, and Ecocity Builders continues to encourage such a project. City

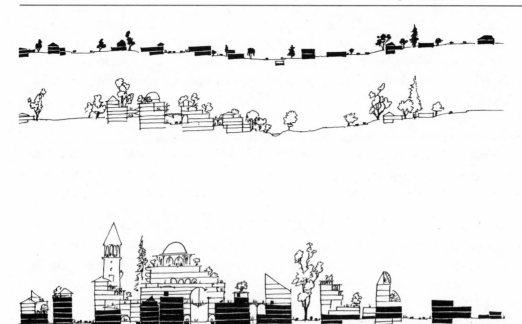

Slice of changing cities. *Here we see a cross section, top, of a typical sprawling low-density development with a car oriented "center" (slightly left of the center of the illustration) becoming a compact town of real diversity of land uses and cultural variety. The buildings in solid black represent two contemporary towns, which become over time, the more centers-oriented ecocity towns in outline.*

Council has endorsed these recommendations and the representatives of the University, owner of most of the real estate that would be involved, have stated that they will seriously consider the recommendations in whatever development they will pursue on the site. Thus the Heart of the City Project is a work in progress. Could its realization be a threshhold event? The world badly needs examples of "integral" or "urban fractal" projects. This could be one. With the Hurricanes Katrina and Rita wakeup call, perhaps Berkeley City and University leaders will — at last — take seriously the notion that we need to "build in balance with nature."

Ecocity Strategies

In terms of a strategy for transforming today's cities to ecocities, this begins to look like a reasonable outline for Berkeley or anywhere else:

- Find your centers and create Heart of the City projects by filling in the missing pieces.
- Create a "vascular system" of bicycle and footpaths, as well as transit, from that heart to the other major nearby centers and to the most appreciated natural features in your area.
- Next or first, whichever sequence can be more easily accomplished, looking

at the whole living system, make an ecocity zoning overlay map to help put in order and connect the proper functions of all of the rest of the city, much of it properly returned to agriculture and nature.

Another approach in this ecocity strategy would be using such tools for transfer of density and financing of the shift as Double TDR. Should local governments be opposed to restoring waterways and opening up agricultural landscapes in areas dependent on cars, then state governments concerned about the collective detriment of car cities, the cost of freeways and air pollution, and the positive benefits of ecological cities could pass preemptive Double TDR legislation that might read something like this: "Any person owning improved property within fifty feet of the historic centerline of a buried or open creek may sell the development rights of his or her property to a developer, by way of a land trust directly, for demolition of improvements and restoration of open space on the sending site and for transfer of those rights to any site within three blocks of a regional rail transit station or intersection of four or more bus lines."

City governments could buy up properties and sell the development rights to developers for this purpose themselves, creating a publicly managed revolving fund or TDR bank for more such projects. Nonprofits could also buy up such land and sell the development rights, thus creating their own revolving

funds for restoring land and waters, or they could become developers of ecological projects themselves. Ecocity Builders is looking into setting up an Ecocity Conservancy Trust for exactly this purpose. If all goes well, we will collect contributions from people and foundations, businesses and government programs to buy houses along the creek culverts, demolish the buildings, carefully recycle the building materials, open up the creek, and either manage the property or donate it to the city to manage. When we sell the development rights to a developer, we will be ready to use our revolving fund all over again, buy more land, remove more buildings, and thus consolidate the land for opening the waters.

The same strategy can be used to expand community gardens and parks. Similarly, when houses in poor condition come up for sale in car-dependent areas, they could be bought and demolished, leaving an empty space between two other houses. The land itself could be sold to one or both of the neighbors, donated to the city as a mini-park or community garden, or retained by the Ecocity Conservancy Trust. Open spaces between houses are not what we are used to, but they could become beautiful places in the neighborhood, source of food, education, camaraderie, or simply visual rest. And the cars that used to park in the driveway there and on the street would be gone. The neighborhood would be quieter by two to six people, while in a transit center two to six

Roll Back Sprawl.
What the world needs now is reversing the asphalt juggernaut, and with it, the domination of cars and oil. It technologically can be easily done and policy tools are available. No waiting for miracle techno fixes, "free energy" or any other rescue from without. Only the ecocity insight and courage to face change has been lacking. Here we see the removal of houses replaced by the apartments and condominiums in the upper right corner, by way of transfer of development rights or other planning tools and strategies.

times as many people would be provided with housing at walkable distances to practically anything urban people might want. I am aware of many such small open spaces in residential areas purchased by city governments not using TDRs that have become neighborhood tot lots and playgrounds. Such open spaces could expand with time, accruing adjacent properties, becoming new parks and agricultural areas, and eventually joining up with creek and ridgeline restorations, profoundly reshaping the city and creating a healthy urban anatomy.

All the tools you have read about here, all the projects, can come into play once a city decides it wants to pursue ecocity building. Basically this is what happened in Curitiba, though a number of the tools offered in this book would be new to that city too. For the strategy to work, we will need to reverse the transportation hierarchy that we currently see in cities to place pedestrians first, and we will need to build the kind of interlinked buildings and public spaces and natural features that become the built expression of the principle of access by proximity. Environmentalists, who already know why to build, and developers, who know how to build and, with regard to the services their buildings provide, have a good idea why, can and must be allies in this effort.

Another part of the overall strategy would be to pursue the International Ecological Rebuilding Program suggested earlier, the ecocity-building equivalent of Al Gore's Global Marshall Plan and Lester Brown's Plan B, with its Departments of Ecological Development, vanishing automobile subsidies, and the like.

Yet another approach would be to challenge a government or large corporation to build experimental ecocities. China, for example, having no fear of density and both a very centralized government and large capitalist corporations, might like to build six experimental cities for one hundred thousand people each — large enough to provide invaluable lessons for us about the basic form and function of particular urban arrangements. I've proposed this to Rusong Wang, convener of the Fifth International Ecocity Conference and a member of the Peoples' Congress of China. Three of these would be located in rural areas but on convenient rail routes to other cities, so that they would function well within the present larger economy. One city would be strictly for pedestrians, very dense and rigorously respecting local bioregional conditions, designed for solar access and technology, organic farming, and so on. No cars, no transit, not even bicycles. A second city would be for pedestrians and bicycles. No cars or transit. A third city would be for pedestrians served by both bicycles and transit but not cars. Three other cities could be built heading for the same goals, but would be existing cities adopting the strategies put forward here to be transformed steadily toward becoming the pedestrian,

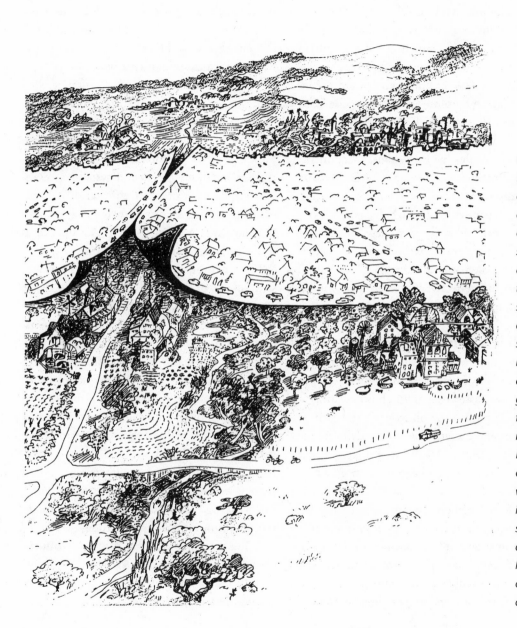

Roll Back Sprawl 2.
Here we extrapolate to a low-density city giving way over decades to the landscape of city centers becoming smaller, taller more diverse and neighborhood centers turning into villages and small towns in agriculture and nature. Something similar needs to happen to metropolitan areas everywhere and is inevitable given collapse of finite fossil fuels but the time it takes to make the transition is crucial. If too long, vast destruction of climate and biodiversity will plague us all. To move intentionally in this direction sooner, rather than later, is a chief means to ecological health and further evolution of hum compassion and creativity.

bicycle/pedestrian, and transit/bicycle/pedestrian cities just mentioned.

Thinking about how to build these six experimental cities, which could be built in any country taking sustainability seriously, and then following through would be as important and useful as anything else conceivable at this time in history, in fact, at this time in evolutionary change on Earth. Since cities are whole, living systems, creations of integral organs or parts, strategies starting from any one part of the whole, if cognizant of the dynamics of the whole, could work equally well. This is good news because it means that we are all, with our different perspectives, talents, and training, in a position to contribute effectively to the ecocity building enterprise.

Building a Culture of Acceptance

Mayor Jaime Lerner knew that Curitiba needed a culture of acceptance for the ecocity, and he very assertively went about creating it. When he put on the uniform of the garbage collector and became one, thus redefining the job as recycler, he became an ordinary citizens engaged in building the ecocity. Everywhere in the world the culture of acceptance can grow as the demonstration projects increase in number, as one incentive after another is passed into law, and as education gets through to ever more people. Good examples and policies will build up and capture the imagination. At some point the imagination of whole cities will kick into gear, we'll get over the threshold of sufficient votes to pass important policies and launch projects, and the latent energy of the whole community will release a cascade of creativity.

To build this culture of acceptance, the strategies I advocate will have to meet the more conventional "better ideas" halfway. Thus we can admire and emulate the cities building new metropolitan light rail and regional medium-weight rail systems around the world, the European cities creating and expanding car-free areas in their ancient cores, the bicycle-promoting cities of Europe, and the developers of traffic-calming streets. When cities calmly reject the next big parking structure, we should celebrate. These decisions should get national urban design prizes until they are so common the local papers don't notice them any more. The New Urbanists take pains to conventionalize their transit villages and even write model ordinances and codes to make their replication faster, the approach even more conventional. And as more and more creek restoration projects, greenbelt laws, and slow-growth initiatives pass, as more natural areas are preserved, we should be happy about it but keep on working until these things are much more common than they are today.

One would hope that the really fundamental experiments like Arcosanti and the ecovillages around the world would become common, too. The same could be said for solar and wind energy technology, "green" architecture, recycling, and nontoxic clean-up products, processes, and technologies as well as for regulations to achieve those ends — fines for polluting, higher car registration and gasoline taxes, and the like. But as we support such endeavors, we can't let our guard down. A billion Chinese are craving cars and almost as many Indians. And Americans want their cars bigger than ever (though the jolt of the 2005 hurricane season has some reconsidering). This is why it is so important to embrace not only the "conventional" improvements but the truly basic changes that come with the use of the tools described in these pages, from the land use foundation up, from the ecocity zoning map to the rooftop garden café. This is why I'm proposing that we acknowledge that we need strategies to reorganize our thoughts, roll back sprawl, and reorganize the city and thus the whole civilization it supports. In times of global warming and approaching peak everything we face immense challenges. But then again, perhaps this is what might wake us up.

Art and Imagination

Can we create afresh our natural and built environments, even ourselves, as we look out onto a hopeful future? Can we, with the strength of our conscience hitched to our reason, make peace between us and our world, pounding the swords of the city on the attack into the plowshares of the city that builds soils and biodiversity? Can we learn to live that way, into a future of ecological health and human creativity and compassion? If we can, we will inevitably build the ecocity. Our home, our garden, our wildness, and nature's wilderness will all spring into full bloom. The ecocity itself will not be Venice or Curitiba, Shaban or an American Southwest Indian pueblo — those precursors are valuable lighthouses to the future — but some new permutation anchored in a sense of our own times and the needs of a profoundly different future.

Building ecocities will be a new version of the ancient art of city building where we are collectively the artist. Each ecocity will be different. The art of city building and the art of living will be combined in this effort, and like all arts, this will be a struggle that promises pain and pleasure, failure and undreamed-of reward. Everything depends on our selection of the field of action, the choice as to what game board we are going to construct, what stage for our performance, what physical community we are going to build. Are we going to continue dismembering our communities and scattering them to the winds? Are we going to continue to pretend that nature is primarily a repository of exploitable resources with no inherent value or rights of its own? Or are we going to set out to make of us and our creations a reflection of our best inner selves, and a celebration of the Earth by whose graces we live?

We have barely begun the ecocity experiment, but concerned citizens can create little pieces of it immediately, giving us all insights into life on the other side of that Ecozoic divide and inspiring us to shift to a healthy way of living. Ultimately we might just build environments that will keep the artist alive in all of us so that we can all live in conscious co-creative coevolution with each other, nature, and the home we build. There is no easy way, no sure way, and each solution will also cause new problems, but there is magic down the path that we have yet to choose and that we have every opportunity to create — the path through ecocities.

Afterword: Rebuilding New Orleans

Will Hurricane Katrina be the wake up call to alert us to rebuilding cities? Will the scenes of Third World USA in collapse or higher gasoline prices do the trick?

Among the flood of articles since the disasters in New Orleans and the Gulf Coast, in none did I notice a writer trace the cause-and-effect chain back through hurricanes on steroids to climate change to the dysfunctional form of the largest creations of our species: car cities humming along outgassing a whole new storm-enhancing, ocean-raising atmosphere for planet Earth. That's only two links in a chain that nobody connected. Naturally, then, no logical conclusion was suggested that therefore, prevention in the future must have something to do with reshaping cities. Neither did anyone *want* to do that thinking because it might imply all us drivers are responsible, if in a slight and widely distributed way, for the catastrophe.

In any case, the conclusion is that we need to rebuild not just New Orleans, but transform car cities into people cities — ecocities. Yes, the levees were neglected there while the government cut taxes for the rich and oil companies made the highest profits in history. Yes, the National Guard was guarding the pipelines in Iraq when it could have been guarding the Gulf Coast. But there are ecological lessons here too.

All cities impact nature and "harness" it for their resources. We've looked at some of that history in these pages. Most are built adjacent to, then spread out across and eliminate much of the very agricultural land needed for their sustenance. But in New Orleans' case, as a major shipping, industrial, food and cultural center for an entire nation, some sort of city is going to be there or nearby, and in that place three ideas emerge from the ecocity perspective that are not yet among the proposals presently circulating. They are mound, labyrinth and shell.

Since these three design elements give us yet another — and especially interesting — way of visualizing cities, I'll wind up this book with suggestions for New Orleans. But first something about the Mississippi Delta.

The great river drains almost one third of the land area of the United States, bringing down millions of tons of silt every year. These muddy waters have for eons spread out on the continental shelf of the Gulf of Mexico, slowed down and dropped their silt, expanding the delta that is Louisiana in a particular direction. As the river grows longer, out across its own deposits, its angle decreases and so does its speed, which increases the rate at which silt falls out of the water. The bed of the river, then, is always rising, water slowing down, breaking through its banks to move out laterally in floods and, where it almost stops on the flood plain, drops yet more silt, raising the level of the whole delta.

After a few hundred years of this, it has a hankering to dash off to lower-lying portions of the delta, taking a shorter, steeper route to the sea. The distances are awesome. The river might want to head 200 miles to the west one century, almost to the edge of Texas, then five hundred years later, it makes a break for a lower lying route back toward the state of Mississippi. Since around 1950, the river has wanted to head out to sea about 120 miles to the west of New Orleans along a branch called the Atchafalaya River.

The US Army Corps of Engineers, however, to maintain the viability of industry, low-land farming, shipping and the city of New Orleans, has built up the banks to pre-vent that from happening — while the Atchafalaya has continued to sink lower and become ever more insistent in saying to the main river, "come this way." Beginning in the 21st century, there is now a set of locks to allow ships to be lowered 15 to 20 feet from the Mississippi to the level of the Atchafalaya, or lifted the other way. In an attempt to deliver silt and slow down the settling along the Atchafalaya, as well as deliver water for a dozen other purposes to agriculture and towns west of New Orleans, the US Army Corps of Engineers is now pouring 30 percent of the flow of the Mississippi over a spillway adjacent the locks and into the smaller, more westerly branch.

One wonders then, that if New Orleans is rebuilt, if it should not be moved too, to a point closer to where the Mississippi would rather spend the next five hundred years being its normal self. But no matter where the city ends up — in exactly the same place or moved anywhere from a short distance to a hundred miles — these same solutions will make sense for any low-lying hurricane coasts.

Mounds were built by the Native Americans up and down the Mississippi. In the Great Mississippi Flood of 1927, animals and people gathered on so-called Indian mounds because they were like islands in the swirling sea of muddy water. The natives built them by carrying basket loads of earth and gradually raising their ceremonial sites and some whole villages above harm's way. The car city covers far too much land area to ever be built upon mounds in that ancient model. It would require astronomical amounts of fill. But with a pedestrian layout and the much smaller land area of a compact city designed for people on foot, it is possible to raise almost any city above the flood. Once again, as in so many other ways described in these pages, the pedestrian layout proves to be the foundation for ecological city solutions.

In some European cities such as hot and windy Nice, France we can see examples of the labyrinth, that is, narrow streets that wind to and fro and sometimes end in "T" intersections or even dead-ends. In Nice, the strong winds are confounded by the twisty streets and become mild breezes of calm air in the "interior" of the Old Town. Similarly, if New Orleans were to have a higher-density core with such a layout, hurricane force winds would be toned down to manageable levels. Broad streets between the towers of the typical downtowns of the car city, along with parking lots, gas stations, freeway interchanges and the like, actually accelerate high winds, which is not a good idea in hurricane country.

If building on mounds and in a configuration of pedestrian streets in a labyrinth-like pattern is accomplished, all that remains is to harden the siding, windows, shutters and rooftops of the city's out-facing buildings, the city's "shell." The "interior" would be high, relatively dry and relatively calm. There would be no need to evacuate.

But where it gets rich and wonderful is when we begin to imagine the full palate of ecocity features coming together. The bare bones of an ecologically tuned New Orleans with the mound, labyrinth and shell as the armature for the whole thing is a start, but we can think through a few other things, too.

Rooftop furniture and smaller trees and bushes could be removed from their vulnerable positions for temporary storage during storms. Though larger trees on rooftops, promoted in this book, are not appropriate to hurricane country, they could be planted in courtyards, atriums and parks deeper in the city, creating an interior "natural" landscape something like the creek daylighting designs I've encouraged for city and neighborhood centers.

The New Orleans streetcar system was much celebrated in the past, and some of it is still there. And who hasn't heard the folk song about the train called the City of New Orleans? Rail transport, along with glass elevators on building exteriors celebrating the city itself with its new-found views — just because it rises up into the views — can hold a larger city together and make it an enormous pleasure to travel from one island neighborhood to another, to downtown and off to the wharf, and out across the rest of the country on the rails that are the ideal transportation partner to the pedestrian city structure itself.

An elevated city rising from the tides provides many new possibilities, such as sloping plazas and parks of the "keyhole" or "view plaza" layout, looking out to the surrounding waters, mangroves, and cypress with their hanging Spanish moss. Staircases and ramps could celebrate the rising and ebbing tides descending into the waters, which could daily inundate portions of these public open spaces as a dedication to keeping the local environment not just in mind, but celebrated in art.

Paths and bridges on floats might last for years leading out to the restored low islands with their wild bayou ecologies. Walks and bicycle rides both short and long could go to gazebos, bandstands, tea houses and art works — all vulnerable to the storms, and once every decade or so, swept away like sandcas-tles to be replenished by new teams of artists and performers.

Will the day come when we can walk such paths through a restored natural world, listening by turns to the songs of people and birds, neither afraid of the other, culture of nature or nature of culture? I'll end this book with these thoughts on the arts of the ecocity in a particular place, just as I discovered my interest in ecological cities more than three decades ago as beautiful places to build and beautiful places to live. But if the idea of ecocities also helps reverse global warming, rolls back sprawl and becomes a healthy response to the disasters of our late hour, so be it. For many reasons, it's time to actually build the ecocity.

New New Orleans center on the edge of water. *Here New Orleans is elevated 30 feet, surrounded by restored marshes, linked by streetcar and pedestrian streets and safe from flooding and high winds. No need to evacuate in hurricanes.*

FURTHER READING

Ecocities — the direct approach:

Register, Richard. *Ecocity Berkeley: Building Cities for a Healthy Future.* North Atlantic Books, 1987.

Good for basic ecocity principles in general, developing a scenario for one set of city transformations based on those principles in Berkeley, California.

Register, Richard and Brady Peeks, eds. *Village Wisdom, Future Cities: The Third International Ecocity and Ecovillage Conference Held in Yoff, Senegal, January 8-12, 1996.* Ecocity Builders, 1997.

Synopsized presentations from the Third International Ecocity Conference, held in Yoff, Senegal, West Africa, in 1996. This book has very broad subject matter and features an international range of efforts to create ecological cities and villages. This book delves into the conference host village's traditions, religion, decision making, and values as well as into ecological technology, agriculture, architecture, and planning from around the world.

Schneider, Kenneth R. *On the Nature of Cities: Toward Enduring and Creative Human Environments.* Jossey-Bass, 1979; and *Autokind vs. Mankind: An Analysis of Tyranny, a Proposal for Rebellion, a Plan for Reconstruction.* Norton, 1971.

Some of the best thinking ever on, as Ken says, the nature of cities as they have functioned in history and as they should/could, if healthy, in the future.

Soleri, Paolo. *Arcology: The City in the Image of Man.* MIT Press, 1969.

First and by far the most fundamental exploration into the urban third-dimension, for the deep future and as a powerful exercise for rethinking cities. This book is like the physics, math, and geometry of the city as a complex organism, with a strong dose of both philosophy about what it might mean and engineering as to how to build it.

Cultural and historical context for ecocities:

Diamond, Jared. *Collapse: How Societies Choose to Fail or Succeed.* Viking, 2005.

A tale told in an exciting and thorough style of societies around the world and throughout history degenerating and partially recovering or collapsing in freefall to annihilation — with lessons for us today about the blinders to ecological and resource realities that make the difference.

Goodman, Percival and Paul. *Communitas: Means of Livelihood and Ways of Life.* University of Chicago

Press, 1947.

A wonderful overview of spatial and cultural city arrangements and the trends of civilization entering the age of sprawl, with perceptive insights into universal and timeless city issues. This book issues many warnings, most thoroughly unheeded! Full of good humor.

Jacobs, Jane. *Cities and the Wealth of Nations: Principles of Economic Life.* Vintage, 1985; and *The Death and Life of Great American Cities.* Random House, 1961.

Her thinking is classic in the field, clearly describing the economics (former title) and social dynamics (latter title) that make cities work, or not work.

Architecture and ecocities:

Rudofsky, Bernard. *Architecture without Architects: A Short Introduction to Non-pedigreed Architecture.* University of New Mexico Press, 1987 [1964]; and *Streets for People: A Primer for Americans.* Doubleday, 1969.

Excellent source books, with many great photos and drawn illustrations, of vernacular architecture from around the world and wonderful urban pedestrian streets.

Land use:

McHarg, Ian. *Design with Nature.* Natural History Press, 1969.

Best mapping system and land use context for the built and natural habitats at the time, leading into Dave Foreman's Wildlands Project mapping system for wild areas and my own ecocity zoning overlay maps for cities.

Pruetz, Rick. *Saved by Development: Preserving Environmental Areas, Farmland and Historic Landmarks with Transfer of Development Rights.* Arje Press, 1997.

Transfer of Development Rights is the most promising tool I have found for reshaping cities on ecological principles — i.e., restoring while rebuilding — and paying for it with profits from the normal real estate development process. One hundred and twenty-five TDR programs around the US are detailed — an excellent source.

Evolution and ecology:

Quammen, David. *The Song of the Dodo: Island Biogeography in an Age of Extinctions.* Scribner, 1996.

An adventure of a book exploring the environments and living creatures that gave us today's ecology and the people and places that gave us the laws of evolution.

Swimme, Brian and Thomas Berry. *The Universe Story: From the Primordial Flaring forth to the Ecozoic Era — A Celebration of the Unfolding of the Cosmos.* HarperSanFrancisco, 1992; and Berry, Thomas. *The Great Work: Our Way into the Future.* Bell Tower, 2000.

Wise and clear overviews of where we stand in cosmic history, with many hints for how cities should develop — and we with them, from the deep past into the deep future.

Wilson, Edward O. *The Diversity of Life.* Norton, 1999 [1992]; and *The Future of Life.* Knopf, 2002.

Evolution education the way it should have been in school: fresh, somehow youthful, and full of enthusiasm and good science. Wilson has a way with words that brings the subject of life to life.

Background information and strategies for ecocity change:

Beatley, Timothy. *Green Urbanism: Learning from European Cities.* Island Press, 2000.

Excellent survey of some of the best features of European cities in rich detail, with clear analysis as to what works and why.

Brown, Lester R. *Plan B: Rescuing a Planet under Stress and a Civilization in Trouble.* Norton, 2003.

Excellent base information on large-scale problems and conventional solutions — but with only one paragraph on the form and function of the city! For arguing the major points of the resources, climate, and biodiversity problems, this is about as good as it gets for the basic information. Lots on wind energy, solar, soil-building, water conservation, etc.

Heinberg, Richard. *The Party's Over: Oil, War and the Fate of Industrial Societies.* New Society Publishers, 2003.

Best overview to my knowledge of the peak oil crisis soon to unfold. Weak on alternatives, but excellent in presenting necessary background for ecocities.

Kenworthy Jeff and Peter Newman. *Sustainability and Cities: Overcoming Automobile Dependence.* Island Press, 1999.

Best readable technical demonstration of relationships between land use, transportation, and the thriving pedestrian environment anywhere, based on studying many cities world-wide. All urban planners should know this work.

The wonder and pleasure of cities:

Bosselaar, Laure-Anne, ed. *Urban Nature: Poems about Wildlife in the City.* Milkweed Editions, 2000.

Truly enjoyable poems with insights into the nature/urban universe — a celebration and mourning by moods and turns.

Calvino, Italo. *Invisible Cities.* Trans. William Weaver. Harcourt Brace Jovanovich, 1974.

The world's most imaginative word-pictures of cities, briefly sketched by the young Marco Polo to Kubla Kahn as he reports on the fantastic cities of the latter's wide realm. An epic exploration, almost in haiku images, of the urban imagination.

NOTES

Preface to the Second Edition

1. Bill Allen, "Editorial," *National Geographic*, September 2004.

2. Jane Jacobs, *Dark Age Ahead* (Random House, 2004).

3. Jared Diamond, *Collapse: How Societies Choose to Fail or Succeed* (Viking, 2005).

Introduction

1. Philip Shabecoff, *Earth Rising: American Environmentalism in the 21st Century* (Island Press, 2000), p. 178.

Chapter 1: As We Build, So Shall We Live

1. Thomas Berry, "The Ecozoic Era," lecture delivered to the E.F. Schumacher Society, Great Barrington, October 19, 1991, online (cited October 22, 2005) at <www.schumacher society.org/publications/berry_91.html>.

2. Lester R. Brown, *Outgrowing the Earth: The Food Security Challenge in the Age of Falling Water Tables and Rising Temperatures* (Norton, 2005), p.190.

3. Paul and Anne Ehrlich, *Healing the Planet* (Addison-Wesley, 1991).

4. I base this very rough estimate on the fact that human beings represent 100 times the biomass of any previous or contempary species in our general size range on Earth.

5. Kenneth R. Schneider, *On the Nature of Cities: Toward Enduring and Creative Human Environments* (Jossey-Bass, 1979).

6. Bernard Rudofsky, *Architecture without Architects: A short Introduction to Non-pedigreed Architecture* (University of New Mexico Press, 1987).

7. Vakhtang Davitaia, in his presentation at the INTERARCH Conference, Sofia, Bulgaria, 1992.

Chapter 2: The City in Evolution

1. Thomas Berry, "The Ecozoic Era," online (cited October 22, 2005) at <www.schumachersociety. org/publications/berry_91.html>.

2. Brian Swimme and Thomas Berry, *The Universe Story: From the Primordial Flaring forth to the Ecozoic Era — A Celebration of the Unfolding of the Cosmos* (HarperSanFrancisco, 1992).

3. Paolo Soleri, *Arcology: The City in the Image of Man*

(MIT Press, 1974), p. 31.

4. Tina Hesman, "Greenhouse Gassed," *Science News*, Vol. 157, No. 13 (March 25, 2000), p. 200. See also online (cited October 22, 2005) at <www.science news.org/articles/20000325/ bob9.asp>.

5. David Quammen, *The Song of the Dodo: Island Biogeography in an Age of Extinctions* (Sribner, 1996).

6. Quammen, p. 117.

7. Mathis Wackernagel and William Rees, *Our Ecological Footprint: Reducing Human Impact on the Earth* (New Society Publishers, 1996).

8. Peter Newman and Jeff Kenworthy, *Sustainability and Cities: Overcoming Automobile Dependence* (Island Press, 1998).

9. James Grier Miller, *Living Systems* (McGraw Hill, 1978), p. 162, 191. I thank Fritjof Capra for calling Miller's book to my atttention.

10. Edward O. Wilson, *The Future of Life* (Knopf, 2002), p. 29.

Chapter 3: The City in Nature

1. Ian McHarg, *Design with Nature* (Natural History Press, 1969), p. 5.

2. Malcolm Margolin, *The Earth Manual: How to Work on Wild Land without Taming It* (Heyday Books, 1975).

3. Kirkpatrick Sale, *The Conquest of Paradise: Christopher Columbus and the Columbian Legacy* (Knopf, 1990); David E. Stannard, *American Holocaust: Columbus and the Conquest of the New World* (Oxford University Press, 1992).

4. Sale, *The Conquest of Paradise* pp. 75-6.

5. Ibid., p. 85.

6. John Perlin, *A Forest Journey: The Role of Wood in the Development of Civilization* (Harvard University Press, 1991).

7. Sale, *The Conquest of Paradise*, p. 78.

8. Ibid., p. 85.

9. David E. Stannard, *American Holocaust* (Oxford University Press, 1992), p. 18.

10. Ibid.

11. Ibid.

12. Ibid., p. 25.

13. Vernon Masayesva, "Conference Report," in Christopher Canfield, ed., *Ecocity Conference 1990: Report of the First International Ecocity Conference, March 29 – April 1, 1990* (Urban Ecology, 1990), p. 28.

14. Jane Jacobs, *Cities and the Wealth of Nations: Principles of Economic Life* (Vintage Books, 1984), p. 29.

15. Lewis Mumford, *The City in History: Its Origins, Its Transformations, and Its Prospects* (Harcourt, Brace and World, 1961), p. 35.

16. Jacobs, *Cities and the Wealth of Nations*, p. 42.

17. Lewis Mumford, *The City in History*, p. 7.

18. Quoted in Jonathan Rabin, *Bad Land: An American Romance* (Pantheon, 1996), p. 60; see also Deborah E. Popper and Frank J. Popper, "The Buffalo Commons as Regional Metaphor and Geographic Method," online (cited October 22, 2005) at <http://www.gprc.org/buffalo_com mons_popper.html>.

19. Deborah Epstein Popper and Frank J. Popper, "The Great Plains: From Dust to Dust," *Planning*, December 1987, p. 12; see also online (cited October 22, 2005) at <www.planning. org/25anniversary/planning/1987dec.htm>.

20. Deborah Epstein Popper and Frank J. Popper, "The Great Plains: From Dust to Dust,"

Planning, December 1987, p. 12; see also online (cited October 22, 2005) at <www.planning.org/25anniversary/planning/1987dec.htm>.

21. Deborah and Frank Popper, "Can We Reinvent the Frontier?", *Wild Earth*, Winter 1999/Spring 2000, p. 32.

22. Ibid.

23. John Reader, *The Rise of Life: The First 3 – 5 Billion Years* (Knopf, 1986), p. 18.

Chapter 4: The City in History

1. Brian Swimme and Thomas Berry, *The Universe Story*, p. 175.

2. Lewis Mumford, *The City in History*, p. 106.

3. Ibid., p. 28.

4. Ibid., p. 19.

5. Jules Henry, *Culture against Man* (Random House, 1963), p. 8.

6. Sybl Moholy-Nagy, *Matrix of Man: An Illustrated History of Urban Environment* (Praeger, 1968), p. 41.

7. Riane Eisler, *The Chalice and the Blade: Our History, Our Future* (Harper and Row, 1987), p. 34.

8. Ibid.

9. Ibid., p. 33.

10. Ibid., p. 32.

11. Christopher Tunnard and Henry Hope Reed, *American Skyline: The Growth and Form of Our Cities and Towns* (New American Library, 1953), p. 122. Many of the facts in this section are gathered from this source and Mumford's *The City in History*

12. See Tunnard and Reed, *American Skyline*, p. 125.

13. Jonathan Kwitny, "The Great Transportation Conspiracy," *This World Magazine, San Francisco Chronicle*, March 1, 1981, p. 19.

14. Kirkpatrick Sale, *The Conquest of Paradise*, p. 79.

15. Kenneth Schneider, *On the Nature of Cities*, pp. 46-7.

16. Ebenezar Howard, *Garden Cities of To-morrow*, ed. F.J. Osborn (Faber and Faber, 1951 [1902]).

17. Lewis Mumford, *The City in History*, p. 522.

Chapter 5: The City Today

1. Kenneth Schneider, *On the Nature of Cities*, p. 255.

2. Ibid., p. 256.

3. Ibid.

4. Speech at a Free Speech Movement demonstration December 2, 1964, on the steps of Sproul Mall, University of California at Berkeley.

5. Council on Environmental Quality, Department of Housing and Urban Development, Environmental Protection Agency, *The Cost of Sprawl* (US Government Printing Office, 1974).

6. E.F. Schuhmacher, *Small Is Beautiful: Economics As If People Mattered* (Blond and Briggs, 1973).

7. Thomas S. Kuhn, *The Structure of Scientific Revolutions* (University of Chicago Press, 1962; 2nd ed., enlarged, 1970).

8. Fritjof Capra, *The Tao of Physics: An Exploration of the Parallels between Modern Physics and Eastern Mysticism* (The Chaucer Press, 1975).

9. Bill Mollison, *Permaculture: A Practical Guide for a Sustainable Future* (Island Press, 1990), p. ix.

10. Declan Kennedy, "Permaculture Workshop: Urban and Rural Approaches to Community in Balance with Nature," in Christopher Canfield, ed., *Ecocity*

Conference 1990: Report of the First International Ecocity Conference, March 29 – April 1, 1990 (Urban Ecology, 1990), p. 23.

11. Peter Calthorpe, "Transforming Suburbia," in Christopher Canfield, ed., *Ecocity Conference 1990: Report of the First International Ecocity Conference, March 29 – April 1, 1990* (Urban Ecology, 1990), p. 36

12. Ibid.

13. *The Charter of the New Urbanism*, pamphlet by the Congress for the New Urbanism, 1994, p.2.

14. Ibid.

15. Peter Calthorpe, "Beyond Solar Suburbia," *Rain*, August/September 1979, p. 12.

16. Ibid.

17. Maria Rosario, interviewed by the author, March 21, 2001.

18. Jamie Lerner, interviewed by Kirstin Miller, February 2, 2000.

19. Michael Kepp, "Curitiba's Creative Solutions: Learning from Lerner," *Choices*, November 1992, p. 24.

20. Ibid., p. 25.

21. Mac Margolis, "A Third World City that Works," *World Monitor*, Vol. 5, No. 3 (March 1992), p. 43.

22. Ibid.

23. Larry Beasley, "Vancouver, British Columbia: New Urban Neighbourhoods in Old Urban Ways," paper distributed at the "Urbanism — New and Other" conference, University of California, Berkeley, February 25, 2000. All subsequent references to Vancouver are from this paper.

24. Richard S. Levine, "Sustainable Development," in Christopher Canfield, ed., *Ecocity Conference 1990: Report of the First International Ecocity Conference,* *March 29 – April 1, 1990* (Urban Ecology, 1990), p. 24.

Chapter 6: Access and Transportation

1. Ivan D. Illich, *Energy and Equity* (Harper and Row, 1974), p. 12.

2. In a conversation with the author, Oakland, California, September 2, 2000.

3. Runzheimer International, "Runzheimer Analyzes Costs for Mid-sized 2006 Vehicles," July 29, 2005.

4. Jeff Silverstein, "189 Die, Hundreds Hurt in Guadalajara Explosions," *San Francisco Chronicle*, April 23, 1992, p.1.

5. John Witelegg, "Do Something Outrageous: Drive a Car Today," *Manchester Guardian*, August 3, 1993.

6. *Utne Reader*, September/October 1993, p. 57.

7. Matthew L. Wald, "A Military-Industrial Alliance Turns Plowshares to Swords," *New York Times*, March 16, 1994, p. 1.

8. Reuters, "Auto Industry Is No. 1 Priority, Says Trade Negotiator," *San Francisco Chronicle*, March 6, 1992.

9. Marshall Berman, quoted in David Engwicht, *Towards an Eco-City: Calming the Traffic* (Envirobook, 1992), p. 75.

10. Peter Newman, "The Truth about Cars and Freeways," in Christopher Canfield, ed., *Ecocity Conference 1990: Report of the First International Ecocity Conference, March 29 – April 1, 1990* (Urban Ecology, 1990), p. 43.

11. *East Bay Express*, mid-1990s.

12. Ed Ayres, "Breaking away," *World Watch*, January 1, 1993, p. 10.

13. Lester Brown and Janet Larson, "World is Turning to Bicycle Mobility and Exercise," at <www.earthpolicy. org/Updates/ Update13.htm>; see also <www. ecocitycleveland.org/transportation/bicycles/globa l_bike_use.html> (cited October 22, 2005).

14. V. Setty Pendakur, "Bicycles: World Class Vehicles," in Christopher Canfield, ed., *Ecocity Conference 1990: Report of the First International Ecocity Conference, March 29 – April 1, 1990* (Urban Ecology, 1990), p. 43.

15. Ibid., p. 43.

16. Ibid.

17. Ed Ayres, "Breaking Away," p. 18.

18. V. Setty Pendakur, "Bicycles: World Class Vehicles," p. 43.

19. Ed Ayres, "Breaking Away," p. 20.

20. Michael David Lipkan, in a letter to the editor, *Permaculture Activist*, 1993.

21. Bernard Rudolfsky, *Streets for People: A Primer for Americans* (Doubleday, 1969), p. 69.

22. Interview with the author, February 1999.

Chapter 7: What to Build

1. Joel H. Crawford, *Carfree Cities* (International Books, 2000).

2. James Kunstler, *The Geography of Nowhere: The Rise and Decline of America's Man-made Landscape* (Simon and Schuster, 1993).

3. Peter van Dresser, *A Landscape for Humans: A Case Study of the Potentials for Ecologically Guided Development in an Uplands Region* (Biotechnic Press, 1972).

4. Anne Whiston Spirn, *The Granite Garden: Urban Nature and Human Design* (Basic Books, 1984).

5. Stewart Brrand, *How Buildings Learn: What Happens after They're Built* (Viking, 1994).

Chapter 8: Plunge On In!

1. Hazel Henderson, *Building a Win-Win World: Life beyond Global Economic Warfare* (Berrett Koehler, 1996), p. 58.

2. Erward O. Wilson, *The Future of Life*, p. 33.

3. Jared Diamond, *Guns, Germs, and Steel: The Fates of Human Societies* (Norton, 1997).

4. Joan Bokaer, "Rebuilding Our Cities in Balance with Nature: A Proposal," (EcoVillage at Ithaca, 1996), p.4.

5. Ibid., p. 4.

6. Ibid.

7. Ibid., p. 5.

8. Quoted in Michael Parenti, *History as Mystery* (City Lights Books, 1999), p. 2.

9. Richard Heinberg, *Powerdown: Options and Actions for a Post-Carbon World* (New Society Publishers, 2004).

Chapter 10: Tools to Fit the Task

1. Rick Pruetz, *Saved by Development: Preserving Environmental Areas, Farmland and Historic Landmarks with Transfer of Development Rights* (Ardje Press, 1998).

2. Al Gore, *Earth in the Balance: Ecology and the Human Spirit* (Houghton Mifflin, 1992), p. 295.

3. Ibid.

4. Ibid., p. 331.

5. Ibid., p. 325.

6. Lester Brown, *Plan B: Rescuing a Planet under Stress and a Civilization in Trouble* (Norton, 2003).

Chaptert 11: What the Fast-Breaking News May Mean

1. Lester R. Brown, *Outgrowing the Earth*, p. *xiv*.

2. Alfred North Whitehead, *Adventures of Ideas* (The Free Press, 1967 [1933]), p. 12.

3. Bill Clinton, *My Life* (Knopf, 2004), p. 933.

4. John Perkings, *Confessions of an Economic Hit Man* (Berrett-Koehler, 2004), p. *xi*.

5. United Nations, *Human Development Report* (United Nations, 1996).

6. Clifford Krauss, "As Arctic Changes, Eskimo's Worries Grow," *International Herald Tribune*, September 6, 2004, p. 2.

7. Interview with John Norris, "Hans Blix: Caught between Iraq and a Hard Place", MTV, online (cited on October 22) at <www.mtv.com/bands/i/iraq/znews_feature_031203/index5.jhtml>

8. Richard Heinberg, *The Party's Over: Oil, War and the Fate of Industrial Societies* (New Society Publishers, 2004), p. 98.

9. Paul R. Ehrlich, *The Population Bomb* (Balantine Books, 1968).

10. Richard Heinberg, "Meditations on Collapse: A Review of Jared Diamond's *Collapse*," *Museletter*, No. 155 (February 2005), online (cited on October 22, 2005) at <www.museletter.com/archive/154.html>.

11. Ibid.

12. Lester R. Brown, *Outgrowing the Earth*, p.177.

13. Ibid., p. 157.

14. Ibid., p. 164.

15. *The End of Suburbia: Oil Depletion and the Collapse of the American Dream*, Barry Silverthorn Producer, Gregory Greene Director, 2004; see <www.endofsuburbia.com> (cited on October 22, 2005).

Chapter 12: Toward Strategies for Success

1. In *Ecological City: A Shared Course*, a video by Tim Alley, 1996.

2. Jaime Lerner, interviewed by Kirstin Miller, *Ecocity Builders Bulletin*, April 2000, p. 6.

3. Ibid.

4. Alfredo Vincente de Castro Trindade, interviewed by the author during the Fourth International Ecocity Conference, Curitiba, Brazil, January 2000.

5. In *Ecological City: A Shared Course*.

6. Jon Jerde, *You Are Here* (Phaidon, 1999).

7. Interview with the author in Jerde's office in Venice, California, in October 1999.

8. See note 105.

Index

Page numbers in italics indicate illustrations

KIRSTEN MILLER

ABOUT THE AUTHOR

RICHARD REGISTER is an urban ecologist, city designer, writer and illustrator. In 1992, he founded Ecocity Builders, of which he is still president. He has been invited to speak on all the continents, and has in his travels visited and photographed many of the best approximation of "pieces of the ecocity" extant.

He is also the author of *Ecocity Berkeley: Building Cities for a Healthy Future,* which was reprinted in Japan in 1993 and China in 2005. The first edition of *Ecocities* was printed in the US, and in China by the Chinese Academy of Science Research Center for Ecological and Environmental Sciences.

Walt Anderson says of Richard that he "belongs to the Ben Franklin tradition of inventors, generalists, and hopeful geniuses. His ideas about ecological cities are a valuable contribution to the great human task of figuring out how to manage the biosphere and enjoy life while we are at it."

Richard lives in Oakland, California; he grew up in New Mexico, but has lived the last 30 years in the San Francisco Bay Area.

If you have enjoyed *Ecocities* you might also enjoy other

BOOKS TO BUILD A NEW SOCIETY

Our books provide positive solutions for people who want to make a difference. We specialize in:

**Sustainable Living • Ecological Design and Planning • Natural Building & Appropriate Technology
New Forestry • Environment and Justice • Conscientious Commerce • Progressive Leadership
Educational and Parenting Resources • Resistance and Community • Nonviolence**

For a full list of NSP's titles, please call 1-800-567-6772 or check out our web site at:

www.newsociety.com

New Society Publishers

ENVIRONMENTAL BENEFITS STATEMENT

New Society Publishers has chosen to produce this book on recycled paper made with 100% post consumer waste, processed chlorine free, and old growth free.

For every 5,000 books printed, New Society saves the following resources:[1]

52	Trees
4,673	Pounds of Solid Waste
5,142	Gallons of Water
6,706	Kilowatt Hours of Electricity
8,495	Pounds of Greenhouse Gases
37	Pounds of HAPs, VOCs, and AOX Combined
13	Cubic Yards of Landfill Space

[1]Environmental benefits are calculated based on research done by the Environmental Defense Fund and other members of the Paper Task Force who study the environmental impacts of the paper industry.

NEW SOCIETY PUBLISHERS